6th Interdisciplinary Conference Progress in Non-equilibrium Green's Functions (PNGF VI)

AA005281

Journal of Physics: Conference Series Volume 696

Lund, Sweden
17 - 21 August 2015

ISBN: 978-1-5108-2280-1
ISSN: 1742-6588

Printed from e-media with permission by:

Curran Associates, Inc.
57 Morehouse Lane
Red Hook, NY 12571

Some format issues inherent in the e-media version may also appear in this print version.

Printed by Curran Associates, Inc. (2016)

For permission requests, please contact the Institute of Physics
at the address below.

Institute of Physics
Dirac House, Temple Back
Bristol BS1 6BE UK

Phone: 44 1 17 929 7481
Fax: 44 1 17 920 0979

techtracking@iop.org

Additional copies of this publication are available from:

Curran Associates, Inc.
57 Morehouse Lane
Red Hook, NY 12571 USA
Phone: 845-758-0400
Fax: 845-758-2633
Email: curran@proceedings.com
Web: www.proceedings.com

Table of contents

Volume 696

Progress in Non-equilibrium Green's Functions (PNGF VI)
17–21 August 2015, Lund, Sweden

Accepted papers received: 25 February 2016
Published online: 12 April 2016

Preface

Papers

Strongly correlated systems

Time-resolved spectroscopy

Finite Systems

Nuclear Matter

Excitons, Lasers

Quantum Transport

Computational Advances

Journal of Physics: Conference Series **696** (2016) 011001 doi:10.1088/1742-6596/696/1/011001

Foreword

The sixth interdisciplinary conference 'Progress in Non-equilibrium Green's Functions' (PNGF6) took place at Lund University, Sweden, on 17-21 August 2015. The conference was attended by 60 scientists, from Europe and overseas, sharing an interest in Green's function methods and/or non-equilibrium phenomena. At the conference, 34 invited and contributed talks were given, together with a poster session with 17 contributions. As its predecessors (Rostock 1999, Dresden 2002, Kiel 2005, Glasgow 2009, Jyväskylä 2012) the conference succeeded in gathering different communities for exchange of recent developments and results.

Among the topics of the conference, we mention approaches for strongly correlated systems, improvements of existing perturbative many-body schemes, electron-phonon/-photon interactions in time-dependent treatments, numerical scalability of NEGF approaches, connections with other non-equilibrium methods and concrete physical applications. For the latter, we mention quantum transport, semiconductor kinetics, multiply excited states in atoms and ions, nuclear reactions, high energy physics, quantum cascade lasers, strongly correlated model systems, graphene-nanostructures, optoelectronics, superconductors, spin-dynamics, photovoltaics, excitations in atoms and ions, and time-resolved spectroscopy.

The present volume contains 20 articles from participants to PNGF6, devoted to these topics. Compared to previous conferences, a completely novel and successful aspect of PNGF6 was the participation of experimentalists among the invited speakers, to establish a connection between emerging experimental techniques (for example, time-dependent spectroscopies) and the theoretical NEGF community. As at the previous PNGF conferences, the atmosphere was friendly and exciting at the same time, favoring vivid and stimulating discussions among experienced scientists, young researchers and students.

The conference would not have been possible without financial support from the Swedish Research Council and from the Christian-Albrechts-Universität-Kiel via Kiel Nano, Surface and Interface Science (KiNSIS). Particular thanks go to Katarina Lindqvist, the administrator at the division of Mathematical Physics in Lund, who provided assistance in many different matters, particularly in making the conference to unfold very smoothly. We also acknowledge the contributors to this volume for their papers and all the referees of the manuscripts.

Finally, we are glad to announce that the seventh conference 'Progress in Nonequilibrium Green's Functions' (PNGF7) is organized by Andrea Marini and Gianluca Stefanucci, and it will take place in August 2018 in Rome, Italy. *Arrivederci in Roma!*

Claudio Verdozzi, Andreas Wacker, Carl-Olof Almbladh,
Lund, February 2016

Michael Bonitz,
Kiel, February 2016

Journal of Physics: Conference Series **696** (2016) 011001 doi:10.1088/1742-6596/696/1/011001

Leo Kadanoff - Obituary

By the Editors of PNGF6

Two months after the PNGF6 conference, the condensed matter and many-body physics communities were saddened by the death of Leo Kadanoff. Kadanoff was an outstanding physicist and teacher, with seminal work in several areas of Physics. His most prominent contribution and legacy arguably remains the introduction and use of the scaling technique in statistical physics; however, for those involved in non-equilibrium Green's function techniques, his name is indissolubly tied to two fundamental, pioneering contributions to the subject, namely one article [1] and one book [2] together with Gordon Baym, both published in the early sixties. These works belong to the very early production of Kadanoff, when he and Baym both were postdoctoral researchers in Copenhagen, investigating conservation laws in theories of transport within the Martin-Schwinger Green's function formalism, and approximations fulfilling such laws. The results, published in a joint paper, were further expanded in a book entitled "Quantum statistical mechanics", where a main topic is the derivation of generalized Boltzmann's equations from Green's functions-based conserving approximations: since then, this short monograph has become a truly classic reference source in the field of non-equilibrium quantum phenomena. It is perhaps worth to add that further advances on conserving Φ-derivable approximations came from subsequent work by Gordon Baym, and that at the same time, largely independently, progress in the same area was being obtained by soviet scientists, primarily by Leonid Keldysh. In a landmark paper, where the notion of round trip contour explicitly plays the central role [3], Keldysh extended the equilibrium standard Feynman-Dyson perturbation (diagrammatic) expansion technique to describe systems far away from equilibrium. It is appropriate to say that by then, the most basic elements of the so-called Martin-Schwinger-Kadanoff-Baym-Keldysh technique for non-equilibrium phenomena, as we know it today, were all in place. These and other historical facts are vividly recollected in the papers by Paul Martin, Gordon Baym, Alexey Abrikosov and Leonid Keldysh, published in the 1st and 2nd volume of the PNGF6 conferences-proceeding series [4,5].

It is interesting to note how Kadanoff himself largely moved away from the topic of NGF for the rest of his career; and yet, in one of his very last papers, he came back one last time to the subject, with a thorough scrutiny of the concept of entropy in non-equilibrium situations and the role played by entanglement ("Entropy is in Flux" arXiv:1403.6162 (2014)). As said before, Kadanoff's contributions cover several different fields; many of these, including

Journal of Physics: Conference Series **696** (2016) 011001 doi:10.1088/1742-6596/696/1/011001

his deep involvement in teaching and mentoring, are mentioned in an article that originally appeared in UChicagoNews after his death, and reproduced here in abridged form by kind permission of the author, Steve Koppes. By us, the Editors of these proceedings, and the rest of the NGF community, Leo Kadanoff will be deeply missed, and remembered as one of the key figures in the beautiful and challenging field of non-equilibrium many-body physics.

Leo Kadanoff, leading figure in theoretical physics, 1937-2015

Adapted from the article published in UChicagoNews by Steve Koppes [6].

Theoretical physicist Leo Kadanoff, who transformed theory and practice across scientific disciplines, died of respiratory failure on October 26 in Chicago. He was 78. "Leo was a prodigious scientist", said his longtime UChicago colleague Sidney Nagel, the Stein-Freiler Distinguished Service Professor in Physics. "His work on statistical mechanics is one of the great achievements of 20th-century theoretical physics. It laid the conceptual and mathematical foundations for some of the most insightful and effective tools on which our modern understanding of nature is based".

Kadanoffs work has applications throughout physics, ranging from condensed matter (liquids and solids) to elementary particles, Nagel said, with the reach of his work extending to mathematics and other sciences. Kadanoff received the 1999 National Medal of Science, the nation's highest science honor, from President Bill Clinton in a White House ceremony. Kadanoff was cited "for leadership in fundamental theoretical research in statistical, solid-state and nonlinear physics, which has led to numerous and important applications in engineering, urban planning, computer science, hydrodynamics, biology, applied mathematics and geophysics."

In 2013, UChicago received a $3.5 million gift from anonymous donors to support the Leo Kadanoff Center for Theoretical Physics. The center brings together physicists who ordinarily work in a specialty such as particle physics, relativity theory or condensed matter theory and encourages them to work on problems of interest in all of those areas.

In the 1960s, Kadanoff made innovative and original contributions to the understanding of phase changes, such as the change of water from liquid to ice. In later years, working in collaboration with students, junior scientists and colleagues, he helped construct a new field of knowledge called soft condensed matter physics, which deals with such phenomena as the flow of fluids and the behavior of granular materials.

Kadanoff was especially interested in how complexity arises from simple phenomena, such as avalanches forming from the forces that are transmitted between grains of sand. A skilled teacher of colleagues, graduate students and undergraduates, one of his contributions was to use and to show others how to use computer models and simplified conceptual models for better understanding the world.

Curator and Cultivator

Kadanoff had been active at the James Franck Institute until a few weeks before his death. One of his particularly important contributions there in his final years was to lead the long-running Computations in Science seminars. "This was easily the most influential seminar connected with the physics department," said longtime UChicago colleague Thomas Witten, the Homer J. Livingston Professor Emeritus in Physics. "Its strategy was to scout actively for the most promising topics that would attract many disciplines, from math and physics to chemistry and geology. Its success was due to Leo's persistent search for the most exciting topics and speakers. He evolved gracefully from his series of landmark discoveries in statistical physics to his last role of curator and cultivator."

Nagel also commented on Kadanoff's ability to attract the attention of scientists from

Journal of Physics: Conference Series **696** (2016) 011001 doi:10.1088/1742-6596/696/1/011001

multiple disciplines.

"Here at the University, he was the center of so much activity that it is difficult to imagine the campus without him," Nagel said. "He had an extraordinary breadth of interests with a keen eye and appreciation for novel and imaginative science of all kinds." Kadanoff championed that work to his colleagues and was central to the culture of collaboration in the Physical Sciences Division. "When he saw an opportunity, he brought many of us together to attack a problem from different perspectives," Nagel noted. "The Leo I knew was all about developing people," Witten said.

Born January 14, 1937 in New York City, Kadanoff received his bachelor's, master's and doctoral degrees from Harvard University. After completing his doctorate in 1960, Kadanoff conducted postdoctoral research at the Bohr Institute for Theoretical Studies in Copenhagen. He joined the faculty of the University of Illinois at Urbana-Champaign in 1962, where he remained until joining the Brown University faculty in 1969.

Distinguished Professorship

Kadanoff became a professor of physics at UChicago in 1978 and was named the John D. and Catherine T. MacArthur Distinguished Service Professor of Physics and Mathematics in 1982. He directed UChicago's Materials Research Center from 1981 to 1984 and from 1994 to 1997. He also had served as a visiting professor at Cambridge University in 1965, and as the Lorentz Professor at the University of Leiden in The Netherlands in 2003. He retired as professor emeritus in 2003 but remained professionally active, serving as president of the American Physical Society in 2007.

Kadanoff had received many honors during his career. These included the Wolf Foundation Prize in Physics, the Grande Medaille d'Or of the Académie des Sciences de l'Institut de France, the American Physical Society's Onsager and Buckley prizes, the Franklin Institute's Elliott Cresson Medal, the Boltzmann Medal of the International Union of Pure and Applied Physics, and the Lorentz Medal of the Royal Dutch Academy of Sciences. He also was a member of the National Academy of Sciences and a fellow of the American Academy of Arts and Sciences, the American Association for the Advancement of Science, of the American Physical Society, and of the Alfred P. Sloan Foundation.

Kadanoff was further recognized for his undergraduate teaching. At UChicago he designed a new undergraduate course, "Chaos, Computers and Physics". Nominated by his students, he received the University's prestigious Quantrell Award for Excellence in Undergraduate Teaching in 1990.

References

[1] Conservation Laws and Correlation Functions, G. Baym and L.P. Kadanoff, Phys. Rev. 124, 237 (1961).

[2] Quantum Statistical Mechanics, G. Baym and L.P. Kadanoff, (New York, W.A. Benjamin, 1962).

[3] Diagram technique for non-equilibrium processes, L. V. Keldysh, Zh. Eksp. Teor. Fiz. 47, 1515 (1964); Sov. Phys. JETP 20, 1018 (1965).

[4] Progress in Nonequilibrium Green's functions, M. Bonitz (Ed.) (World Scientific, Singapore, 2000).

[5] Progress in Nonequilibrium Green's functions II, M. Bonitz and D. Semkat (Eds.) (World Scientific, Singapore, 2003).

[6] http://news.uchicago.edu/source/steve-koppes

Journal of Physics: Conference Series **696** (2016) 011001 doi:10.1088/1742-6596/696/1/011001

General Information

International Advisory Board: M. Bonitz, P. Danielewicz, F. Jahnke, A.-P. Jahuo, R. van Leeuwen.

Organising Committee: C. Verdozzi (Chair), A. Wacker, C.-O. Almbladh.

Conference Secretary: Katarina Lindqvist.

Support: The organisers acknowledge support from the Swedish Research Council and from the Christian-Albrechts-Universität-Kiel via Kiel Nano, Surface and Interface Science (KiN-SIS).

Workshop Venue: The Department of Physics, University of Lund
Postal address: Box 118, 221 00 LUND, Sweden

Journal of Physics: Conference Series **696** (2016) 011001 doi:10.1088/1742-6596/696/1/011001

List of Talks

I=Invited talk, C= Contributed talk

A. P. Jauho,
Green's function techniques applied to large scale nanostructuring of graphene (I)
M. Galperin,
Transport and optical response in a nanoscale device (I)
T. Kubis,
Contact self-energies: More than simple boundary conditions (I)
M. Garny,
Renormalization out of equilibrium in relativistic quantum field theory (I)
P. Danielewicz,
Progress in Application of Nonequilibrium Green's Functions to Nuclear Reactions (I)
K. Thygesen,
Many-body GW calculations for molecular transport junctions (I)
M. Schüler,
Time propagation of coupled fermionic-bosonic Kadanoff-Baym equations for plasmon-assisted double photoemission (C)
M. Sentef,
Theoretical investigations of laser-driven superconductivity (C)
S. Ristinmaa Sörensen,
Double ionization: investigating mechanisms and time scales (I)
L. Reining,
A direct approach to the calculation of many-body Green's functions (I)
I. Knezevic,
Coupling Electrons, Phonons, and Photons at the Nanoscale: Challenges in Nonequilibrium Transport Simulation (I)
H. S. Köhler,
Nuclear Response Functions by two-time Green's functions (C)
R. van Leeuwen,
Kadanoff-Baym equations for time-dependent coupled electron-boson systems (I)
P. Lipavský,
Formation of superconducting condensate under non-equilibrium conditions (I)
D. Winge,
Second Harmonic Generation in Quantum Cascade Lasers Simulated using Green's Functions (C)
M. Hopjan,
Real-time dynamics of Hubbard-type model systems via a combination of the Kadanoff-Baym formalism with adiabatic DFT (C)
A. L'Huillier,
Ultrafast Atomic Physics using Attosecond Pulses (I)
S. Hermanns,
Nonequilibrium Green's functions - Going beyond standard approximations for extended systems with state-of-the-art computing (I)
K. Balzer,
Solving the Kadanoff-Baym equations via the auxiliary Hamiltonian representation: method and applications (I)
M. Pereira,
Nonequilibrium Green's Functions Theory for Transport and Optics of TERA-MIR Materials

and Devices (I)
R. Citro,
Spin pumping through an interacting quantum dot (I)
P. Werner,
Nonequilibrium dynamics of electron-phonon systems
M. Potthoff,
Non-equilibrium self-energy-functional theory (I)
E. Arrigoni,
Steady-state dynamical mean-field theory within an auxiliary master equation approach (C)
H. Strand,
Real-time dynamics of lattice bosons in high dimensions from nonequilibrium dynamical mean-field theory (C)
E. Lindroth,
Study of attosecond dynamics in atomic many-body systems using complex scaling (I)
E. K. U. Gross,
Some thoughts on the electron-phonon interaction (I)
G. Stefanucci,
NEGF approach to pump-probe photoabsorption spectroscopy (I)
J. Wang,
Current conserving theory on shot noise (I)
F. Jahnke,
Semiconductor nanolaser with superradiant light emission connecting carrier and photon correlations (I)
K. Kaasbjerg,
Non-equilibrium Green's function approach to light emission from plasmonic contacts (C)
N. Schlünzen,
Nonequilibrium Green's functions approach to transport and diffusion in strongly coupled finite quantum systems (C)
U. Aeberhard,
Nonequilibrium Green's function theory of nanostructure solar cells (I)
Y. Pavlyukh,
Non-equilibrium approach to scattering theory and to plasmon losses in photoemission (I)

Journal of Physics: Conference Series **696** (2016) 011001 doi:10.1088/1742-6596/696/1/011001

List of Posters

BAUCH, Sebastian
The time-dependent generalized-active-space configuration interaction approach to correlated ionization dynamics
BONITZ, Michael
Path integral Monte Carlo for correlated fermions - ways around the sign problem
BOSTRÖM, Emil
Time-Resolved Spectroscopy and Dissociation Dynamics of Atoms at Surfaces: Insight from a Model-System Approach
DAMTIE, Fikeraddis Ahmed
Time dependent study of multiple exciton generation (MEG) in nanocrystal QDS
DORDA, Antonius
Spectral and transport properties of the nonequilibrium Anderson impurity model: auxiliary master equation approach within matrix product states
FRANCKIÉ, Martin
Non-Equilibrium Green's Functions as a Validation Tool for QCL Modeling
HERRMANN, Andreas
Non-Equilibrium Dynamical Cluster Approximation of the Falicov-Kimball Model
HINZ, Christopher
Towards a Unified Numerical Framework for Quantum Mechanical Simulations
HYRKÄS, Markku
Computing Exact Self-Energies with Polynomial Expansion
KARLSSON, Daniel
NEGF for studying strongly interacting systems in quantum transport
LANI, Giovanna
The SCE functional in the time domain: insights into its formal properties
RIDLEY, Michael
Generalised Theory of Current Noise for an Arbitrary Time-Dependent Bias
SHAHID, Nazish
Different forms of the Kadanoff-Baym equations in quantum statistical mechanics
TANG, Gaomin
Full-counting statistics of charge and spin transport in the transient regime
TITVINIDZE, Irakli
Non-equilibrium inhomogeneous DMFT for correlated Heterostructures
TUOVINEN, Riku
Extending the time-dependent Landauer-Büttiker formalism to superconducting junctions and arbitrary temperatures
YANG, Kaike
Dynamical correction of thermoelectric coefficients for strongly interacting electrons in the Coulomb blockade regime

Journal of Physics: Conference Series **696** (2016) 011001 doi:10.1088/1742-6596/696/1/011001

List of Participants

AEBERHARD, Urs; *u.aeberhard@fz-juelich.de*, IEK-5: Photovoltaik, Forschungszentrum Juelich, 52425 Juelich, Germany.

ALMBLADH, Carl-Olof; *Carl-Olof.Almbladh@ask.teorfys.lu.se*, Lund University, Box 118, 22100 Lund, Sweden.

ARRIGONI, Enrico; *arrigoni@tugraz.at*, Institute of Theoretical and Computational Physics, Graz University of Technology, Petersgasse 16, 8010 Graz, Austria.

BALZER, Karsten; *balzer@rz.uni-kiel.de*, Computing Center, High-Performance-Computing Division, Christian-Albrechts-Universität zu Kiel, Ludewig-Meyn-Strasse 4, 24118 Kiel, Germany.

von BARTH, Ulf; *barth@ask.teorfys.lu.se*; Lund University, Box 118, 221 00 Lund, Sweden.

BAUCH, Sebastian; *bauch@theo-physik.uni-kiel.de*, Institute for Theoretical Physics and Astrophysics, University of Kiel, Leibnizstrasse 15, 24098 Kiel, Germany.

BONITZ, Michael; *bonitz@physik.uni-kiel.de*, ITAP, CAU Kiel, Leibnizstrasse 15, 24098 Kiel, Germany.

BOSTRÖM, Emil; *emil.bostrom@teorfys.lu.se*, Lund University, Box 118, 22100 Lund, Sweden.

CITRO, Roberta; *citro@sa.infn.it*, Department of Physics "E.R. Caianiello", University of Salerno, Via Giovanni Paolo II, 132 -84084 Fisciano (Sa), Italy.

DAMTIE, Fikeraddis Ahmed; *fikeraddis.damtie@teorfys.lu.se*, Lund University, Box 118, 221 00 Lund, Sweden.

DANIELEWICZ Pawel; *danielewicz@nscl.msu.edu*, National Superconducting Cyclotron Laboratory, Michigan State University, 640 South Shaw Lane, T7/Rm 7, East Lansing, MI 48824, USA.

DORDA, Antonius; *dorda@tugraz.at*, Graz University of Technology, Petersgasse 16, 8010 Graz, Austria.

FRANCKIE, Martin; *martin.franckie@gmail.com*, Lund University, Box 118, 22100 Lund, Sweden.

GALPERIN, Michael; *migalperin@ucsd.edu*, University of California - San Diego, 9500 Gilman Dr., La Jolla, CA 92093-0340, USA.

GARNY, Mathias; *mathias.garny@cern.ch*, CERN Theory Group, CERN, CH 1211 Geneva, Switzerland.

GROSS, Eberhard; *office.theory@mpi-halle.mpg.de*, Max Planck Institute of Microstructure Physics, Weinberg 2, 06120 Halle, Germany.

HERMANNS, Sebastian; *hermanns@theo-physik.uni-kiel.de*, Institut für theoretische Physik und Astrophysik, Christian-Albrechts-Universität zu Kiel, Leibnizstr. 15, 24098 Kiel, Germany.

HERRMANN, Andreas; *andreas.herrmann@unifr.ch*, University of Fribourg, Ch.du Musée 3, 1700 Fribourg, Switzerland.

HINZ, Christopher; *hinz@theo-physik.uni-kiel.de*, Institut für Theoretische Physik und Astrophysik, CAU Kiel, Leibnizstrasse 15, 24098 Kiel, Germany.

HOFMANN, Felix; *fhofmann@physik.uni-hamburg.de*, University of Hamburg,

Jungiusstrasse 9, 20355 Hamburg, Germany.

HOPJAN, Miroslav; *miroslav.hopjan@teorfys.lu.se*; Lund University, Box 118, 22100 Lund, Sweden.

HYRKÄS, Markku; *markku.hyrkas@jyu.fi*, University of Jyväskylä, Yliopistonkatu 20 B 44, Finland.

JAHNKE, Frank; *jahnke@itp.uni-bremen.de*, University of Bremen, PO Box 330 440, 28334 Bremen, Germany.

JAUHO, Antti-Pekka; *antti-pekka.jauho@nanotech.dtu.dk*, CNG, DTU Nanotech, DTU, Bldg. 345E, Orsteds Plads, 2800 Kgs Lyngby, Denmark.

KAASBJERG, Kristen; *cosby@fys.ku.dk*, Center for Nanostructured Graphene, Technical University of Denmark, Örsteds Plads, bygning 345Ø2800 Kgs. Lyngby, Denmark.

KARLSSON, Daniel; *daniel.l.e.karlsson@jyu.fi*, Nanoscience Center, P.O. Box 35 FI-40014, University of Jyväskylä, Finland.

KNEZEVIC, Irena; *irena.knezevic@wisc.edu*, University of Wisconsin - Madison, Electrical and Computer Engineering, 1415 Engineering Dr, Madison, WI 53706, USA.

KUBIS, Tillmann; *tkubis@purdue.edu*, Network for Computation Nanotechnology, Purdue University, 207 S Martin Jischke Drive, West Lafayette, Indiana 47907-1971, USA.

KÖHLER, H. Sigurd; *kohler@physics.arizona.edu*, Physics Department, University of Arizona,Tucson AZ 85721, USA.

L'HUILLIER, Anne; *anne.lhuillier@fysik.lth.se*; Lund University, Box 118, 22100 Lund, Sweden.

LANI, Giovanna; *g.lani@vu.nl*,Vrije Universiteit Amsterdam, De Boelelaan 1105, 1081 HV Amsterdam, The Netherlands.

LEE, Geunsik; *leegeunsik@gmail.com*, Ulsan National Institute of Science and Technology, Republic of Korea, Ulsan.

LINDROTH, Eva; *Eva.Lindroth@fysik.su.se*, Department of Physics, Stockholm University, SE-106 91 Stockholm, Sweden

LIPAVSKY, Pavel; *lipavsky@karlov.mff.cuni.cz*, Faculty of Mathematics and Physics, Ke Karlovu 3, 121 16, Prague 2, Czech Republic.

MACKINNON, Angus; *a.mackinnon@imperial.ac.uk*, Imperial College London, Blackett Laboratory, South Kensington campus, London SW7 2AZ, United Kingdom.

PAVLYUKH, Yaroslav; *yaroslav.pavlyukh@physik.uni-halle.de*, Institut für Physik, Martin-Luther-Universität Halle-Wittenberg, Karl-Freiherr-von-Fritsch Str. 3, D-06120 Halle (Saale), Germany.

PEREIRA, Mauro; *m.pereira@shu.ac.uk*, Materials and Engineering Research Institute, Sheffield Hallam University, Howard Street S1 1WB, Sheffield, United Kingdom.

POTTHOFF, Michael; *michael.potthoff@physik.uni-hamburg.de*, I. Institute for Theoretical Physics, University of Hamburg, Jungiusstrasse 9, D-20355 Hamburg, Germany.

RAMMER, Jörgen; *jorgen.rammer@matfys.lth.se*, Lund University, Box 118, 22100 Lund, Sweden.

REINING, Lucia; *lucia.reining*@polytechnique.fr, CNRS, LSI, Ecole Polytechnique, 91128 Palaiseau,

France.

RIDLEY, Michael; *michael.ridley12@imperial.ac.uk*, Department of Physics, The Blackett Laboratory, Imperial College London, South Kensington Campus, London SW7 2AZ, United Kingdom.

RISTINMAA SÖRENSEN, Stacey; *stacey.sorensen@sljus.lu.se*, Lund University, Box 118, 221 00 Lund, Sweden.

SCHLÜNZEN, Niclas; *schluenzen@theo-physik.uni-kiel.de*, Institut für theoretische Physik und Astrophysik, Christian-Albrechts-Universität zu Kiel, Leibnizstr. 15, 24098 Kiel, Germany.

SCHÜLER, Michael; *michael.schueler@physik.uni-halle.de*, Martin Luther University, Karl-Freiherr-von-Fritsch-Strasse 3, 06120 Halle (Saale), Germany.

SENTEF, Michael; *michael.sentef@mpsd.mpg.de*, Max-Planck Institute for the Structure and Dynamics of Matter, Luruper Chaussee 149, 22761 Hamburg, Germany.

SHAHID, Nazish; *nash_shhd@hotmail.com*, Department of Mathematics, Forman Christian College, Ferozepur Road, Lahore 54300, Pakistan.

STEFANUCCI, Gianluca; *gianluca.stefanucci@roma2.infn.it*, University of Rome Tor Vergata, via della Ricerca Scientifica 1, Italy.

STRAND, Hugo; *hugo.strand@unifr.ch*, University of Fribourg, Chemin du Musée 3, CH-1700 Fribourg, Switzerland.

TANG, Gaomin; *gmtang@hku.hk*, Department of Physics, Rm418, CYM physics building, The University of Hong Kong, Pokfulam Road, Hong Kong.

THYGESEN, Kristian; *thygesen@fysik.dtu.dk*, Technical University of Denmark, Fysikvej 1, Denmark.

TITVINIDZE, Irakli; *irakli.titvinidze@tugraz.at*, Institute of theoretical and computational physics, Graz University of Technology, Petersgasse 16, 8010 Graz, Austria.

TUOVINEN, Riku; *riku.m.tuovinen@jyu.fi*, Department of Physics, Nanoscience Center, FI-40014, University of Jyväskylä, Finland.

VAN LEEUWEN, Robert; *robert.vanleeuwen@jyu.fi*, University of Jyväskylä, Survontie 9 40500, Finland.

VERDOZZI, Claudio; *claudio.verdozzi@teorfys.lu.se*, Lund University, Box 118, 221 00, Lund, Sweden.

WACKER, Andreas; *Andreas.Wacker@fysik.lu.se*, Lund University, Box 118, 221 00 Lund, Sweden.

WANG, Jian; *jianwang@hku.hk*, Department of physics, University of Hong Kong, Pokfulam Road, Hong Kong.

WERNER, Philipp; *philipp.werner@unifr.ch*, University of Fribourg, Chemin du Musée 3, Switzerland.

WINGE, David; *david.winge@teorfys.lu.se*; Lund University, Box 118, 221 00 Lund, Sweden.

YANG, Kaike; *g.kkyang@gmail.com*, University of the Basque Country, UPV/EHU, Nano-Bio Spectroscopy Group, Departamento de Fisica de Materiales, Universidad del Pais Vasco UPV/EHU, Avenida de Tolosa 72, 20018 San Sebastian, Spain.

ÅBERG, Sven; *Sven.Aberg@matfys.lth.se* , Lund University, Box 118, 221 00 Lund, Sweden.

Journal of Physics: Conference Series **696** (2016) 011001 doi:10.1088/1742-6596/696/1/011001

Conference Photographs

Visit at Kulturen in Lund. From the left: Frank Jahnke, Angus MacKinnon, Andreas Wacker, Gaomin Tang, Jian Wang, Miroslav Hopjan, Eva Lindroth, Gianluca Stefanucci, Lucia Reining, Kaike Yang, Carl-Olof Almbladh, Pawel Danielewicz, Hardy Gross, Daniel Karlsson, Markku Hyrkäs, Claudio Verdozzi, Robert van Leeuwen, Giovanna Lani, Karsten Balzer

Visit at Kulturen in Lund. From the left: Andreas Wacker, Michael Galperin, Emil Boström, Yaroslav Pavlyukh, Claudio Verdozzi, Daniel Karlsson, Lucia Reining, Karsten Balzer, Eva Lindroth, Nazish Shahid, Hardy Gross, Angus MacKinnon, Michael Ridley.

Journal of Physics: Conference Series **696** (2016) 011001 doi:10.1088/1742-6596/696/1/011001

Poster Session: Kaike Yang (left), Riku Tuovinen (center), Pawel Danielewicz (right)

Poster Session: Hardy Gross (left) and Robert van Leeuwen (right)

Journal of Physics: Conference Series **696** (2016) 011001 doi:10.1088/1742-6596/696/1/011001

The PNGF6 organizers. From the left: Claudio Verdozzi, Carl-Olof Almbladh, Andreas Wacker

Journal of Physics: Conference Series **696** (2016) 011001 doi:10.1088/1742-6596/696/1/011001

Conference Photo Legend

(1) U. von Barth; (2) G. Stefanucci; (3) K. Thygesen; (4) K. Lindqvist; (5) P. Danielewicz; (6) G. Lani; (7) R. van Leeuwen; (8) G. Lee; (9) E. Boström; (10) F. A. Damtie; (11) A. Herrmann; (12) R. Tuovinen; (13) M. Pereira; (14) E. Gross; (15) M. Franckié; (16) H. Strand; (17) K. Yang; (18) N. Shahid; (19) C.-O. Almbladh; (20) R. Citro; (21) F. Brange; (22) K. Balzer; (23) C. Hinz; (24) L. Reining; (25) G. Tang; (26) J. Wang; (27) A. Dorda; (28) U. Aeberhard; (29) M. Schüler; (30) M. Hyrkäs; (31) D. Karlsson; (32) E. Arrigoni; (33) M. Ridley; (34) F. Hofmann; (35) S. Hermanns; (36) S. Köhler; (37) T. Kubis; (38) A.-P. Jauho; (39) A . Wacker; (40) A. Mackinnon; (41) M. Hopjan; (42) Y. Pavlyukh; (43) F. Jahnke; (44) I. Titvinidze; (45) M. Galperin; (46) E. Lindroth; (47) M. Potthoff; (48) A. L'Huillier; (49) S. Bauch; (50) C. Verdozzi; (51) D. Winge;(52) N. Schlünzen; (53) M. Sentef; (54) P. Lipavsky.

Peer review statement

All papers published in this volume of *Journal of Physics: Conference Series* have been peer reviewed through processes administered by the proceedings Editors. Reviews were conducted by expert referees to the professional and scientific standards expected of a proceedings journal published by IOP Publishing.

Journal of Physics: Conference Series **696** (2016) 012001 doi:10.1088/1742-6596/696/1/012001

The auxiliary Hamiltonian approach and its generalization to non-local self-energies

Karsten Balzer

Rechenzentrum, Christian-Albrechts-Universität zu Kiel, Ludewig-Meyn-Strasse 4, 24118 Kiel, Germany

E-mail: balzer@rz.uni-kiel.de

Abstract. The recently introduced auxiliary Hamiltonian approach [Balzer K and Eckstein M 2014 *Phys. Rev. B* **89** 035148] maps the problem of solving the two-time Kadanoff-Baym equations onto a noninteracting auxiliary system with additional bath degrees of freedom. While the original paper restricts the discussion to spatially local self-energies, we show that there exists a rather straightforward generalization to treat also non-local correlation effects. The only drawback is the loss of time causality due to a combined singular value and eigen decomposition of the two-time self-energy, the application of which inhibits one to establish the self-consistency directly on the time step. For derivation and illustration of the method, we consider the Hubbard model in one dimension and study the decay of the Néel state in the weak-coupling regime, using the local and non-local second-order Born approximation.

1. Introduction

Many numerical approaches which are based on the Keldysh formalism require the solution of the two-time Kadanoff-Baym equations (or the nonequilibrium Dyson equation) which are equations of motions for the one-particle nonequilibrium Green's function defined on the L-shaped Keldysh contour [1, 2, 3]. The particular double-time and non-Markovian form of these integro-differential equations generally renders their integration a non-trivial and challenging task. Moreover, big efforts are needed to reach either long times, handling a large memory kernel, or to simulate quantum systems with many degrees of freedom and especially spatially inhomogeneous systems.

Despite these obstacles, much computational progress has been made over the recent decades, driven by the widespread success of nonequilibrium Green's functions in various fields, including nuclear [4, 5, 6], plasma [7, 8], semiconductor [9, 10, 11], condensed matter [12, 13] and atomic and molecular physics [14, 15]. Following the pioneering work of Refs. [4, 5], which focuses on the numerical simulation of heavy-ion collisions, time propagation schemes of the Kadanoff-Baym equations (KBE) have been developed in different contexts to study the correlation build-up in homogeneous fermion systems [16, 17]. A general Fortran code is presented in Ref. [18] which efficiently evaluates the collision integrals (convolutions) by using fast Fourier transforms of the momentum resolved Green's function. Concerning the treatment of initial correlations different pathways have been pursued which range from the use of adiabatic switching [19, 20] and the embedding of additional collision terms derived from the equilibrium pair correlation function [21] to the extension of the real-time Green's function onto the imaginary branch of the Keldysh contour [22, 23]. With the advance of computer power, also finite and inhomogeneous systems have become accessible. A corresponding well recognized predictor-corrector-based time

Journal of Physics: Conference Series **696** (2016) 012001 doi:10.1088/1742-6596/696/1/012001

propagation code is described in Ref. [24] and is used in many applications. Further progress is due to adapted basis sets which have been deployed to simplify the interaction (Coulomb) matrix elements, e.g., [25], and due to efficient parallelization strategies [26, 27, 28], where the Green's function is distributed over the memory of multiple compute nodes in such a way that the collision integrals can be calculated locally and a minimum communication is required to perform the time stepping.

A feature, that is common to virtually all time propagation codes, is the direct evaluation of the memory kernel, i.e., the computation of the collision integrals which cover the whole time history. Together with the double-time structure of the Green's function, this leads to the typical n_t^3-scaling of the numerical algorithms in the number of time steps. Only further approximations can overcome this unfavorable scaling, such as the application of the generalized Kadanoff-Baym ansatz [20, 29, 30], which reconstructs the two-time Green's function from its time-diagonal value and is expected to be a reasonable option to extend state-of-the art simulations, e.g., [31, 32, 33].

The prevailing similarity of the time propagation schemes raises the question whether there exist alternative methods to solve the two-time KBE. Aside from matrix inversion techniques, e.g. [34], a promising idea has been triggered by nonequilibrium dynamical mean-field theory (DMFT) [13], where Hamiltonian-based impurity solvers have recently become available through a two-time decomposition of the hybridization function [35, 36, 37, 38]. In Ref. [39] it has been shown that a similar decomposition of the self-energy can be used to derive an auxiliary Hamiltonian representation of the Kadanoff-Baym equations, where the one-particle nonequilibrium Green's function of the interacting many-body problem under consideration is obtained from a *noninteracting* auxiliary system which couples to additional bath orbitals the number of which is finite. Although progress has been made recently by showing that a nonequilibrium self-energy has a unique Lehmann representation [40], the auxiliary Hamiltonian method has so far been proven to be effective only for a local self-energy, which represents an approximation becoming exact in the limit of infinite spatial dimensions (and is just the central approximation in DMFT).

In this contribution, we show that the method can be extended rather easily to treat also non-local self-energies and that it is thus applicable more generally. Particularly, it can improve the n_t^3-scaling in the same manner as in Ref. [39] as long as the computation of the self-energy itself represents a sufficiently simple task. Note that this is a rather general prerequisite of the approach, independent of the self-energy's spatial form. We start the discussion with a brief review of the auxiliary Hamiltonian approach for systems with a local self-energy (Sec. 2.1). Thereafter, we derive an extended auxiliary system for one of the most simple test cases, the (two-site) Hubbard dimer, and show how the additionally emerging bath parameters follow from a singular value decomposition of the offdiagonal components of the self-energy (Sec. 2.2.1). After numerical validation of the scheme in Sec. 2.2.2, we then generalize the approach to an arbitrary number of sites and hence to a general quantum system (Sec. 2.3). Based on further numerical results, which are presented in Sec. 3, we finally discuss additional simplifications of the method.

2. Theory

For the discussion of the auxiliary Hamiltonian approach and its extension to non-local self-energies, we consider the single-band Fermi-Hubbard model,

$$H = \sum_{ij\sigma} T_{ij} c^\dagger_{i\sigma} c_{j\sigma} + U \sum_i \left(n_{i\uparrow} - \frac{1}{2} \right) \left(n_{i\downarrow} - \frac{1}{2} \right) , \tag{1}$$

where T_{ij} are hopping matrix elements connecting the lattice sites i and j, U is the local Coulomb repulsion, $c^\dagger_{i\sigma}$ and $c_{i\sigma}$ are creation and annihilation operators for particles of spin $\sigma \in \{\uparrow, \downarrow\}$,

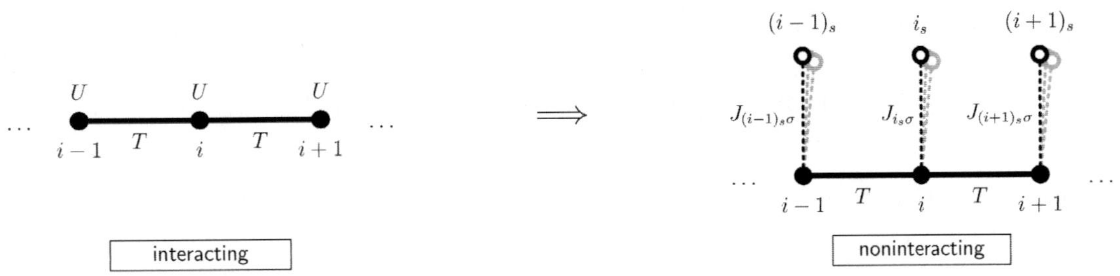

Figure 1. Construction of the auxiliary Hamiltonian in the case of a local self-energy for a Hubbard chain with nearest-neighbor hopping T and on-site interaction U. While the filled circles denote the original lattice sites, the open circles mark the bath orbitals i_s which are coupled independently of one another to a lattice site i by a time- and (in general) spin-dependent hopping matrix element $J_{i_s\sigma}(t)$.

and $n_{i\sigma} = c^\dagger_{i\sigma}c_{i\sigma}$ is the density. Using the nonequilibrium Green's function $G_{ij\sigma}(t,t') = -\mathrm{i}\langle T_{\mathcal{C}} c_{i\sigma}(t)c^\dagger_{j\sigma}(t')\rangle$, with $\langle T_{\mathcal{C}}\ldots\rangle = \mathrm{tr}[T_{\mathcal{C}}\exp(S)\ldots]/\mathrm{tr}[T_{\mathcal{C}}\exp(S)]$, action $S = -\mathrm{i}\int_{\mathcal{C}} \mathrm{d}s\, H(s)$ and contour-ordering operator $T_{\mathcal{C}}$, the time evolution of the system (1) is given by the Kadanoff-Baym equation

$$\sum_k [\mathrm{i}\partial_t\delta_{ik} - T_{ik}]G_{kj\sigma}(t,t') = \delta_{\mathcal{C}}(t,t')\delta_{ij} + \sum_k \int_{\mathcal{C}} \mathrm{d}s\, \Sigma_{ik\sigma}(t,s)G_{kj\sigma}(s,t'), \tag{2}$$

which is accompanied by its adjoint equation with $t \leftrightarrow t'$. For simplicity, we incorporate the Hartree potential $V_i(t) = U(\langle n_{i\bar{\sigma}}(t)\rangle - \frac{1}{2})$ into the hopping matrix, thus we define as $\Sigma_{ij\sigma}(t,t')$ only the correlation part of the self-energy. As the unit of energy (time) we choose the (inverse) hopping T.

2.1. Auxiliary Hamiltonian for a local self-energy

Following Ref. [39], the Kadanoff-Baym equations (2) can be mapped onto a *noninteracting* auxiliary system when the self-energy is local in space, i.e., if $\Sigma_{ij\sigma}(t,t') \propto \delta_{ij}\Sigma_{i\sigma}(t,t')$. In this case the retardation effects described by $\Sigma_{i\sigma}$ can be mimicked by sets $\mathcal{S}_i$ of bath orbitals which are coupled to each individual lattice site. The corresponding auxiliary Hamiltonian is given by

$$\begin{aligned} H_{\mathrm{aux}}(t) = &\sum_{ij\sigma} T_{ij}c^\dagger_{i\sigma}c_{j\sigma} + \sum_{i\sigma}\sum_{s} \epsilon_{i_s\sigma}(t)a^\dagger_{i_s\sigma}a_{i_s\sigma} \\ &+ \sum_{i\sigma}\sum_{s\in\mathcal{S}_i}\left(J_{i_s\sigma}(t)a^\dagger_{i_s\sigma}c_{i\sigma} + J^*_{i_s\sigma}(t)c^\dagger_{i\sigma}a_{i_s\sigma}\right), \end{aligned} \tag{3}$$

where $a^\dagger_{i_s\sigma}$ and $a_{i_s\sigma}$ describe the creation and annihilation of particles with spin σ in a bath orbital i_s with energy $\epsilon_{i_s\sigma}(t)$ that is assigned to the lattice site i, see Fig. 1. By comparing the KBE of the auxiliary system (3) and the original lattice system (1) [Eq. (2)], one finds that the auxiliary system has the same Green's function $G_{ij\sigma}(t,t')$ (with i,j being lattice indices) provided that for all times t and t' on the Keldysh contour the bath parameters can be determined from

$$\Sigma_{i\sigma}(t,t') = \sum_{s\in\mathcal{S}_i} J_{i_s\sigma}(t)g(\epsilon_{i_s\sigma};t,t')J^*_{i_s\sigma}(t'), \tag{4}$$

Journal of Physics: Conference Series **696** (2016) 012001 doi:10.1088/1742-6596/696/1/012001

where $g(\epsilon; t, t')$ is the Green's function of an isolated bath orbital,

$$g(\epsilon; t, t') = \mathrm{i}[f_\beta(\epsilon(0)) - \theta_\mathcal{C}(t, t')]\mathrm{e}^{-\mathrm{i}\int_t^{t'} \mathrm{d}s\, \epsilon(s)}\,, \tag{5}$$

with $f_\beta(\epsilon) = 1/(\mathrm{e}^{\beta\epsilon} + 1)$ being the Fermi-Dirac distribution for an inverse temperature β, and $\theta_\mathcal{C}$ the Heavyside step function on the contour.

Analyzing Eq. (4) regarding the location of the time arguments on the contour, one can distinguish two classes $\mathcal{S}_i^{(1)} \cup \mathcal{S}_i^{(2)} = \mathcal{S}_i$ of bath orbitals that model different dynamical effects:

(i) For the mixed components ($\Sigma_{i\sigma}^{\rceil}$) of the self-energy, the bath orbitals in $\mathcal{S}_i^{(1)}$ describe the time evolution of correlations that are present in the initial state. As these contributions typically decay as function of time, the corresponding bath orbitals become decoupled ($J_{i_s\sigma}(t) \to 0$) in the limit $t \to \infty$. In general, the bath parameters of this class depend on the generalized spectral function

$$C_{i\sigma}^{\rceil}(t, \epsilon) = \frac{1}{2\pi}\left(\Sigma_{i\sigma}^{\rceil}(t, \epsilon + \mathrm{i}0) - \Sigma_{i\sigma}^{\rceil}(t, \epsilon - \mathrm{i}0)\right)\,. \tag{6}$$

Details on the construction of the parameters are given in Ref. [35].

(ii) For the real-time greater ($\Sigma_{i\sigma}^{>}$) and lesser ($\Sigma_{i\sigma}^{<}$) components of the self-energy, the bath orbitals in $\mathcal{S}_i^{(2)}$ describe the dynamical build-up of correlations. If we choose the initial bath energies $\epsilon_{i_s\sigma}(0)$ [at time $t = 0$] such that the Fermi-Dirac distribution $f_\beta(\epsilon_{i_s\sigma}(0))$ is either 0 or 1, the Green's function g evaluates to $g^{\gtrless}(\epsilon_{i_s\sigma}; t, t') = \mp\mathrm{i}$, and one can decompose both real-time components separately,

$$-\mathrm{i}\Sigma_{i\sigma}^{<}(t, t') = \sum_{s \in \mathcal{S}_i^{<}} J_{i_s\sigma}(t) J_{i_s\sigma}^*(t')\,, \tag{7}$$

$$\mathrm{i}\Sigma_{i\sigma}^{>}(t, t') = \sum_{s \in \mathcal{S}_i^{>}} J_{i_s\sigma}(t) J_{i_s\sigma}^*(t')\,, \tag{8}$$

where $\mathcal{S}_i^{<} \cup \mathcal{S}_i^{>} = \mathcal{S}_i^{(2)}$.

The form of Eqs. (7) and (8) reveals that the bath parameters $J_{i_s\sigma}(t)$ in $\mathcal{S}_i^{(2)}$ can be obtained directly from a matrix decomposition of the two-time self-energy. As the left-hand sides represent Hermitian *and* positive definite matrices (in t and t'), one can either apply an eigenvalue[1] or a Cholesky decomposition, see [35, 39] for details. We emphasize that an exact representation requires the number of bath orbitals to be equal to the number of time steps. To drastically reduce the size of the baths one however can use low-rank representations, e.g., by performing the sums in Eqs. (7) and (8) only over the important eigenvalues or by using a finite number of bath orbitals to get an exact Cholesky decomposition on the first time steps while fitting the time dependency at later times [35]. Below, we consider only bath orbitals which fall into class (ii), i.e., we start either from a noninteracting initial state, a mean-field state or the atomic limit. In all these cases, the Matsubara and mixed components of the self-energy vanish ($\Sigma_{i\sigma}^{\mathrm{M}} = \Sigma_{i\sigma}^{\lceil} = 0$), and consequently the bath orbitals in $\mathcal{S}_i^{(1)}$ are redundant.

As a result of the mapping procedure, the solution of the interacting many-body problem (1) has been replaced by the determination of the Green's function $G_{ij\sigma}^{\mathrm{aux}}(t, t')$ of the *noninteracting* auxiliary system. The only assumption, at this stage, is the DMFT approximation of a local

[1] In case of an eigenvalue decomposition, the square roots of the real eigenvalues are integrated into the hopping parameters.

self-energy. A self-consistent solution is further prepared by iteration, starting, e.g., from the noninteracting Green's function. Moreover, we can use the causal property of the Cholesky decomposition (where an increase of the matrix dimension does not alter the decomposition at earlier times) to get the self-consistent bath parameters stepwise directly on the time step, see in particular Ref. [35]. In practice, we also do not need to store the Green's function. Instead it is sufficient to save the self-energy which enters the construction of the bath parameters. For the numerical computation of the auxiliary Green's function, we have developed a Krylov-based time-propagation scheme (see Appendix of Ref. [39]), where the greater ($G^>_{ij\sigma}(t_s,t)$) and lesser component ($[G^<_{ij\sigma}(t,t_s)]^\dagger$) of the Green's function are calculated for all times $t \leq t_s$ on the time step t_s, i.e., on the edge of the lower and upper propagation triangle, compare, e.g., with Ref. [24]. In addition, OpenMP-based parallelization of the code is achieved by performing the time stepping for all vectors $\mathbf{G}^>_{j\sigma}(t_s,t)$ with components $G^>_{ij\sigma}(t_s,t)$ [for all i] in parallel, thus for an infinitesimal time step δt and times $t \leq t_s$ (and analogously for the lesser component), we evaluate

$$\mathbf{G}^>_{j\sigma}(t_s+\delta t,t) = |\mathbf{G}^>_{j\sigma}(t_s,t)| \exp\left[-\mathrm{i}\mathbf{h}^{\mathrm{aux}}_\sigma(t)\delta t\right] \frac{\mathbf{G}^>_{j\sigma}(t_s,t)}{|\mathbf{G}^>_{j\sigma}(t_s,t)|}, \tag{9}$$

where $\mathbf{h}^{\mathrm{aux}}_\sigma(t)$ denotes the one-particle Hamiltonian of the auxiliary system $H_{\mathrm{aux}}(t)$. We note that the normalization of the vector $\mathbf{G}^>_{j\sigma}$ on the right-hand side of Eq. (9) is a precondition to apply the unitary time-evolution operator within the Krylov method [41].

2.2. Generalization of the approach to non-local self-energies

From the first point of view, it seems to be not a straightforward endeavor to take into account arbitrary non-local self-energies in the auxiliary Hamiltonian approach. This is due to the fact that any non-local, i.e., offdiagonal component of the self-energy does not describe retardation effects on a single site which is in clear correspondence to a noninteracting site coupled to a bath as discussed in Sec. 2.1. Moreover, offdiagonal components of Σ_σ are less symmetric, which raises the question whether it is at all possible to construct an auxiliary system such that the conservation laws for particle number, momentum and energy (which come with a conserving approximation) are obeyed. Nevertheless, it is intuitive to imagine that correlation effects, which originate from non-local parts of the self-energy, should be representable by additional particle motion that connects different sites of the lattice. In the following, we show that a valid mapping exists for the Hubbard dimer. How the findings can be generalized to more extended systems is discussed afterwards (see Sec. 2.3).

2.2.1. Hubbard dimer The simplest lattice model, in which a non-local self-energy appears, is the Hubbard dimer [42, 43], consisting of two connected sites 1 and 2. In order to account for diagonal *and* offdiagonal components of the self-energy we extend the auxiliary Hamiltonian of Eq. (3) by a set of bath orbitals [with energies $\epsilon_{3_s\sigma}(t)$] which exchange particles with both sites of the dimer. If we refer to the corresponding hopping matrix elements as $C_{s\sigma}(t)$ and $D_{s\sigma}(t)$ and define $J_{1_s\sigma}(t) = A_{s\sigma}(t)$ and $J_{2_s\sigma}(t) = B_{s\sigma}(t)$ (compare with Fig. 2), the ansatz for the mapping is given by

$$\begin{aligned} H_{\mathrm{aux}}(t) = &-T\sum_\sigma \left(c^\dagger_{1\sigma}c_{2\sigma} + c^\dagger_{2\sigma}c_{1\sigma}\right) + \sum_{i\in\{1,2,3\}\sigma}\sum_{s\in S_i} \epsilon_{i_s\sigma}(t) a^\dagger_{i_s\sigma}a_{i_s\sigma} \\ &+\sum_s \left(A_{s\sigma}(t)a^\dagger_{1_s\sigma}c_{1\sigma} + B_{2_s\sigma}(t)b^\dagger_{s\sigma}c_{2\sigma} + C_{s\sigma}(t)a^\dagger_{3_s\sigma}c_{1\sigma} + D_{s\sigma}(t)a^\dagger_{3_s\sigma}c_{2\sigma} + \mathrm{H.c.}\right), \end{aligned} \tag{10}$$

where in the last term s runs only over existing bath sites. To determine the parameters of the baths 1_s, 2_s and 3_s in the presence of all components of the self-energy, we formulate the

Figure 2. Construction of the auxiliary Hamiltonian representation of the KBE for the Hubbard dimer, including non-local correlation effects.

equations of motions for the one-particle Green's function of the auxiliary system (10) and compare them to the KBE of the Hubbard dimer ($i, j \in \{1, 2\}$),

$$[\mathrm{i}\partial_t + T\delta_{|i-j|1}]G_{ij\sigma}(t,t') - \delta_{ij}\delta_{\mathcal{C}}(t,t') = \sum_{k\in\{1,2\}} \int_{\mathcal{C}} \mathrm{d}s\, \Sigma_{ik\sigma}(t,s)G_{kj\sigma}(s,t') \,. \tag{11}$$

For the auxiliary system we have

(i) the *lattice components* of the auxiliary Green's functions, which evolve in time according to

$$\begin{aligned} \mathrm{i}\partial_t G^{\mathrm{aux}}_{1j\sigma} - \delta_{1j}\delta_{\mathcal{C}}(t,t') &= -TG^{\mathrm{aux}}_{2j\sigma}(t,t') + \sum_s \left[A_{s\sigma}(t)G^{\mathrm{aux}}_{1_sj\sigma}(t,t') + C_{s\sigma}(t)G^{\mathrm{aux}}_{3_sj\sigma}(t,t')\right] \,, \\ \mathrm{i}\partial_t G^{\mathrm{aux}}_{2j\sigma} - \delta_{2j}\delta_{\mathcal{C}}(t,t') &= -TG^{\mathrm{aux}}_{1j\sigma}(t,t') + \sum_s \left[B_{s\sigma}(t)G^{\mathrm{aux}}_{2_sj\sigma}(t,t') + D_{s\sigma}(t)G^{\mathrm{aux}}_{3_sj\sigma}(t,t')\right] \,, \end{aligned} \tag{12}$$

where $j \in \{1, 2\}$, and

(ii) the *mixed bath-lattice terms*, which obey

$$\begin{aligned} [\mathrm{i}\partial_t - \epsilon_{1_s\sigma}(t)]G^{\mathrm{aux}}_{1_sj\sigma} &= A^*_{s\sigma}(t)G^{\mathrm{aux}}_{1j\sigma}(t,t') \,, \\ [\mathrm{i}\partial_t - \epsilon_{2_s\sigma}(t)]G^{\mathrm{aux}}_{2_sj\sigma} &= B^*_{s\sigma}(t)G^{\mathrm{aux}}_{2j\sigma}(t,t') \,, \\ [\mathrm{i}\partial_t - \epsilon_{3_s\sigma}(t)]G^{\mathrm{aux}}_{3_sj\sigma} &= \left[C^*_{s\sigma}(t)G^{\mathrm{aux}}_{1j\sigma}(t,t') + D^*_{s\sigma}(t)G^{\mathrm{aux}}_{2j\sigma}(t,t')\right] \,. \end{aligned} \tag{13}$$

Analogously to Ref. [39], the mixed bath-lattice terms can be determined by using the Green's function $g(\epsilon_{i_s\sigma}; t, t')$ of an isolated bath orbital (cf. Eq. (5)):

$$\begin{aligned} G^{\mathrm{aux}}_{1_sj\sigma}(t,t') &= \int_{\mathcal{C}} \mathrm{d}s\, g(\epsilon_{1_s\sigma}; t, s)A^*_{s\sigma}(s)G^{\mathrm{aux}}_{1j\sigma}(s,t') \,, \\ G^{\mathrm{aux}}_{2_sj\sigma}(t,t') &= \int_{\mathcal{C}} \mathrm{d}s\, g(\epsilon_{2_s\sigma}; t, s)B^*_{s\sigma}(s)G^{\mathrm{aux}}_{2j\sigma}(s,t') \,, \\ G^{\mathrm{aux}}_{3_sj\sigma}(t,t') &= \int_{\mathcal{C}} \mathrm{d}s\, g(\epsilon_{3_s\sigma}; t, s)\left[C^*_{s\sigma}(s)G^{\mathrm{aux}}_{1j\sigma}(s,t') + D^*_{s\sigma}(s)G^{\mathrm{aux}}_{2j\sigma}(s,t')\right] \,. \end{aligned} \tag{14}$$

Inserting the results of Eq. (14) into Eq. (12) and comparing term by term to the Kadanoff-Baym equation (11), we find that the offdiagonal components of the self-energy can indeed be represented by the bath 3_s. However, the expressions for the diagonal components do not only

Journal of Physics: Conference Series **696** (2016) 012001 doi:10.1088/1742-6596/696/1/012001

involve the hopping matrix elements $A_{s\sigma}(t)$ and $B_{s\sigma}(t)$. Instead, we obtain

$$\begin{aligned}
\Sigma_{11\sigma}(t,t') &= \sum_{s\in\mathcal{S}_1} A_{s\sigma}(t) g(\epsilon_{1_s\sigma};t,t') A^*_{s\sigma}(t') + \sum_{s\in\mathcal{S}_3} C_{s\sigma}(t) g(\epsilon_{3_s\sigma};t,t') C^*_{s\sigma}(t')\,, \\
\Sigma_{22\sigma}(t,t') &= \sum_{s\in\mathcal{S}_2} B_{s\sigma}(t) g(\epsilon_{2_s\sigma};t,t') B^*_{s\sigma}(t') + \sum_{s\in\mathcal{S}_3} D_{s\sigma}(t) g(\epsilon_{3_s\sigma};t,t') D^*_{s\sigma}(t')\,, \\
\Sigma_{12\sigma}(t,t') &= \sum_{s\in\mathcal{S}_3} C_{s\sigma}(t) g(\epsilon_{3_s\sigma};t,t') D^*_{s\sigma}(t')\,, \\
\Sigma_{21\sigma}(t,t') &= \sum_{s\in\mathcal{S}_3} D_{s\sigma}(t) g(\epsilon_{3_s\sigma};t,t') C^*_{s\sigma}(t')\,.
\end{aligned} \tag{15}$$

In passing, we note that the result (15) can also be obtained by integrating out the bath orbitals in a path integral, which can be done exactly if the bath orbitals appear only up to quadratic order [44]. Furthermore, the right-hand sides of the last two lines in Eq. (15) are in line with the Hermitian symmetry of the real-time quantities, i.e.,

$$\begin{aligned}
\Sigma^{\gtrless}_{12\sigma}(t,t') = -[\Sigma^{\gtrless}_{21}(t',t)]^* &\stackrel{\text{def.}}{=} -\sum_{s\in\mathcal{S}^{\gtrless}} D^*_{s\sigma}(t')[g^{\gtrless}(\epsilon_{3_s\sigma};t',t)]^* C_{s\sigma}(t) \\
&= \sum_{s\in\mathcal{S}^{\gtrless}} C_{s\sigma}(t) g^{\gtrless}(\epsilon_{3_s\sigma};t,t') D^*_{s\sigma}(t')\,,
\end{aligned} \tag{16}$$

where we used $g^{\gtrless}(\epsilon;t,t') = -[g^{\gtrless}(\epsilon;t',t)]^*$ in the last equality.

For the discussion of the result (15) we assume that the initial state of the Hubbard dimer is not correlated, i.e., $\Sigma^{\mathrm{M}} = \Sigma^{\lceil} = 0$. To fix the bath parameters, it is instructive to begin with the decomposition of the offdiagonal components of the self-energy. This will yield the time-dependent hopping parameters $C_{s\sigma}(t)$ and $D_{s\sigma}(t)$ and, subsequently, allows us to determine the parameters $A_{s\sigma}(t)$ and $B_{s\sigma}(t)$ from the first two lines in Eq. (15). As in Sec. 2.1, we choose the energies of the bath orbitals such that the initial occupations of the bath orbitals are either 0 (for the greater component of Σ) or 1 (for the lesser component of Σ), which leads to

$$\pm\mathrm{i}\Sigma^{\gtrless}_{12\sigma}(t,t') = \sum_{s\in\mathcal{S}^{\gtrless}} C_{s\sigma}(t) D^*_{s\sigma}(t')\,. \tag{17}$$

As function of t and t', the matrices $\mathrm{i}\Sigma^{>}_{12\sigma}$ and $-\mathrm{i}\Sigma^{<}_{12\sigma}$ are neither Hermitian nor positive definite which generally rules out an eigenvalue and a Cholesky decomposition. This is in contrast to the diagonal components of the self-energy discussed in Sec. 2.1 (Eqs. (7) and (8)). However, one can perform a singular value decomposition of the objects as

$$\pm\mathrm{i}\Sigma^{\gtrless}_{12\sigma}(t,t') = \sum_{s\in\mathcal{S}_3^{\gtrless}} U_{s\sigma}(t) S_{s\sigma} V_{s\sigma}(t')\,, \tag{18}$$

where U_σ and V_σ denote unitary matrices and S_σ is a real vector containing the non-negative singular values. As the singular values $S_{s\sigma}$ typically decay rapidly in magnitude, we can further use a low-rank decomposition by choosing the number of bath orbitals smaller than the number of time steps. Incorporating the singular values symmetrically into the matrices U_σ and V_σ, we obtain the first set of bath parameters as

$$C_{s\sigma}(t) = \sqrt{S_{s\sigma}} U_{s\sigma}(t)\,, \qquad D_{s\sigma}(t) = \sqrt{S_{s\sigma}} V^*_{s\sigma}(t)\,. \tag{19}$$

As a next step we can use the result of Eq. (19) to determine the bath parameters $A_{s\sigma}(t)$ and $B_{s\sigma}(t)$. Taking into account the initial occupations, the first two lines of Eq. (15) can be rewritten as ($i = 1, 2$)

$$\Delta^{\gtrless}_{i\sigma}(t,t') = \pm \mathrm{i}\Sigma^{\gtrless}_{ii\sigma}(t,t') - \Gamma^{\gtrless}_{i\sigma}(t,t')\,, \tag{20}$$

with

$$\begin{aligned}
\Delta^{\gtrless}_{1\sigma}(t,t') &= \sum_{s\in\mathcal{S}^{\gtrless}_1} A_{s\sigma}(t)A^*_{s\sigma}(t')\,, &\qquad \Gamma^{\gtrless}_{1\sigma}(t,t') &= \sum_{s\in\mathcal{S}^{\gtrless}_3} C_{s\sigma}(t)C^*_{s\sigma}(t')\,,\\
\Delta^{\gtrless}_{2\sigma}(t,t') &= \sum_{s\in\mathcal{S}^{\gtrless}_2} B_{s\sigma}(t)B^*_{s\sigma}(t')\,, &\qquad \Gamma^{\gtrless}_{2\sigma}(t,t') &= \sum_{s\in\mathcal{S}^{\gtrless}_3} D_{s\sigma}(t)D^*_{s\sigma}(t')\,.
\end{aligned} \tag{21}$$

While the matrices $\pm \mathrm{i}\Sigma^{\gtrless}_{ii\sigma}$ and $\Gamma^{\gtrless}_{i\sigma}$ are both Hermitian and positive definite by construction, the positive definiteness does not hold in general for their difference $\Delta^{\gtrless}_{i\sigma}$, Eq. (20). For this reason, the partitioning of $\Delta^{\gtrless}_{i\sigma}$ into the bath parameters $A_{s\sigma}(t)$ and $B_{s\sigma}(t)$, respectively, cannot be performed through a Cholesky decomposition. Instead, we use an eigenvalue decomposition,

$$\begin{aligned}
\Delta^{\gtrless}_{1\sigma}(t,t') &= \sum_{s\in\mathcal{S}^{\gtrless}_1} P_{s\sigma}(t)E_{s\sigma}P^*_{s\sigma}(t')\,,\\
\Delta^{\gtrless}_{2\sigma}(t,t') &= \sum_{s\in\mathcal{S}^{\gtrless}_2} Q_{s\sigma}(t)F_{s\sigma}Q^*_{s\sigma}(t')\,,
\end{aligned} \tag{22}$$

where the eigenvectors form the rows of the matrices P_σ and Q_σ and the corresponding eigenvalues are contained in the vectors E_σ and F_σ. Comparing Eqs. (21) and (22), we thus obtain the remaining bath parameters as

$$A_{s\sigma}(t) = \sqrt{E_{s\sigma}}P_{s\sigma}(t)\,, \qquad B_{s\sigma}(t) = \sqrt{F_{s\sigma}}Q_{s\sigma}(t)\,. \tag{23}$$

2.2.2. Numerical validation The derivation presented in the previous section proofs that there exists a procedure which maps the full Kadanoff-Baym equations of the Hubbard dimer onto a noninteracting auxiliary model. In this section, we demonstrate the validity of the mapping also numerically. To this end, we prepare the Hubbard dimer (at time $t = 0$) in the perfect Néel state $|\Psi_0\rangle = c^\dagger_{2\downarrow}c^\dagger_{1\uparrow}|0\rangle$ and use the second-order Born approximation of the self-energy,

$$\Sigma^{2\mathrm{B},\gtrless}_{ij\sigma}(t,t') = U^2 G^{\gtrless}_{ji\sigma}(t,t')G^{\gtrless}_{ji\bar\sigma}(t,t')G^{\lessgtr}_{ij\bar\sigma}(t',t)\,, \tag{24}$$

to monitor the decay of the local staggered magnetization $m_i(t) = \langle n_{i\uparrow}(t)\rangle - \langle n_{i\downarrow}(t)\rangle$ in the weak-coupling regime. For convenience, we compute the Green's function only for one spin direction. The opposite component follows from the symmetry $G_{ij\bar\sigma}(t,t') = G_{(L+1-i)(L+1-j)\sigma}(t,t')$, where $L = 2$ is the spatial dimension of the dimer.

Fig. 3a shows the self-consistent second Born result for an on-site interaction $U = 0.5$, where the KBE has been solved with a time step of size $\delta t = 0.01$ (on a time window $[0, 6]$) including $n_{\mathrm{aux}} = 20$ orbitals for each bath i_s. The result is further compared to the Hartree dynamics (given by the black dashed line) and to the exact dynamics (see the black solid line), which has been obtained by exact diagonalization. On the short time scale, unlike the mean-field result, correlations induce a damping of the oscillatory magnetization dynamics in the dimer. If we

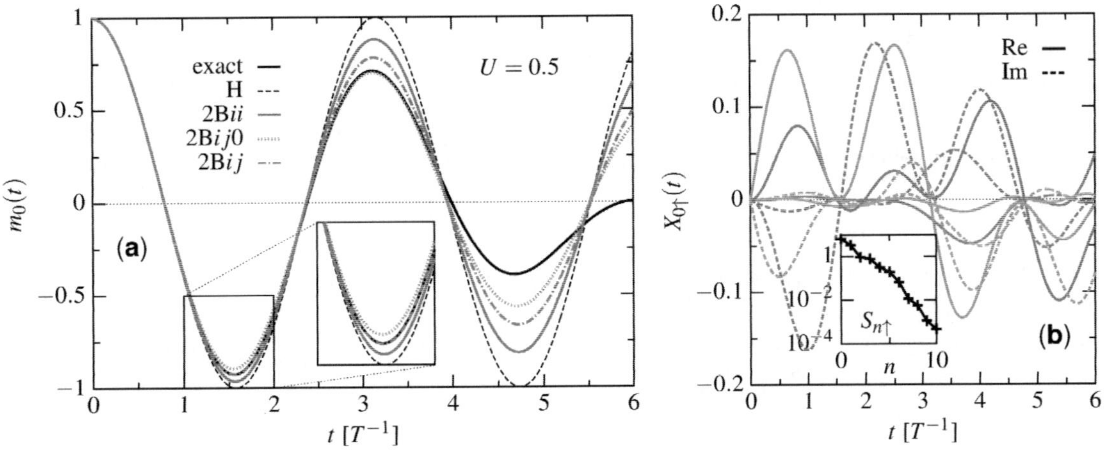

Figure 3. Hubbard dimer prepared in the Néel state at $t = 0$. (a) Time evolution of the staggered magnetization $m_0(t) = \langle n_{0\uparrow}(t)\rangle - \langle n_{0\downarrow}(t)\rangle$ on the left site of the dimer for $U = 0.5$ in Hartree (H) and second-order Born approximation (2B), as computed from the self-consistent auxiliary Green's function $G^{\rm aux}_{ij\sigma}(t,t')$. The line labeled 2B$ij$ (2Bii) denotes the result for a non-local (local) self-energy. The curve denoted as 2Bij0 shows the result where $\Gamma_{i\sigma}$ is additionally set to zero in Eq. (20). (b) Time-dependent hopping parameters $A_{0\sigma}(t)$ (red lines), $B_{0\sigma}(t)$ (orange lines), $C_{0\sigma}(t)$ (blue lines) and $D_{0\sigma}(t)$ (green lines). The inset shows the singular values of the decomposition (18) of the non-local self-energy $\Sigma_{12\sigma}(t,t')$.

consider nothing but the local parts of the self-energy, i.e., if we set to zero the bath parameters $C_{s\sigma}(t)$ and $D_{s\sigma}(t)$ in the auxiliary system (Eq. (10)), the time evolution of the magnetization follows the blue solid line (labeled 2Bii). The inclusion of the non-local self-energy improves this result as expected, see the red dash-dotted line referred to as 2Bij. Particularly, we find good agreement with the exact dynamics for times $t \lesssim 2.5$. As additional result (denoted 2Bij0), we present the green dotted curve, where in Eq. (21) the quantities $\Gamma^{\gtrless}_{1\sigma}$ and $\Gamma^{\gtrless}_{2\sigma}$ have been neglected in the determination of the bath parameters $A_{s\sigma}(t)$ and $B_{s\sigma}(t)$. We note that such an incomplete solution also leads to an acceptable dynamics, although there are discrepancies at short times. Moreover, we emphasize that all schemes conserve the particle number as function of time.

To further illustrate the method, Fig. 3b shows the time dependence of the complex hopping parameters $A_{0\sigma}(t)$ to $D_{0\sigma}(t)$, i.e., for the first bath orbital $s = 0$ in each set $\mathcal{S}_i^<$. Starting from zero (at time $t = 0$ where the self-energy $\Sigma^{\gtrless}_{ij\sigma}$ vanishes), the time evolution of the bath parameters is rather complicated despite the simplicity of the Hubbard dimer. Moreover, the inset in Fig. 3b displays the singular values $S_{n\uparrow}$ of Eq. (18) which decay exponentially in magnitude and thus allow one to use a low-rank representation of the bath 3_s. Concerning the baths 1_s and 2_s, we have included the 20 most important eigenvalues of each self-energy decomposition, i.e., those which have the largest absolute values. Note that an exact representation of the self-energy using a time step of $\delta = 0.01$ would require the dimension of each individual bath to be $n_{\rm aux} = 600$ on the considered time interval.

2.3. Generalization to an arbitrary lattice system

For a lattice system with more than two sites, the self-energy should have a similar auxiliary representation as discussed above in Sec. 2.2.1. Without giving the proof, we state that the

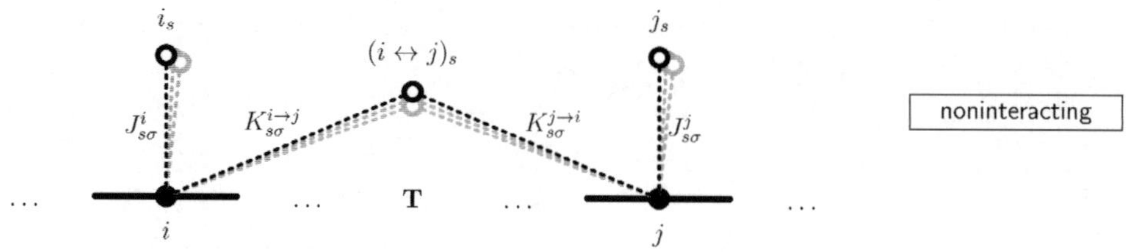

Figure 4. Generalization of the auxiliary Hamiltonian approach to an extended lattice system with a hopping matrix $\mathbf{T}$ and an arbitrary non-local self-energy $\Sigma^{\gtrless}_{ij\sigma}(t,t')$. Together with the dashed lines, the open circles show the layout of the bath and how it is connected to the lattice sites (filled circles) by complex hopping matrix elements $K^{i\to j}_{s\sigma}(t)$, $K^{j\to i}_{s\sigma}(t)$, $J^i_{s\sigma}(t)$ and $J^j_{s\sigma}(t)$.

offdiagonal components of the self-energy can always be expressed in the form ($i \neq j$)

$$\pm \mathrm{i}\Sigma^{\gtrless}_{ij\sigma} = \sum_{s\in\mathcal{S}^{\gtrless}_{i\leftrightarrow j}} K^{i\to j}_{s\sigma}(t)[K^{j\to i}_{s\sigma}(t')]^* \,, \tag{25}$$

with time-dependent hopping parameters $K^{i\to j}_{s\sigma}(t)$ connecting a lattice site i and a bath orbital s. In the course of this, it is important to remark that the parameters $K^{i\to j}_{s\sigma}(t)$ are different from $K^{j\to i}_{s\sigma}(t)$. However, both sets follow from a single singular value decomposition of the left-hand side of Eq. (25). On the other hand, the auxiliary representation for the *diagonal* components of the self-energy is given by

$$\pm\Sigma^{\gtrless}_{ii\sigma} = \sum_{s\in\mathcal{S}^{\gtrless}_i} J^i_{s\sigma}(t)[J^i_{s\sigma}(t')]^* + \sum_{j\neq i}\sum_{s\in\mathcal{S}^{\gtrless}_{i\leftrightarrow j}} K^{i\to j}_{s\sigma}(t)[K^{i\to j}_{s\sigma}(t')]^* \,, \tag{26}$$

where the bath parameters $J^i_{s\sigma}(t) \equiv J_{i_s\sigma}(t)$ are defined as in Sec. 2.1 and can be obtained from an eigenvalue decomposition applied to Eq. (26). For an illustration of the resulting auxiliary system see Fig. 4.

In summary, the one-particle auxiliary Hamiltonian which replaces the full Kadanoff-Baym equation of Eq. (2) for an arbitrary non-local self-energy is of the form

$$\mathbf{h}^{\mathrm{aux}}_\sigma(t) = \begin{pmatrix} \mathbf{T} & \mathbf{J}_\sigma & \mathbf{K}_\sigma \\ \mathbf{J}^\dagger_\sigma & \mathbf{0} & \mathbf{0} \\ \mathbf{K}^\dagger_\sigma & \mathbf{0} & \mathbf{0} \end{pmatrix} \,, \tag{27}$$

where the matrix elements of $\mathbf{T}$ are the hopping parameters T_{ij} of the original Hubbard model (1), the matrix $\mathbf{J}$ is diagonal involving the bath parameters $J^i_{s\sigma}(t)$, and the matrix $\mathbf{K}$ includes the bath parameters $K^{i\to j}_{s\sigma}(t)$ and $K^{j\to i}_{s\sigma}(t)$. If we use n_{aux} bath orbitals for the representation of each component $\Sigma_{ij\sigma}$ (with $i \leq j$) of the self-energy and consider a Hubbard model with L sites, the dimension of the auxiliary Hamiltonian $\mathbf{h}^{\mathrm{aux}}_\sigma$ is given by

$$\mathcal{D}_{\mathrm{aux}} = L + 2n_{\mathrm{aux}}L + n_{\mathrm{aux}}(L^2 - L) = (n_{\mathrm{aux}}+1)L + L^2 \,, \tag{28}$$

i.e., the size of the generalized auxiliary system scales linearly with the size of the bath and quadratically with the number of lattice sites. Moreover, we emphasize that the low-rank representation of Eqs. (25) and (26) [with fixed $n_{\mathrm{aux}} \ll n_t$] generically leads to a time propagation scheme which scales quadratically with the number of time steps n_t. This is one of the main advantages of the method and allows the scheme to outperform standard KBE solvers.

3. Application of the method

To validate the generalization of the auxiliary Hamiltonian approach for a quantum system which is more complex than the Hubbard dimer discussed in Sec. 2.2.2, we now consider a Hubbard chain with $L = 10$ sites. For this system it is still easy to compute the dynamics exactly, and therefore it is possible to judge upon the quality of the second-order Born approximation, cf. Eq. (24). Similarly to the analysis performed on exact grounds in Ref. [45], we study the time evolution of the Néel state, i.e., the initial state is given by the one-particle density matrix

$$\rho_{ij\sigma}(0) = -\mathrm{i}G^{<}_{ij\sigma}(0,0) = -\mathrm{i}\delta_{ij}\langle\Psi_0|n_{i\sigma}|\Psi_0\rangle\,, \tag{29}$$

where

$$|\Psi_0\rangle = \prod_{i\in A} c^{\dagger}_{i\downarrow} \prod_{j\in B} c^{\dagger}_{j\uparrow}|0\rangle\,, \tag{30}$$

and A and B denote the two sublattices. To perform the simulations, we choose $n_{\mathrm{aux}} = 30$, $\delta t = 0.01$ and iterate until self-consistency is reached on the time interval $[0, 6]$.

We consider two cases, first a Hubbard chain with open boundary conditions and second a closed chain, respectively, a ring. Figure. 5 summarizes the results for the former system, showing the time evolution of the average magnetization $m(t) = \frac{2}{L}\sum_{i\in A}(\langle n_{i\uparrow}(t)\rangle - \langle n_{i\downarrow}(t)\rangle)$ for different approximation levels of the self-energy and different strengths of the interaction U. Generally, we find that the antiferromagnetic order parameter melts on the time scale of a few hopping times at weak interaction. In the limit $U \to 0$ and large L (see Fig. 5a), we observe a damped oscillation of $m(t)$ which is of the form of a zeroth-order Bessel function $J_0(\gamma t)$ with some constant γ, compare, e.g., with Ref. [46]. At small $U \lesssim 1$, the second-order Born approximation well describes the relaxation of the magnetization, whereas the Hartree approximation (H) leads to larger oscillations of $m(t)$. These deficiencies of the mean-field results particularly increase for larger interactions. On the contrary, the second Born approximation (also for larger U) correctly captures the time scale on which the order parameter relaxes towards $m = 0$, see, e.g., Fig. 5f. Furthermore, the inclusion of non-local parts of the second-order Born self-energy (see the lines labeled 2Bij in comparison to 2Bii) leads to results which agree better with the exact data. This is of course expected because a DMFT approximation generally performs inadequately in low-dimensional systems.

In Fig. 6 we show the simulations for the closed chain. Here, the oscillations of the Hartree results are even more dominant for larger values of the interaction ($U \gtrsim 1.5$), and correlations are essential to describe the dynamics of the magnetization. It is not our intention, in this paper, to discuss the physics behind the melting process. The time evolution is basically driven by the fact that local magnetic moments decay along with the long-range order at sufficiently small interactions, whereas in the Mott insulating regime (large U) energy is transferred from charge excitations to the spin background while local moments persist [38]. By means of nonequilibrium DMFT simulations for the Bethe lattice it has further been shown in Ref. [38] that the dynamics at weak interaction is governed by residual quasiparticles, which generate oscillations in the offdiagonal components of the momentum distribution.

In both types of systems, we finally observe that the conservation of particles, which is an essential test for the numerical implementation, is not destroyed when the bath parameters $K^{i\to j}_{s\sigma}(t)$ and $K^{j\to i}_{s\sigma}(t)$ are *not* taken into account in the decomposition of the diagonal parts of the self-energy (Eq. (26)). Instead, the corresponding time evolution of the magnetization (see the green dotted lines) is relatively close to the full second-order Born approximation where these terms are included, compare with the red dash-dotted lines. We can therefore identify another time propagation scheme where the diagonal and offdiagonal parts of the self-energy are decomposed independently. This means an additional simplification of the method, because

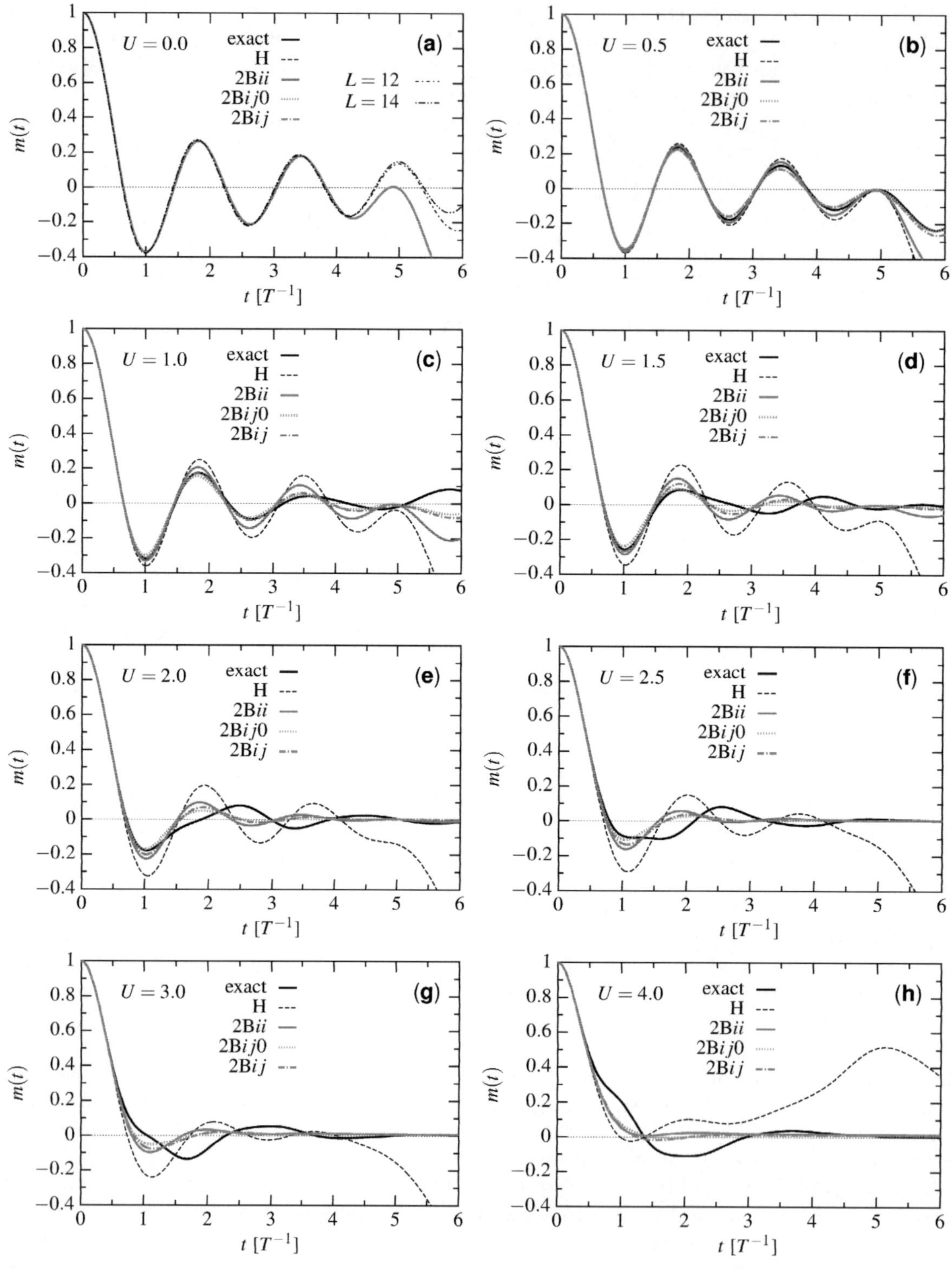

Figure 5. (a)-(h): Self-consistent results for the decay of the magnetization $m(t)$ in an open 10-site Hubbard chain with nearest-neighbor hopping T and on-site interaction U, which ranges from zero to 4.0. While the black dashed (solid) line shows the Hartree (exact) result, the colored lines correspond to different levels of the second-order Born approximation, recall Fig. 3 in Sec. 2.2.2. In panel (a), we also include exact results for a chain with $L = 12$ and 14 sites.

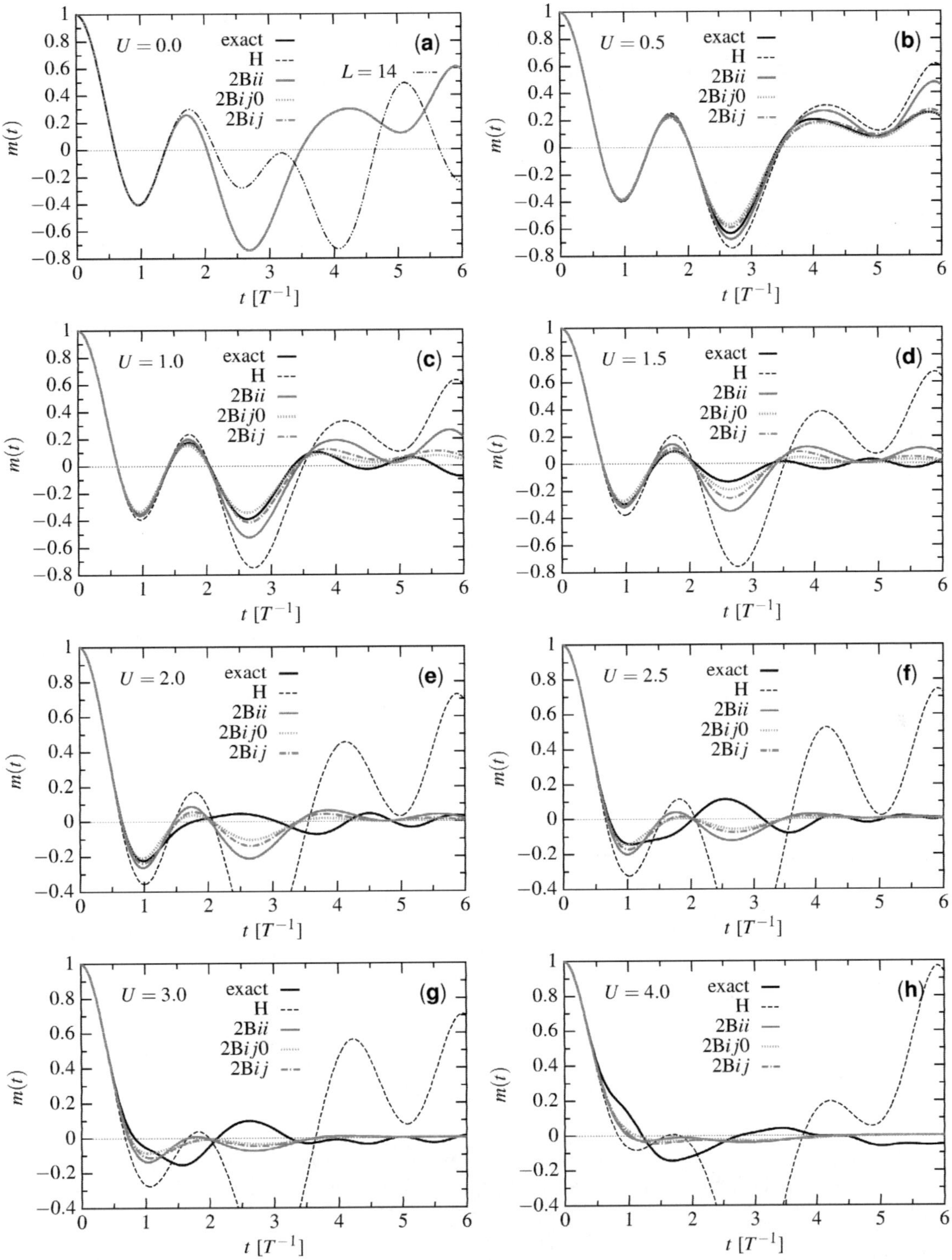

Figure 6. Same as in Fig. 5 but for a closed Hubbard chain, i.e., a ring with 10 sites where fermions can additionally hop between sites 1 and 10.

the associated matrices become Hermitian and positive definite again. However, we note that one still looses the causal structure in the time propagation as a singular value decomposition is required to treat Eq. (25).

4. Conclusion

The intriguing possibility to map *any* KBE with a given self-energy onto a noninteracting auxiliary system means that there exists a general alternative scheme to solve two-time quantum kinetic equations. This scheme does not at all require the evaluation of the memory kernel and can bypass the n_t^3-scaling by representing the two-time self-energies in terms of their most important "modes". These modes appear as time-dependent couplings to bath orbitals which are connected to either one or two (lattice) coordinates and thus mimic the retardation effects of the interacting system which are described by the local and non-local parts of the self-energy. In the course of this, all the properties which come with the self-energy are retained by the scheme, i.e., in particular, a Φ-derivable approximation automatically leads to an auxiliary system in which particle number, energy and momentum are preserved. In general, the Hamiltonian representation can exploit its full potential when the calculation of the self-energy scales essentially better than n_t^3 which is the case for the second Born approximation. However, the method should be efficiently applicable also for some more advanced self-energies, e.g., for the GW approximation, where a similar two-time decomposition of the polarization could be used to self-consistently determine the screened interaction.

Acknowledgements

We thank E. Arigoni and C. Verdozzi for stimulating discussions and acknowledge M. Eckstein for carefully reading the manuscript.

References

[1] Kadanoff L P and Baym G 1962 Quantum Statistical Mechanics (W.A. Benjamin, Inc. New York)
[2] Keldysh L V 1964 *Zh. Eksp. Teor. Fiz.* **47** 1515 [1965 *Sov. Phys. JETP* **20** 1018]
[3] Stefanucci G and van Leeuwen R 2013 Nonequilibrium Many-Body Theory of Quantum Systems: A Modern Introduction (Cambridge University Press, Cambridge)
[4] Danielewicz P 1984 *Annals of Phys.* **152** 239
[5] Danielewicz P 1984 *Annals of Phys.* **152** 305
[6] Köhler H S and Morawetz K 2001 *Phys. Rev. C* **64** 024613
[7] Kwong N-H and Bonitz M 2000 *Phys. Rev. Lett.* **84** 1768
[8] Kremp D, Schlanges M, and Kraeft W-D 2005 Quantum Statistics of Nonideal Plasmas (Springer, Berlin)
[9] Kwong N-H, Bonitz M, Binder R, and Köhler H S 1998 *phys. stat. sol. (b)* **206** 197
[10] Gartner P, Bányai L, and Haug H 1999 *Phys. Rev. B* **60** 14234
[11] Lorke M, Nielsen T R, Seebeck J, Gartner P, and Jahnke F 2006 *Phys. Rev. B* **73** 085324
[12] Freericks J K, Turkowski V M, and Zlatić V 2006 *Phys. Rev. Lett.* **97** 266408
[13] Aoki H, Tsuji N, Eckstein M, Kollar M, Oka T, and Werner P 2014 *Rev. Mod. Phys.* **86** 779
[14] Dahlen N E and van Leeuwen R 2007 *Phys. Rev. Lett.* **98** 153004
[15] Perfetto E, Uimonen A-M, van Leeuwen R, and Stefanucci G 2015 *Phys. Rev. A* **92** 033419
[16] Morawetz K and Köhler H S 1999 *Eur. Phys. J. A* **4** 291
[17] Kremp D, Bornath T, Bonitz M, Kraeft W D, and Schlanges M 2000 *Phys. Plasmas* **7** 59
[18] Köhler H S, Kwong N H, and Yousif H A 1999 *Comp. Phys. Comm.* **123** 123
[19] Rios A, Barker B, Buchler M, and Danielewicz P 2011 *Annals of Phys.* **326** 1274
[20] Hermanns S, Balzer K, and Bonitz M 2012 *Physica Scripta* **T151** 014036
[21] Semkat D, Kremp D, and Bonitz M 1999 *Phys. Rev. E* **59** 1557
[22] Morozov V G and Röpke G 1999 *Annals of Phys.* **278** 127
[23] Dahlen N E and van Leeuwen R 2005 *J. Chem. Phys.* **122** 164102
[24] Stan A, Dahlen N E, and van Leeuwen R 2009 *J. Chem. Phys.* **130** 224101
[25] Balzer K, Bauch S, and Bonitz M 2010 *Phys. Rev. A* **81** 022510
[26] Garny M and Müller M M 2010 High Performance Computing in Science and Engineering, Garching/Munich 2009 (Springer, Berlin)

[27] Balzer K, Bauch S, and Bonitz M 2010 *Phys. Rev. A* **82** 033427
[28] Balzer K and Bonitz M 2013 *Lect. Notes Phys.* **867**
[29] Lipavský P, Špička V, and Velický B 1986 *Phys. Rev. B* **34** 6933
[30] Hermanns S, Balzer K, and Bonitz M 2013 *J. Phys.: Conf. Ser.* **427** 012008
[31] Latini S, Perfetto E, Uimonen A-M, van Leeuwen R, and Stefanucci G 2014 *Phys. Rev. B* **89** 075306
[32] Balzer K, Hermanns S, and Bonitz M 2013 *J. Phys.: Conf. Ser.* **427** 012006
[33] Hermanns S, Schlünzen N, and Bonitz M 2014 *Phys. Rev. B* **90** 125111
[34] Freericks J K 2008 *Phys. Rev. B* **77** 075109
[35] Gramsch C, Balzer K, Eckstein M, and Kollar M 2013 *Phys. Rev. B* **88** 235106
[36] Wolf F A, McCulloch I P, and Schollwöck U 2014 *Phys. Rev. B* **90** 235131
[37] Balzer K, Li Z, Vendrell O, and Eckstein M 2015 *Phys. Rev. B* **91** 045136
[38] Balzer K, Wolf F A, McCulloch I P, Werner P, and Eckstein M 2015 *Phys. Rev. X* **5** 031039
[39] Balzer K and Eckstein M 2014 *Phys. Rev. B* **89** 035148
[40] Gramsch C and Potthoff M 2015 *Phys. Rev. B* **92** 235135
[41] Hochbruck M and Lubich C 1997 *SIAM J. Numer. Anal.* **34** 1911
[42] Jafari S A 2008 *Iranian J. Phys. Res.* **8** 113
[43] Carrascal D J, Ferrer J, Smith J C, and Burke K 2015 *J. Phys.: Cond. Mat.* **27** 393001
[44] Negele J W and Orland H 1998 Quantum Many-Particle Systems (Westview Press, Boulder, CO)
[45] Bauer A, Dorfner F, and Heidrich-Meisner F 2015 *Phys. Rev. A* **91** 053628
[46] Barmettler P, Punk M, Gritsev V, Demler E, and Altman E 2009 *Phys. Rev. Lett.* **102** 130603

Journal of Physics: Conference Series **696** (2016) 012002 doi:10.1088/1742-6596/696/1/012002

Nonequilibrium variational-cluster approach to real-time dynamics in the Fermi-Hubbard model

Felix Hofmann[1], Martin Eckstein[2] and Michael Potthoff[1]

[1] I. Institut für Theoretische Physik, Universität Hamburg, Jungiusstraße 9, 20355 Hamburg, Germany

[2] Max Planck Research Department for Structural Dynamics at the University of Hamburg, CFEL, Notkestraße 85, 22607 Hamburg, Germany

E-mail: fhofmann@physik.uni-hamburg.de

Abstract. The nonequilibrium variational-cluster approach is applied to study the real-time dynamics of the double occupancy in the one-dimensional Fermi-Hubbard model after different fast changes of hopping parameters. A simple reference system, consisting of isolated Hubbard dimers, is used to discuss different aspects of the numerical implementation of the approach in the general framework of nonequilibrium self-energy functional theory. Opposed to a direct solution of the Euler equation, its time derivative is found to serve as numerically tractable and stable conditional equation to fix the time-dependent variational parameters.

1. Introduction

Over the past decade, considerable progress in the field of ultracold gases has given access to experimentally simulating prototypical many-body models, e.g., known from condensed-matter physics, with a high degree of dynamic control [1–3]. Accompanied by theoretical work, this opened up the possibility to study their emergent phases and nonequilibrium dynamics induced by fast changes of the model parameters [4,5]. As *the* paradigmatic model for strongly correlated fermions on a lattice, we consider the Fermi-Hubbard model [6–8]. Not only the bosonic variant [9–12] but also the Fermi-Hubbard model has been realized by recent optical-lattice experiments [13–15]. This promises to provide insight into fundamental questions related to the Mott transition, to collective order or even to non-Fermi-liquid physics.

Using standard notations, the Fermi-Hubbard Hamiltonian is given by

$$H(t) = \sum_{\langle ij \rangle,\sigma} T_{ij}(t) c^\dagger_{i\sigma} c_{j\sigma} + U(t) \sum_i n_{i\uparrow} n_{i\downarrow} \, , \tag{1}$$

where a fermion at site i and with spin projection $\sigma = \uparrow, \downarrow$ is annihilated (created) by $c^{(\dagger)}_{i\sigma}$, and where $n_{i\sigma} = c^\dagger_{i\sigma} c_{i\sigma}$ is the density operator. Tunneling between neighboring sites $\langle ij \rangle$ is described by the hopping amplitude $T_{ij}(t)$. On the same site, fermions are subjected to the repulsive Hubbard interaction $U(t)$. A nonequilibrium state and nontrivial real-time dynamics of observables can be initiated by fast changes of the time-dependent model parameters $T_{ij}(t)$ and $U(t)$.

Despite its conceptual simplicity, the Hubbard model has challenged theoreticians for more than five decades, and has stimulated the development of a large number of different methods.

Many approaches which are based on the language of Green's functions [16] have successfully been extended to describe real-time phenomena [17–25], using the Keldysh formalism [26]. Of particular interest in this context are quantum-cluster theories [27], which are formulated in the thermodynamic limit, but nevertheless account for nonlocal correlations on a length scale defined by the linear extension of a reference cluster, which can have a profound influence on the nonequilibrium dynamics [28, 29]. A simple but conceptually appealing quantum-cluster approach is the cluster-perturbation theory (CPT) [30–33]. Within the CPT, the infinite lattice is tiled into small clusters, which can be solved exactly by numerical means. This solution is then extended to the full lattice by infinite-order perturbation theory with respect to the inter-cluster hopping but neglecting vertex corrections [32]. This amounts to approximate the self-energy of the full model by the self-energy of the reference system of disconnected clusters. A major drawback of the CPT consists in the fact that, even for a given geometrical tiling of the lattice, the partitioning of the hopping part and hence the choice of the reference system, i.e., the starting point of the perturbative expansion is not unique. However, the nonuniqueness of the CPT construction can be turned into an advantage, if one can find a variational prescription for finding the *optimal* starting point for the cluster-perturbation theory. In fact, this is achieved with the variational-cluster approach (VCA) [34, 35] which is best understood in the general framework of self-energy functional theory [36, 37].

The main purpose of the present paper is to discuss different practical issues related to the application of the *nonequilibrium* VCA. While the main theoretical concept is essentially the same as for the equilibrium VCA, we demonstrate that the numerical implementation of the nonequilibrium variant of the approach is by far more complex and requires new techniques. As a proof of principle, we present a first numerical implementation of the nonequilibrium VCA, based on simple two-site reference clusters, and study two types of parameter quenches (or fast ramps) in the one-dimensional Fermi-Hubbard model with alternating (dimerized) hopping amplitudes which have attracted experimental attention recently [38, 39].

The paper is organized as follows: In Section 2 we briefly review the concept of nonequilibrium Green's functions used for the nonequilibrium extension of the self-energy functional theory, as described in Sec. 3. We derive the central Euler equation in Sec. 4 and discuss its numerical implementation in Sec. 5. Results for the dimerized Hubbard model are presented in Sec. 6. A summary is given in Sec. 7.

2. Nonequilibrium Green's functions

The self-energy functional theory relies on functionals that are formally defined by means of all-order perturbation theory and thus on the concept of (nonequilibrium) Green's functions [16,26]. Throughout this paper, we make use of the general theory provided by Refs. [40] but in first place closely follow the formal setup by Wagner [41].

For a general fermionic lattice model with Hamiltonian

$$H_{\boldsymbol{T},\boldsymbol{U}}(t) = \sum_{\alpha\beta} T_{\alpha\beta}(t) c_\alpha^\dagger c_\beta + \frac{1}{2} \sum_{\alpha\beta\gamma\delta} U_{\alpha\beta\delta\gamma}(t) c_\alpha^\dagger c_\beta^\dagger c_\gamma c_\delta \,, \tag{2}$$

which is specified by the parameters $\boldsymbol{T}$ and $\boldsymbol{U}$, and which initially (at time $t = t_0$) is prepared in a thermal state with inverse temperature β and chemical potential μ, the elements of the contour-ordered Green's function $\boldsymbol{G_{T,U}}$ are defined as

$$iG_{\boldsymbol{T},\boldsymbol{U}}(1,2) = \left\langle \mathcal{T}_\mathcal{C} c_\mathcal{H}(1) c_\mathcal{H}^\dagger(2) \right\rangle_{\boldsymbol{T},\boldsymbol{U}} \,. \tag{3}$$

Here, greek indices combine all one-particle orbitals, and $c_\mathcal{H}^{(\dagger)}(i)$ is the annihilator (creator) in the Heisenberg picture with respect to $\mathcal{H}(z) = H(z) - \mu N$, where N is the total particle-number

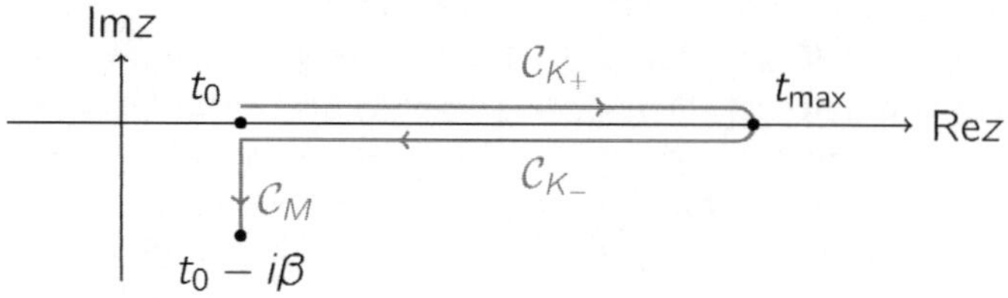

Figure 1. The three-branched Keldysh-Matsubara contour $\mathcal{C}$ in the complex time plane, extending up to time $t_{\max}$; $\mathcal{C}_{K_\pm}$ denotes the upper/lower branch and $\mathcal{C}_M$ the Matsubara branch.

operator. We use the short-hand notation $i \equiv (\alpha_i, z_i)$, where z_i marks an arbitrary point on the Keldysh-Matsubara contour $\mathcal{C}$ (see Fig. 1). $\mathcal{T}_\mathcal{C}$ is the contour time ordering. Further, $\langle \cdots \rangle = \mathrm{tr}(\rho \cdots)$ denotes the expectation value in the initial state, where ρ is density operator, $\rho = \exp(-\beta \mathcal{H}_{\text{ini}})/Z$, and $Z = \mathrm{tr}\exp(-\beta \mathcal{H}_{\text{ini}})$ with $\mathcal{H}_{\text{ini}} = H(t_0) - \mu N$ is the partition function.

From the Heisenberg equation of motion for the annihilator, we find the equation of motion

$$(i\partial_z - (\boldsymbol{T}(z) - \mu))\, \boldsymbol{G}_{\boldsymbol{T},\boldsymbol{U}}(z, z') = \mathbf{1}\delta_\mathcal{C}(z, z') + (\boldsymbol{\Sigma}_{\boldsymbol{T},\boldsymbol{U}} \circ \boldsymbol{G}_{\boldsymbol{T},\boldsymbol{U}})\,(z, z')\,, \tag{4}$$

where $\delta_\mathcal{C}$ is the contour delta-function and $\boldsymbol{\Sigma}_{\boldsymbol{T},\boldsymbol{U}}$ is the self-energy. The circle $\circ$ on the right-hand side stands for an integration along $\mathcal{C}$ and an implicit summation over all orbital indices. By setting $\boldsymbol{U} = 0$ in Eq. (4), we obtain the inverse of the "free" Green's function $\boldsymbol{G}_{\boldsymbol{T},0}$ as

$$G^{-1}_{\boldsymbol{T},0;\alpha\alpha'}(z, z') = \delta_\mathcal{C}(z, z') \left(\delta_{\alpha\alpha'} i\partial_{z'} - (T_{\alpha\alpha'}(z') - \mu\delta_{\alpha\alpha'})\right)\,. \tag{5}$$

Using this, we can rewrite Eq. (4) as Dyson's equation:

$$\boldsymbol{G}_{\boldsymbol{T},\boldsymbol{U}} = \boldsymbol{G}_{\boldsymbol{T},0} + \boldsymbol{G}_{\boldsymbol{T},0} \circ \boldsymbol{\Sigma}_{\boldsymbol{T},\boldsymbol{U}} \circ \boldsymbol{G}_{\boldsymbol{T},\boldsymbol{U}}\,. \tag{6}$$

3. Self-energy functional theory

The (nonequilibrium) self-energy functional theory is based on the following functional of the nonequilibrium self-energy (see Refs. [37, 42] for details):

$$\widehat{\Omega}_{\boldsymbol{T},\boldsymbol{U}}[\boldsymbol{\Sigma}] = \frac{1}{\beta}\,\mathrm{Tr}\ln\left(\boldsymbol{G}^{-1}_{\boldsymbol{T},0} - \boldsymbol{\Sigma}\right)^{-1} + \widehat{F}_{\boldsymbol{U}}[\boldsymbol{\Sigma}]\,. \tag{7}$$

Here, the first term on the right-hand side is a simple and explicit functional of $\boldsymbol{\Sigma}$ and depends on the system's one-particle parameters $\boldsymbol{T}$. The second term is the Legendre transform of the Luttinger-Ward functional $\widehat{\Phi}_{\boldsymbol{U}}[\boldsymbol{G}]$ [43]. The latter can be defined by means of all-order perturbation theory or, nonperturbatively, within a path-integral formalism analogous to the equilibrium case [42]. Note, that functionals are indicated by a hat, and that the trace is defined as $\mathrm{Tr}\,\boldsymbol{A} = \sum_\alpha \int_\mathcal{C} dz\, A_{\alpha\alpha}(z, z^+)$, where z^+ is infinitesimally later than z on $\mathcal{C}$.

There are two important properties of the self-energy functional (7): First, if $\widehat{\Omega}_{\boldsymbol{T},\boldsymbol{U}}[\boldsymbol{\Sigma}]$ is evaluated at the physical self-energy $\boldsymbol{\Sigma} = \boldsymbol{\Sigma}_{\boldsymbol{T},\boldsymbol{U}}$, one obtains the physical grand potential of the system in its initial thermal state. All contributions from the upper and the lower Keldysh branch cancel in this case. Second, the self-energy functional is stationary at the physical self-energy, i.e.,

$$\left.\frac{\delta\widehat{\Omega}_{\boldsymbol{T},\boldsymbol{U}}[\boldsymbol{\Sigma}]}{\delta\boldsymbol{\Sigma}}\right|_{\boldsymbol{\Sigma}=\boldsymbol{\Sigma}_{\boldsymbol{T},\boldsymbol{U}}} = 0\,. \tag{8}$$

In fact, using the well-known properties of the Luttinger-Ward functional and its Legendre transform, it is easy to see that Eq. (8) is equivalent with Dyson's equation (6). Eq. (8) constitutes a general variational principle by which one could determine the self-energy (as

Journal of Physics: Conference Series **696** (2016) 012002 doi:10.1088/1742-6596/696/1/012002

well as the grand potential) of a system of correlated fermions. However, the explicit form of the functional $\widehat{F}_{\boldsymbol{U}}[\boldsymbol{\Sigma}]$ is generally unknown.

An important observation is that the Luttinger-Ward functional, and thus also $\widehat{F}_{\boldsymbol{U}}[\boldsymbol{\Sigma}]$ is universal, i.e., it depends (besides the argument) on the interaction parameters $\boldsymbol{U}$ only, as made explicit by the subscript. Hence, the functional form of $\widehat{F}_{\boldsymbol{U}}[\boldsymbol{\Sigma}]$ remains unchanged for any reference system with the same interaction $\boldsymbol{U}$ as the original system but with a different set of one-particle parameters $\boldsymbol{\lambda}'$, i.e., for any reference system with a Hamiltonian of the form $H' \equiv H_{\boldsymbol{\lambda}',\boldsymbol{U}}$.

Let us choose a reference system in a particular subspace of one-particle parameters $\boldsymbol{\lambda}'$, which has a sufficiently simple structure so that its one-particle Green's function, its self-energy, and its grand potential of the initial thermal state at β and μ can be computed exactly for any $\boldsymbol{\lambda}'$ of the subspace. (Here and in the following, primed quantities will refer to the reference system.) We can write down the self-energy functional of the reference system,

$$\widehat{\Omega}_{\boldsymbol{\lambda}',\boldsymbol{U}}[\boldsymbol{\Sigma}] = \frac{1}{\beta}\,\mathrm{Tr}\ln\left(\boldsymbol{G}^{-1}_{\boldsymbol{\lambda}',0} - \boldsymbol{\Sigma}\right)^{-1} + \widehat{F}_{\boldsymbol{U}}[\boldsymbol{\Sigma}]\,, \tag{9}$$

and, due to the mentioned universality, eliminate $\widehat{F}_{\boldsymbol{U}}[\boldsymbol{\Sigma}]$ by combining Eqs. (7) and (9):

$$\widehat{\Omega}_{\boldsymbol{T},\boldsymbol{U}}[\boldsymbol{\Sigma}] = \widehat{\Omega}_{\boldsymbol{\lambda}',\boldsymbol{U}}[\boldsymbol{\Sigma}] + \frac{1}{\beta}\,\mathrm{Tr}\ln\left(\boldsymbol{G}^{-1}_{\boldsymbol{T},0} - \boldsymbol{\Sigma}\right)^{-1} - \frac{1}{\beta}\,\mathrm{Tr}\ln\left(\boldsymbol{G}^{-1}_{\boldsymbol{\lambda}',0} - \boldsymbol{\Sigma}\right)^{-1}. \tag{10}$$

This expression for the self-energy functional, which is still exact, can be evaluated in practice for a certain subclass of trial self-energies, namely for the physical self-energies of the reference system. Inserting $\boldsymbol{\Sigma} = \boldsymbol{\Sigma}_{\boldsymbol{\lambda}',\boldsymbol{U}}$, we have

$$\widehat{\Omega}_{\boldsymbol{T},\boldsymbol{U}}[\boldsymbol{\Sigma}_{\boldsymbol{\lambda}',\boldsymbol{U}}] = \Omega_{\boldsymbol{\lambda}',\boldsymbol{U}} + \frac{1}{\beta}\,\mathrm{Tr}\ln\left(\boldsymbol{G}^{-1}_{\boldsymbol{T},0} - \boldsymbol{\Sigma}_{\boldsymbol{\lambda}',\boldsymbol{U}}\right)^{-1} - \frac{1}{\beta}\,\mathrm{Tr}\ln\left(\boldsymbol{G}_{\boldsymbol{\lambda}',\boldsymbol{U}}\right), \tag{11}$$

where we used $\widehat{\Omega}_{\boldsymbol{\lambda}',\boldsymbol{U}}[\boldsymbol{\Sigma}_{\boldsymbol{\lambda}',\boldsymbol{U}}] = \Omega_{\boldsymbol{\lambda}',\boldsymbol{U}}$ and Dyson's equation for the reference system, see Eq. (6). Using the stationarity principle Eq. (8), we can optimize the self-energy over the restricted subspace of trial self-energies spanned by the subspace of variational parameters $\boldsymbol{\lambda}'$ via

$$\left.\frac{\delta\widehat{\Omega}_{\boldsymbol{T},\boldsymbol{U}}[\boldsymbol{\Sigma}_{\boldsymbol{\lambda}',\boldsymbol{U}}]}{\delta\boldsymbol{\lambda}'(z)}\right|_{\boldsymbol{\lambda}'(z)=\boldsymbol{\lambda}'_{\mathrm{opt}}(z)} = 0\,. \tag{12}$$

This Euler equation yields the optimal parameters $\boldsymbol{\lambda}'_{\mathrm{opt}}(z)$, the optimal initial-state grand potential $\widehat{\Omega}_{\boldsymbol{T},\boldsymbol{U}}[\boldsymbol{\Sigma}_{\boldsymbol{\lambda}'_{\mathrm{opt}},\boldsymbol{U}}]$, the optimal self-energy $\boldsymbol{\Sigma}_{\boldsymbol{\lambda}'_{\mathrm{opt}},\boldsymbol{U}}$, and the related one-particle Green's function

$$\boldsymbol{G}^{\mathrm{SFT}} \equiv (\boldsymbol{G}^{-1}_{\boldsymbol{T},0} - \boldsymbol{\Sigma}_{\boldsymbol{\lambda}'_{\mathrm{opt}},\boldsymbol{U}})^{-1}\,. \tag{13}$$

The type of approximation is specified by the choice of the reference system. Typically, this is defined by cutting the lattice, which underlies the original model, into disconnected clusters consisting of a small number of sites L_c each. For Hubbard-type systems with local interactions, this generates simple reference systems with a small Hilbert space, which can be solved, e.g., by exact diagonalization techniques. The resulting approximation is called the variational-cluster approximation (VCA) [34, 35, 37]. A concrete example will be given in Sec. 6. To enlarge the number of variational degrees of freedom locally, a number L_b of uncorrelated "bath sites" can be coupled via a hybridization term to each of the correlated ones. One may formally re-derive the (nonequilibrium) dynamical mean-field theory (DMFT) [18, 19, 44] by choosing a reference system of decoupled correlated sites ($L_c = 1$) hybridizing with a continuum of bath sites ($L_b = \infty$). The reference system is given as a set of decoupled single-impurity Anderson models in this case. The VCA, the DMFT, and other approximations defined in this way are nonperturbative by construction and can be improved systematically by enlarging the space of variational parameters.

Self-energy functional theory in principle provides approximations which respect the macroscopic conservation laws resulting from the invariance of the original system under continuous U(1) and SU(2) symmetry groups. This is similar to the perturbatively defined "conserving" approaches in the sense of Baym and Kadanoff [45, 46] which are obtained from the standard recipe by deriving the self-energy from a truncated Luttinger-Ward functional. The conserving nature of the nonperturbative approximations generated within the SFT framework is ensured by a formal proof [37]. This makes use of the fact that variations of the one-particle parameters of the reference system comprise gauge transformations of the form $\boldsymbol{\lambda}' \mapsto \bar{\boldsymbol{\lambda}}'$ with

$$\boldsymbol{\varepsilon}'(z) \mapsto \bar{\boldsymbol{\varepsilon}}'(z) = \boldsymbol{\varepsilon}'(z) - \partial_z \boldsymbol{\chi}(z) \,, \qquad \boldsymbol{T}'(z) \mapsto \bar{\boldsymbol{T}}'(z) = e^{i\boldsymbol{\chi}(z)} \boldsymbol{T}'(z) e^{-i\boldsymbol{\chi}(z)} \,, \tag{14}$$

where $\boldsymbol{\varepsilon}'$ denotes the (spatially) diagonal part of $\boldsymbol{\lambda}'$ and $\boldsymbol{T}'$ its off-diagonal part, and where $\chi_{ij,\sigma\sigma'}(z)$ is a spatially diagonal contour function. One shows that stationarity with respect to those gauge transformations implies that conservation laws, as expressed by continuity equations, are proliferated from the reference system, where they hold exactly, to the original system where they are expressed in terms of the approximate quantities. In fact, this is unproblematic in practice for the case of the total particle number and of the z-component of the total spin. As discussed in Ref. [37], however, respecting total-energy conservation requires a continuum of variational degrees of freedom. In simple approximations, such as the VCA, one therefore has to tolerate a violation of total-energy conservation or use this as a measure for the quality of the approximation.

4. Euler equation

To proceed to a practical computation, one first has to note that the trial self-energy depends on variational parameters $\boldsymbol{\lambda}'(z)$ which can be different on the two Keldysh branches and we therefore distinguish $\boldsymbol{\lambda}'_+(t)$ and $\boldsymbol{\lambda}'_-(t)$ on $\mathcal{C}_{K_+}$ and $\mathcal{C}_{K_-}$ (cf. Fig. 1). In the end, we are only interested in physical solutions of the Euler equation (12), namely solutions which satisfy $\boldsymbol{\lambda}'_+(t) = \boldsymbol{\lambda}'_-(t)$ and hence specify a *physical* Hamiltonian. To make this more explicit, we perform a simple rotation in parameter space and define $\boldsymbol{\lambda}'_{\text{phys}}(t) = \frac{1}{2}(\boldsymbol{\lambda}'_+(t) + \boldsymbol{\lambda}'_-(t))$ and $\boldsymbol{\lambda}'_{\text{trans}}(t) = \frac{1}{2}(\boldsymbol{\lambda}'_+(t) - \boldsymbol{\lambda}'_-(t))$. The physical parameter manifold is given by $\boldsymbol{\lambda}'_{\text{trans}}(t) = 0$. Now, it is easy to see that the self-energy functional is trivially always stationary with respect to physical variations when evaluated at a physical parameter set:

$$\left. \frac{\delta \widehat{\Omega}_{\boldsymbol{T},\boldsymbol{U}}[\boldsymbol{\Sigma}_{\boldsymbol{\lambda}',\boldsymbol{U}}]}{\delta \boldsymbol{\lambda}'_{\text{phys}}(t)} \right|_{\boldsymbol{\lambda}'_{\text{trans}}(t)=0} = 0 \,. \tag{15}$$

Hence, stationarity with respect to physical variations simply places the solution *on* the physical manifold. To fix the solution *within* the physical manifold a second condition is needed, namely stationarity with respect to the transverse variations:

$$\left. \frac{\delta \widehat{\Omega}_{\boldsymbol{T},\boldsymbol{U}}[\boldsymbol{\Sigma}_{\boldsymbol{\lambda}',\boldsymbol{U}}]}{\delta \boldsymbol{\lambda}'_{\text{trans}}(t)} \right|_{\boldsymbol{\lambda}'_{\text{trans}}(t)=0} = 0 \,. \tag{16}$$

This is in fact the central equation of the nonequilibrium SFT. Hence, variations necessarily go off the physical manifold and thus, opposed to the standard strategy in the equilibrium case (see

Journal of Physics: Conference Series **696** (2016) 012002 doi:10.1088/1742-6596/696/1/012002

e.g. Ref. [47]), it is convenient to carry out the functional derivative analytically before solving the resulting equation numerically on the physical parameter manifold.

To compute the functional derivative in Eq. (12) or (16), we use the chain rule:

$$\frac{\delta\widehat{\Omega}_{\boldsymbol{T},\boldsymbol{U}}[\boldsymbol{\Sigma}_{\boldsymbol{\lambda}',\boldsymbol{U}}]}{\delta\lambda'_{\alpha_1\alpha_2}(z)} = \mathrm{Tr}\left(\frac{\delta\widehat{\Omega}_{\boldsymbol{T},\boldsymbol{U}}[\boldsymbol{\Sigma}_{\boldsymbol{\lambda}',\boldsymbol{U}}]}{\delta\boldsymbol{\Sigma}_{\boldsymbol{\lambda}',\boldsymbol{U}}} \circ \frac{\delta\boldsymbol{\Sigma}_{\boldsymbol{\lambda}',\boldsymbol{U}}}{\delta\lambda'_{\alpha_1\alpha_2}(z)}\right). \tag{17}$$

The first factor is given by $\beta\delta\widehat{\Omega}_{\boldsymbol{T},\boldsymbol{U}}[\boldsymbol{\Sigma}_{\boldsymbol{\lambda}',\boldsymbol{U}}]/\delta\boldsymbol{\Sigma}_{\boldsymbol{\lambda}',\boldsymbol{U}} = \left(\boldsymbol{G}_{\boldsymbol{T},0}^{-1} - \boldsymbol{\Sigma}_{\boldsymbol{\lambda}',\boldsymbol{U}}\right)^{-1} - \boldsymbol{G}_{\boldsymbol{\lambda}',\boldsymbol{U}}$ and can be rewritten in a more convenient way. With the inter-cluster hopping $\boldsymbol{V}(z) \equiv \boldsymbol{T}(z) - \boldsymbol{\lambda}'(z)$ we immediately have $G_{\boldsymbol{T},0}^{-1}(1,2) = G_{\boldsymbol{\lambda}',0}^{-1}(1,2) - \delta_{\mathcal{C}}(z_1,z_2)V_{\alpha_1\alpha_2}(z_2)$, see Eq. (5). Plugging this into the definition of the SFT Green's function, Eq. (13), and using Dyson's equation for the reference system, we get

$$\boldsymbol{G}^{\mathrm{SFT}} = \boldsymbol{G}_{\boldsymbol{\lambda}',\boldsymbol{U}} + \boldsymbol{G}_{\boldsymbol{\lambda}',\boldsymbol{U}}\boldsymbol{V} \circ \boldsymbol{G}^{\mathrm{SFT}}. \tag{18}$$

This is actually the central equation of nonequilibrium cluster-perturbation theory [32]. With the associated Lippmann-Schwinger equation $\boldsymbol{G}^{\mathrm{SFT}} = \boldsymbol{G}_{\boldsymbol{\lambda}',\boldsymbol{U}} + \boldsymbol{G}_{\boldsymbol{\lambda}',\boldsymbol{U}} \circ \boldsymbol{Y}_{\boldsymbol{\lambda}',\boldsymbol{T},\boldsymbol{U}} \circ \boldsymbol{G}_{\boldsymbol{\lambda}',\boldsymbol{U}}$, where $\boldsymbol{Y}_{\boldsymbol{\lambda}',\boldsymbol{T},\boldsymbol{U}}(z_1,z_2) = \boldsymbol{V}(z_1)\delta_{\mathcal{C}}(z_1,z_2) + \boldsymbol{V}(z_1)\boldsymbol{G}^{\mathrm{SFT}}(z_1,z_2)\boldsymbol{V}(z_2)$ is the corresponding T-matrix, we eventually obtain

$$\frac{\delta\widehat{\Omega}_{\boldsymbol{T},\boldsymbol{U}}[\boldsymbol{\Sigma}_{\boldsymbol{\lambda}',\boldsymbol{U}}]}{\delta\boldsymbol{\Sigma}_{\boldsymbol{\lambda}',\boldsymbol{U}}} = \frac{1}{\beta}\boldsymbol{G}_{\boldsymbol{\lambda}',\boldsymbol{U}} \circ \boldsymbol{Y}_{\boldsymbol{\lambda}',\boldsymbol{T},\boldsymbol{U}} \circ \boldsymbol{G}_{\boldsymbol{\lambda}',\boldsymbol{U}} \tag{19}$$

for the first factor in Eq. (17). To evaluate the second factor, we write $\boldsymbol{\Sigma}_{\boldsymbol{\lambda}',\boldsymbol{U}} = \boldsymbol{G}_{\boldsymbol{\lambda}',0}^{-1} - \boldsymbol{G}_{\boldsymbol{\lambda}',\boldsymbol{U}}^{-1}$. Apart from the $\boldsymbol{\lambda}'$-dependence of the inverse free Green's function, see Eq. (5), we have to consider the derivative of the inverse of the full Green's function: From the identity $\delta(\boldsymbol{G}_{\boldsymbol{\lambda}',\boldsymbol{U}} \circ \boldsymbol{G}_{\boldsymbol{\lambda}',\boldsymbol{U}}^{-1})/\delta\boldsymbol{\lambda}' = 0$ we deduce $\delta\boldsymbol{G}_{\boldsymbol{\lambda}',\boldsymbol{U}}^{-1}/\delta\boldsymbol{\lambda}' = -\boldsymbol{G}_{\boldsymbol{\lambda}',\boldsymbol{U}}^{-1} \circ \delta\boldsymbol{G}_{\boldsymbol{\lambda}',\boldsymbol{U}}/\delta\boldsymbol{\lambda}' \circ \boldsymbol{G}_{\boldsymbol{\lambda}',\boldsymbol{U}}^{-1}$. Finally, making use of the properties of the time ordering operator and the exponential function, a straightforward calculation yields:

$$\frac{\delta G_{\boldsymbol{\lambda}',\boldsymbol{U}}(3,4)}{\delta\lambda'_{\alpha_1\alpha_2}(z_1)} = G_{\boldsymbol{\lambda}',\boldsymbol{U}}(3,4)\, G_{\boldsymbol{\lambda}',\boldsymbol{U}}(2,1^+)\big|_{z_2=z_1} - G^{(2)}_{\boldsymbol{\lambda}',\boldsymbol{U}}(3,2,1^+,4)\Big|_{z_2=z_1}, \tag{20}$$

where $\boldsymbol{G}^{(2)}_{\boldsymbol{\lambda}',\boldsymbol{U}}$ is the two-particle Green's function of the reference system. Collecting the results we find:

$$-\beta\frac{\delta\widehat{\Omega}_{\boldsymbol{T},\boldsymbol{U}}[\boldsymbol{\Sigma}_{\boldsymbol{\lambda}',\boldsymbol{U}}]}{\delta\lambda'_{\alpha_1\alpha_2}(z_1)} = \iint d3d4\, Y_{\boldsymbol{\lambda}',\boldsymbol{T},\boldsymbol{U}}(4,3)\; L_{\boldsymbol{\lambda}',\boldsymbol{U}}(3,2,1^+,4)\big|_{z_2=z_1} =: K^{(0)}[\boldsymbol{\lambda}']_{\alpha_2\alpha_1}(z_1), \tag{21}$$

where $L_{\boldsymbol{\lambda}',\boldsymbol{U}}(1,2,3,4) = G_{\boldsymbol{\lambda}',\boldsymbol{U}}(2,4)G_{\boldsymbol{\lambda}',\boldsymbol{U}}(1,3) - G_{\boldsymbol{\lambda}',\boldsymbol{U}}(1,4)G_{\boldsymbol{\lambda}',\boldsymbol{U}}(2,3) + G^{(2)}_{\boldsymbol{\lambda}',\boldsymbol{U}}(1,2,3,4)$ is the two-particle (four-point) vertex function with external legs. Therewith, we have the Euler equation of the nonequilibrium SFT:

$$\boldsymbol{K}^{(0)}[\boldsymbol{\lambda}'](t)\Big|_{\boldsymbol{\lambda}'=\boldsymbol{\lambda}'_{\mathrm{opt}}} = 0, \tag{22}$$

which will be the starting point for the numerical determination of the optimal parameters. Note that those parameters depend on the real time variable t only rather than on the contour variable z, because after the transverse variations have been carried out analytically we can restrict the search for optimal parameters to the physical manifold, see Eq. (16).

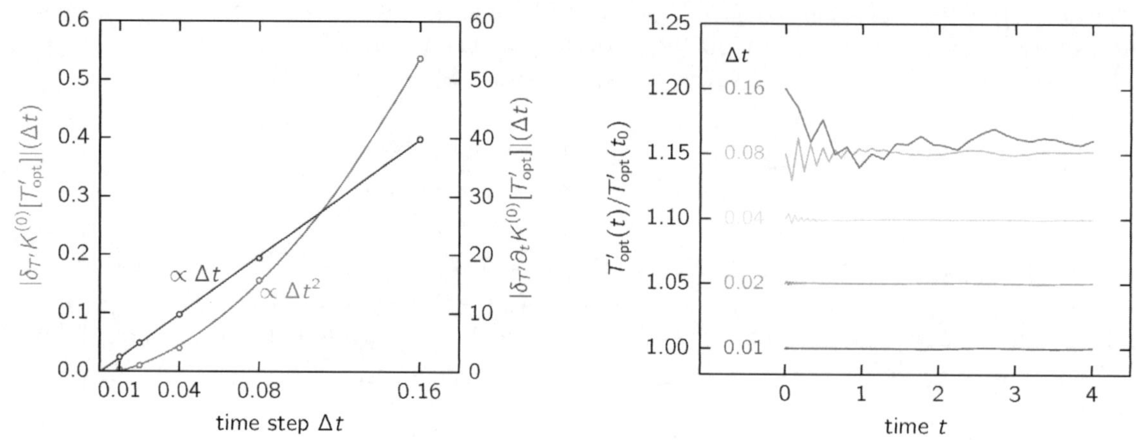

Figure 2. Test of the numerical implementation for the time-propapagation of a system in equilibrium. The system is a one-dimensional chain of 100 sites at $\beta = 6$ and $U = 4$. As a reference system a two-site cluster is used. *Left*: Δt dependence of the Jacobian of $\boldsymbol{K}^{(0)}[\boldsymbol{\lambda}'](t)$ (red, left ordinate axis) and of $\partial_t \boldsymbol{K}^{(0)}[\boldsymbol{\lambda}'](t)$ (blue, right ordinate axis). Note, that the ordinate axes' scales differ by two orders of magnitude. Results are independent of the time t. *Right*: Time dependence of the optimal parameters. VCA results obtained as the roots of $\partial_t \boldsymbol{K}^{(0)}[\boldsymbol{\lambda}'](t) = 0$ [Eq. (24)] for different Δt. The ordinate axis refers to the result for $\Delta t = 0.01$; results for larger Δt are constantly shifted by multiples of 0.05.

5. Numerical implementation

With the *time-dependent* Euler equation (22) we are facing a profound root-finding problem of formally infinite dimensions. However, the SFT variational principle is inherently causal which allows us to determine the optimal parameters at a given time on a discrete time grid, without affecting the results at previous grid times, and then proceed to the next time. This causality is most easily seen when discussing Eq. (21): The integrals over z_3 and z_4 extend over the entire contour $\mathcal{C}$ but can be confined to times which are (physically) earlier than $t_{\max} = t_1$ (cf. Fig. 1). If one or both times are located beyond t_1, the later time can be shifted from the upper to the lower branch (or vice versa) without altering the contour ordering, and the respective contributions to the integral cancel. Thus, at time t_1, all quantities in the Euler equation (22) and therewith the parameter $\boldsymbol{\lambda}'_{\rm opt}(t_1)$ itself depend on parameters at earlier times only.

Unfortunately, a straightforward numerical solution of Eq. (22) turns out to be impossible. The optimal parameters, determined by standard root-finding techniques, quickly accumulate a large error after a few time steps only such that a reasonable solution cannot be found in this way. Generally, for a functional depending parametrically on time, which is evaluated on a finite time grid and the parameters of which are varied only at the very last instant of time, we expect the explicit dependence of the variation of the functional on the time step Δt to scale as $(\Delta t)^n$ with $n \geq 1$, since for an infinitesimally fine grid variations at a single instant of time would reduce to variations on a null set, and thus the variation of the functional must vanish.

In the present case, if one aims to solve Eq. (22) at a given time t, keeping the solution $\boldsymbol{\lambda}'_{\rm opt}$ fixed at earlier times, the variation $\delta_{\boldsymbol{\lambda}'(t)} \boldsymbol{K}^{(0)}[\boldsymbol{\lambda}'_{\rm opt}](t)$ of the parameters at the last time-step will scale as $(\Delta t)^2$, if the whole implementation is based on an equidistant time-grid Δt. This is verified numerically in Fig. 2 (left) for a special case (described in the figure caption). This scaling is independent of the quadrature rules used in solving the time integrals (if the accuracy of those scales as $(\Delta t)^2$ or better). Hence the observed $(\Delta t)^2$ scaling is in fact due to the Jacobian

$\delta_{\boldsymbol{\lambda}'(t)}\boldsymbol{K}^{(0)}[\boldsymbol{\lambda}'_{\rm opt}](t)$ itself and turned out to be detrimental for the stability of the algorithm for practically reasonable choices of Δt.

Fortunately, this problem can be overcome by requiring the *time derivative* of $\boldsymbol{K}^{(0)}[\boldsymbol{\lambda}'](t)$ [see Eq. (21)] to vanish for all times $t > t_0$ and by fixing the initial conditions at t_0 by Eq. (22). Hence, instead of solving the Euler equation (22) for all times, we rather consider

$$\boldsymbol{K}^{(0)}[\boldsymbol{\lambda}'](t)\Big|_{\boldsymbol{\lambda}'=\boldsymbol{\lambda}'_{\rm opt}} = 0\,, \qquad \text{for } t = t_0\,, \tag{23a}$$

$$\partial_t\boldsymbol{K}^{(0)}[\boldsymbol{\lambda}'](t)\Big|_{\boldsymbol{\lambda}'=\boldsymbol{\lambda}'_{\rm opt}} = 0\,, \qquad \text{for } t > t_0\,. \tag{23b}$$

As can be seen in Fig. 2 (left), the dependence on Δt is only linear in this case. This greatly improves the accuracy of the parameter optimization and allows us to trace the optimal variational parameters as a function of time. As a simple numerical check, one may consider the *equilibrium* problem. The expected time *in*dependence of the optimal variational parameters is indeed found for sufficiently small Δt, see Fig. 2 (right).

According to Eq. (21), the time derivative $\partial_t\boldsymbol{K}^{(0)}[\boldsymbol{\lambda}'](t)$ can be obtained from the corresponding equations of motion for the four-point vertex function $\boldsymbol{L}_{\boldsymbol{\lambda}',\boldsymbol{U}}$. There are two qualitatively different dependencies on t: The first one results from the boundary terms due to the time-ordering operator comprised by all Green's functions. Contributions from this one cancel out when taking the time derivative. The second one is the explicit dependence of the annihilator and the creator on the external time t_1. This is governed by the Heisenberg equation of motion. Commuting the operators with the one-particle part of the Hamiltonian simply results in matrix products with $\boldsymbol{\lambda}'$ while commuting with the interacting part gives rise to higher-order products of annihilators and creators which we denote by ψ or $\psi^\dagger$, respectively: $\left[c(1), H'_1(1)\right] \equiv \psi(1)$ and $\left[H'_1(1), c^\dagger(1)\right] \equiv \psi^\dagger(1)$. Thus, the time derivative of $\boldsymbol{K}^{(0)}[\boldsymbol{\lambda}'](t)$ acquires the form:

$$\partial_t\boldsymbol{K}^{(0)}[\boldsymbol{\lambda}'](t) = \left[\boldsymbol{K}^{(0)}[\boldsymbol{\lambda}'](t), \boldsymbol{\lambda}'(t)\right] + \boldsymbol{K}^{(1)}[\boldsymbol{\lambda}'](t)\,, \tag{24}$$

where we have used Eq. (21) and where

$$K^{(1)}[\boldsymbol{\lambda}']_{\alpha_2\alpha_1}(z_1) = \iint d3d4\; Y_{\boldsymbol{\lambda}',\boldsymbol{T},\boldsymbol{U}}(4,3) M_{\boldsymbol{\lambda}',\boldsymbol{U}}(3,2,1^+,4)\big|_{z_2=z_1}\,, \tag{25}$$

with

$$M_{\boldsymbol{\lambda}',\boldsymbol{U}}(3,2,1^+,4) = L_{\boldsymbol{\lambda}',\boldsymbol{U}}(3,2,1^+_\psi,4) - L_{\boldsymbol{\lambda}',\boldsymbol{U}}(3,2_\psi,1^+,4)\,. \tag{26}$$

The subscript ψ indicates that c and $c^\dagger$ are replaced by ψ and $\psi^\dagger$ in the respective correlation function. For example, $iG_{\boldsymbol{\lambda}',\boldsymbol{U}}(1_\psi,2) = \langle \mathcal{T}_{\mathcal{C}}\psi(1)c^\dagger(2)\rangle_{\boldsymbol{\lambda}',\boldsymbol{U}}$.

The algorithm for the numerical implementation of the VCA in the general nonequilibrium case [Eq. (23)] is sketched in Fig. 3. The following steps are performed at each time step:

(i) At a given time t and for certain guessed parameters $\boldsymbol{\lambda}'_{\rm g}(t)$, we propagate $c^\dagger_\alpha(t)$ and $c_\alpha(t)$ for all relevant orbitals α by a time step Δt and store their respective representations in the occupation-number basis. For small cluster sizes, this is straightforward. Therewith, arbitrary correlation functions can be calculated for arbitrary times on the contour up to the time t. While the single particle Green's function $\boldsymbol{G}'$ can be updated from time step to time step and kept in the storage, the higher correlation functions $\boldsymbol{L}'$ and $\boldsymbol{M}'$ have to be recalculated on-the-fly for *any* time step and after any update of $\boldsymbol{\lambda}'_{\rm g}(t)$, because of the external time t. Symmetries can be exploited to reduce the actual number of elements that have to be calculated.

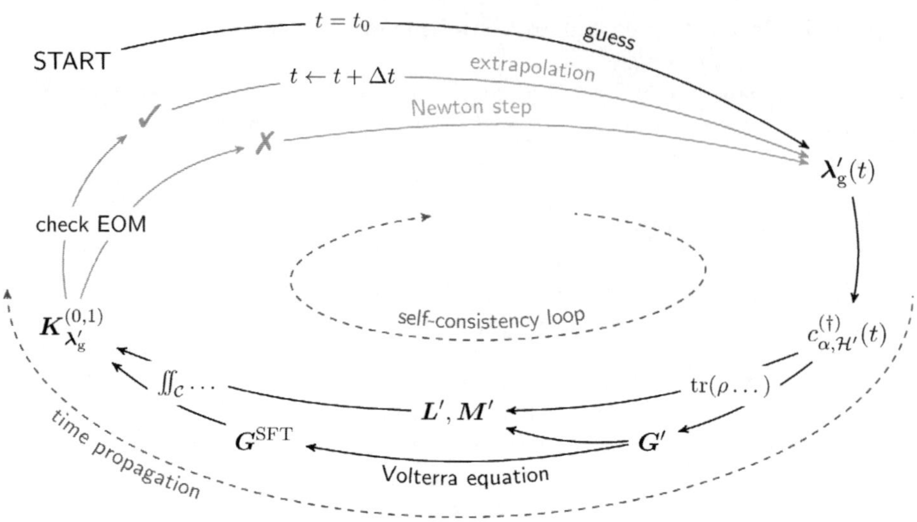

Figure 3. Sketch of the propagation algorithm for the optimal parameters. See text for discussion.

(ii) The SFT Green's function $\boldsymbol{G}^{\rm SFT}$ is obtained from the CPT equation (18), which can be cast in the form of a Volterra equation of second kind: $(\mathbf{1} - \boldsymbol{G}_{\boldsymbol{\lambda}',\boldsymbol{U}}\boldsymbol{V}) \circ \boldsymbol{G}^{\rm SFT} = \boldsymbol{G}_{\boldsymbol{\lambda}',\boldsymbol{U}}$. The latter is solved up to time t by standard techniques (cf. Refs. [48–50]). Here, we exploit the translational symmetries of $\boldsymbol{V}$ and of $\boldsymbol{G}^{\rm SFT}$ by Fourier transformation with respect to the super-lattice.

(iii) The correlation functions $\boldsymbol{L}'$, $\boldsymbol{M}'$ and $\boldsymbol{G}^{\rm SFT}$, are then used to calculate both $\boldsymbol{K}^{(0)}[\boldsymbol{\lambda}'](t)$ and $\boldsymbol{K}^{(1)}[\boldsymbol{\lambda}'](t)$ via Eqs. (21) and (25). For the initial time t_0, only those integrations contribute where both times are on the Matsubara branch. For any later time, the mixed and Keldysh integrations have to be carried out, too. Results for the integrals from earlier time steps cannot be recycled, due to the external time t appearing in the correlation functions $\boldsymbol{L}'$ and $\boldsymbol{M}'$. For both, the integrations involved in the Euler equation (23) as well as in the Volterra equation, we use high-order integration schemes, like the Newton-Cotes rules or the Gregory rules [50, 51], to allow for large $\Delta\tau$ and large Δt steps [52].

(iv) Generally, the initial guess $\boldsymbol{\lambda}'_{\rm g}(t)$ will not be a root of Eqs. (23), and must therefore be updated. The standard way of doing this, is to apply Newton's method. This, however, requires the knowledge of the (inverse of the) functional's Jacobian $\delta_{\boldsymbol{\lambda}'(t_0)}\boldsymbol{K}^{(0)}[\boldsymbol{\lambda}'](t_0)$ or $\delta_{\boldsymbol{\lambda}'(t)}\partial_t\boldsymbol{K}^{(0)}[\boldsymbol{\lambda}'](t)$ respectively. Both, the implementation of an analytical expression for the Jacobian or a direct numerical evaluation via finite difference methods are rather costly options and thus not feasible. We thus employ Broyden's method [53, 54], which provides increasingly improved updates for the (inverse of the) Jacobian during the course of the Newton iteration, starting from an initial guess for both, the root as well as the Jacobian itself. However, at the initial time we evaluate the Jacobian numerically by finite differences at some guessed parameter. This improves the success and the speed of the method significantly. At later times, we extrapolate both the optimal parameter as well as the inverse Jacobian of $\partial_t\boldsymbol{K}^{(0)}[\boldsymbol{\lambda}'](t)$ to again start the Broyden iteration from an accurate initial guess - with this, we reliably find the roots of the respective functional after only one to three iterations per time step.

Finally, we would like to add an important remark regarding the solution of Eq. (23b). As Eq. (24) contains $\boldsymbol{\lambda}'(t)$ explicitly, a first naive approach would be to guess an "optimal" value

Journal of Physics: Conference Series **696** (2016) 012002 doi:10.1088/1742-6596/696/1/012002

$\boldsymbol{\lambda}'_{\rm g}(t)$, calculate $\boldsymbol{K}^{(0)}[\boldsymbol{\lambda}'_{\rm g}](t)$ and $\boldsymbol{K}^{(1)}[\boldsymbol{\lambda}'_{\rm g}](t)$, plug these into Eq. (24), solve the resulting equation $0 = \left[\boldsymbol{K}^{(0)}[\boldsymbol{\lambda}'_{\rm g}](t), \boldsymbol{\lambda}'(t)\right] + \boldsymbol{K}^{(1)}[\boldsymbol{\lambda}'_{\rm g}](t)$ for a new $\boldsymbol{\lambda}'(t)$, update both functionals and iterate this procedure until convergence. However, there are two complications. First, close to the optimal point, $\boldsymbol{K}^{(0)}[\boldsymbol{\lambda}'](t)$, and hence also $\boldsymbol{K}^{(1)}[\boldsymbol{\lambda}'](t)$ must vanish and therefore solving Eq. (24) for $\boldsymbol{\lambda}'(t)$ will get more and more unstable. Second, for a guessed parameter it is unlikely that the corresponding functionals will be compatible with Eq. (24), i.e., that a solution is existing at all. For this to be the case, e.g., $\boldsymbol{K}^{(1)}[\boldsymbol{\lambda}'](t)$ has to be trace-less, since it otherwise could not equal a commutator of two other matrices. Indeed, in our numerics this requirement is not met in general. More generally, Eqs. (24) and (23) can be understood as a special case of Sylvester's equation, namely $\boldsymbol{AX} - \boldsymbol{XB} = \boldsymbol{C}$, which has a unique solution for any matrix $\boldsymbol{C}$, if and only if the matrices $\boldsymbol{A}$ and $\boldsymbol{B}$ have distinct spectra (see Refs. [55–57]). In our case, this is clearly not the case, which is why we cannot expect a unique solution (or any solution at all), for some $\boldsymbol{K}^{(1)}[\boldsymbol{\lambda}'](t)$ as obtained for an arbitrarily guessed optimal parameter. A way out would be to not solve Eq. (23) directly, but minimize the norm of the left-hand side, which then, in any case, again suffers from the first complication.

6. Application of the nonequilibrium variational-cluster approach

In the following, two concrete examples for the application of the nonequilibrium variational-cluster approach are presented to discuss some of its characteristics in detail. We consider the one-dimensional Hubbard model at half-filling and at different interaction strengths and study quenches (or fast ramps) of the hopping parameters. Both examples have been in the focus of recent experiments where the redistribution of antiferromagnetic correlations between different bonds and for different ramp times [38] as well as the topological properties of Bloch bands in optical lattices [39] for the uncorrelated variant of the model [58, 59] were studied.

Figure 4 provides an illustration of the initial and of the final states as well as of the reference systems: In both cases, the system is initially in the thermal state of a dimerized Hubbard model specified by some inverse temperature β. In case (a), this state is generated by an initial Hamiltonian which consists of decoupled two-site clusters. Hence, the initial state is a simple valence-bond state with nearest-neighbor correlations and reduced translational symmetries. The intra-cluster hopping $T_{\rm intra} = 1$ fixes energy and time units. In case (b), the clusters with $T_{\rm intra} = 1$ are weakly coupled by an inter-cluster hopping $T_{\rm inter} = 0.2$.

In case (a), the final-state dynamics after the quench is governed by the full Hubbard Hamiltonian where the initially disconnected clusters are linked by a final inter-cluster hopping $T_{\rm inter} = 1$. This is a highly nontrivial example where the system should build up longer-ranged nonlocal correlations and entanglement in the course of time. In case (b) the final-state

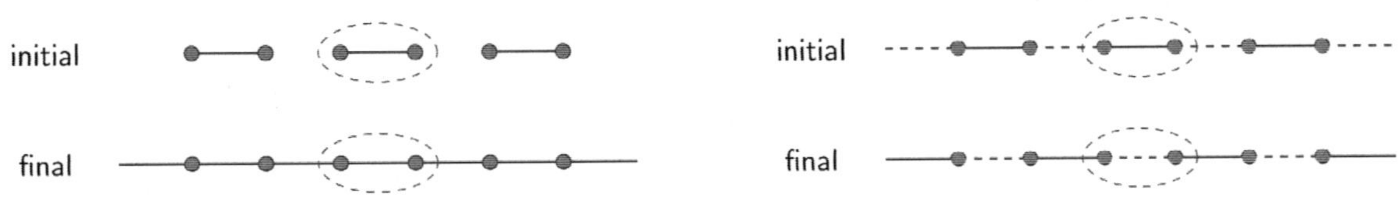

(a) dimerized lattice → homogeneous lattice (b) dimerization change

Figure 4. Illustration of initial and final states for both (a) the quench from a dimerized configuration to a homogeneous lattice as well as (b) change of the dimerization. Black solid lines indicate a nearest-neighbor intra- or inter-cluster hopping, $T_{\rm intra} = 1$ or $T_{\rm inter} = 1$. Black dashed lines: $T_{\rm inter} = 0.2$. Correlated sites are represented by red filled dots. The reference system used for the VCA calculations is indicated by a representative two-site reference cluster highlighted as the blue dashed ellipse.

Journal of Physics: Conference Series **696** (2016) 012002 doi:10.1088/1742-6596/696/1/012002

Hamiltonian is basically the same as the Hamiltonian specifying the initial state but with the important difference that the nearest-neighbor hopping $T_{\text{inter}} = 1$ now connects clusters that are shifted by one lattice constant. This example is also highly nontrivial as it corresponds to a sudden switch between Hamiltonians describing states with well-developed but incompatible valence bonds where the entanglement and the spin correlations must reorganize between two different local situations.

The VCA, as a cluster mean-field approximation, can only partly cover the expected final-state dynamics. Clearly, the quality of the approximation decisively depends on the size of the cluster used in the reference system. Note that in both cases one in principle needs clusters of infinite size to recover the exact solution. Here, however, our main intention is to discuss some numerical issues and to demonstrate that the VCA can be implemented successfully and yields reasonable results (which can in principle be improved systematically by going to larger cluster sizes). To this end, we have choosen a simple reference system, namely a system of disconnected clusters consisting of two sites each, indicated by the blue dashed lines in Fig. 4. No additional bath degrees of freedom have been added. Thus, the only possible variational parameters are the intra-cluster hopping T' and the on-site energies ε'_i in the reference system. The latter are fixed by the symmetries of the model at half-filling and, therefore, only the intra-cluster hopping T' is left for optimization.

We have tested the computer code in several trivial limits. Furthermore, for the equilibrium case our data for different U and β and for a *homogeneous* lattice are fully consistent with those obtained previously for $\beta \to \infty$ in Ref. [60] using a completely different algorithm.

Calculations for the systems sketched in Fig. 4 have been performed at inverse temperature $\beta = 10$ which, on the level of the approximation employed, is already representative of the zero-temperature limit as has been checked by varying β. Using fifth-order integration schemes on the Matsubara branch, converged results are obtained with $\Delta\tau = 0.1$. On the Keldysh branch we are limited to the trapezoidal rule only since the implementation of higher-order schemes for the evaluation of the time-propagation operator gives rise to numerical instabilities (see Sec. 5 and comment [52]). Converged results are obtained for a time step of $\Delta t = 0.02$, and a maximum propagation time of $t_{\max} = 10$ is easily achieved with a desktop computer. To get converged results with respect to the spatial extension of the one-dimensional lattice, it is sufficient to consider systems with $L = 100$ sites (using periodic boundary conditions). Rather than a sudden parameter quench we assume a finite (but short) ramp within a time interval $\Delta t_{\text{ramp}} = 0.5$, using a ramp profile $r(t) = (1 - \cos(\pi t/\Delta t_{\text{ramp}}))/2$ with continuous first-order derivatives at the joints at $t = 0$ and $t = \Delta t_{\text{ramp}}$. This has been recognized to stabilize the algorithm significantly. It is also numerically advantageous to hold the original system in its initial equilibrium state for a few time steps, before starting with the ramp at $t = t_0 = 0$ to build up the desired integration order.

Results are shown in Fig. 5. We first discuss case (a). For $t = 0$, the optimal value of the variational parameter is found as $T'_{\text{opt}} = T_{\text{intra}} = 1$ (dotted line). In fact, this had to be expected since, for the given problem, the self-energy of the reference system equals the full self-energy [cf. Eq. (13)] and thus the VCA is exact in the initial state. This represents another nontrivial check of our algorithm.

For $t > 0$ we find that T'_{opt} becomes time-dependent, i.e., the reference system adjusts itself to the parameter ramp in a time-dependent way to optimally describe the dynamics of the original system. In the limit of infinitely large clusters where VCA formally becomes exact, one would expect $T'_{\text{opt}}(t)$ to become constant after a certain relaxation time. As seen in the figure, this relaxation of the optimal variational parameter, and also of the double occupancy, takes place on a short time scale given by one or two inverse hoppings. With a reference cluster of two sites only, however, some finite-size effects must be tolerated. These show up indeed as oscillations of $T'_{\text{opt}}(t)$ around an average value after the relaxation process. For weak U, where the physics is

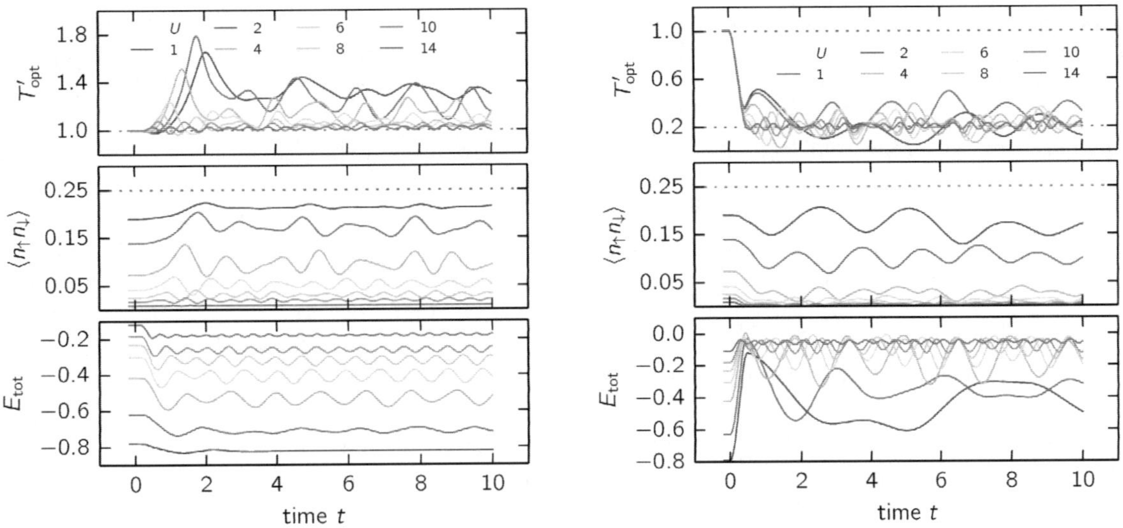

(a) dimerized lattice → homogeneous lattice
(b) dimerization change

Figure 5. Time dependencies of the optimal hopping T'_{opt} of the two-site reference cluster, the double occupancy and the total energy $E_{\text{tot}}(t) = \langle H(t) \rangle$ for different U and for the two different ramps of the hopping parameters of the original system (see Fig. 4, a and b).

governed by the hopping part of the Hamiltonian, this average is by about 30% higher than the initial value $T_{\text{intra}} = 1$. With increasing U, the final average value of $T'_{\text{opt}}(t)$ is seen to decrease and approaches unity for $U \to \infty$. This is plausible since for weak U the fermion system is more itinerant and thus an increased hopping parameter in the reference system is necessary to (at least partially) compensate for the missing inter-cluster hopping in the reference system while for strong U this is less important as the physics is more local.

Similar arguments can be used to explain the time dependence of the double occupancy. For $t = 0$, there is a strong U dependence of $\langle n_\uparrow n_\downarrow \rangle$, which is exactly reproduced by the VCA. For $t > 0$, the double occupancy quickly relaxes to a higher value (apart from significant finite-size oscillations) as the system becomes more itinerant due to the additional physical inter-cluster hopping. The effect is strongest for weak U.

For case (b), finite-size effects are generally somewhat stronger, see Fig. 5 (right). Still the main trends are clear and plausible: As the inter-cluster hopping is weak, the initial values of T'_{opt} and of $\langle n_\uparrow n_\downarrow \rangle$ are close to those of case (a). For $t > 0$, the relaxation process takes place on essentially the same time scale of one or two inverse hoppings. However, one now expects that the optimal intra-cluster hopping adjusts to a value close to $T_{\text{inter}} = 0.2$, i.e., close to the physical hopping parameter (see final state in Fig. 4(b)). This is in fact seen (dotted line in Fig. 5, right). The decrease of the final average value of the double occupancy with increasing U is understood in the same way as in case (a).

Note that the geometrical structure of the reference system is the same for the initial and the final state, see Fig. 4. Given this, we expect that the description of the initial state is better than that of the final state in the case of the dimerization change. We have also performed calculations using a reference system that is shifted by one lattice constant for both, the initial and the final state. In this case, one expects that the VCA description of the final-state dynamics is more accurate than for the initial state. The calculations (not shown) yield unphysical results in this case with a diverging optimal intra-cluster hopping after one to two inverse hoppings, depending

Journal of Physics: Conference Series **696** (2016) 012002 doi:10.1088/1742-6596/696/1/012002

somewhat on U. This demonstrates the crucial importance of an accurate description of the initial state for the subsequent real-time dynamics. A sudden switch of the geometrical structure of the reference system at $t = 0$, which follows the dimerization change of the original system, would be a legitimate choice and would result in a superior approximation. This, however, requires a substantially higher numerical effort as more than a single two-site cluster must be considered as a building block in the calculation which is beyond the scope of the present work.

In both cases, (a) and (b), there is a fairly good conservation of the total energy right after the ramp (recall that $\Delta t_{\text{ramp}} = 0.5$) with some remaining finite-size oscillations but no long-time drift. Note that, *a priori*, this could not have been expected as a matter of course, since strict energy conservation within SFT can only be ensured by variations *nonlocal* in time, which would correspond to the optimization of infinitely many bath-degrees of freedom (see Ref. [37] for a detailed discussion).

In case (a) and for all interaction strengths, we find that the total energy decreases after the ramp (see Fig. 5a). This implies that the increase of the total energy due to the heating of the system during the ramp is overcompensated by the energy decrease that is induced by the coupling of the different isolated clusters via $H_{T_{\text{inter}},0}$ and the corresponding lowering of the kinetic energy. In case (b) the total energy increases for all U after the ramp (see Fig. 5b). Since the initial and the final Hamiltonians are identical apart from a translation by one lattice constant, they have the same ground-state energies. The observed increase of the total energy must therefore be exclusively due to the heating of the system during the ramp.

7. Summary

The self-energy functional theory (SFT) provides a unifying framework for different types of cluster and impurity approximations and has recently been extended to nonequilibrium cases [37]. In the present work, we have discussed its numerical implementation in detail. An important observation is that a straightforward solution of the central SFT Euler equation is impossible in practice as it suffers from numerical instabilities which could traced back to an inherent quadratic dependence of the Jacobian on the time step Δt. Fortunately, the Euler equation can equivalently be replaced by its time derivative (and the appropriate initial condition). The corresponding Jacobian shows a numerically much more favorable linear-in-Δt scaling. By using high-order integration schemes and Broyden's method we have put forward and have implemented a stable propagation algorithm for the optimal parameters of the reference system.

As a concrete example, the variational cluster approach (VCA) has been applied to study the dynamics initiated by two different ramps of the hopping parameters of an initially dimerized one-dimensional Hubbard model. The time-evolution of the optimal parameters, the double occupancy and the total energy has been studied at different interaction strengths. These calculations have been performed with a two-site reference cluster and demonstrate that plausible and consistent results can be obtained in fact. Note that, as compared to plain CPT, the strengths of the VCA are expected to show up when considering situations involving (dynamical) phase transitions or in the possibility to construct approximations involving bath degrees of freedom. This, however, is beyond the scope of the present paper.

Despite the simple two-site reference system used here, the resulting real-time dynamics is in fact completely different from a mere superposition of oscillations with frequencies that are characteristic for the finite reference system. Namely, the variational embedding of the cluster rather allows to describe the relaxation of the system to a new stationary final state. Depending on the system, on the type of process studied and on the model parameters, however, the final state does show some unphysical oscillations which are caused by the small size of the reference system and which must be tolerated at the given approximation level. The SFT framework allows to systematically improve the approximation by consideration of larger clusters, clusters with more variational parameters or by attaching uncorrelated bath sites. We expect that the remaining finite-size oscillations can be systematically controlled in this way.

In practice, calculations for larger clusters are computationally expensive. Since the exact intra-cluster 4-point correlation function $L(3, 2, 1^+, 4)$ is required, the practically accessible cluster size is *eventually* limited by the exponential growth of the cluster Hilbert-space dimension. However, for small clusters the computational bottleneck of the algorithm actually consists in the necessity to repeatedly calculate the correlation functions $\boldsymbol{K}^{(0)}[\boldsymbol{\lambda}'](t)$ (and $\boldsymbol{K}^{(1)}[\boldsymbol{\lambda}'](t)$) at each instant of time and for each Newton step via a double contour integration and a double sum over the intra-cluster orbitals (see Eqs. (21) and (25)). With the present implementation, clusters with 6 or more sites are clearly beyond our capabilities.

Approximations generated within the SFT framework do respect macroscopic conservation laws in general [37]. Regarding total-energy conservation, however, this would require the optimization of parameters which are nonlocal in time, i.e., the allocation of infinitely many bath sites. Since this has not been considered here, we also have to tolerate superimposed finite-size oscillations on the total energy. Apart from those, however, conservation of the total energy is described fairly well, in particular, there is no long-time drift. However, to answer the question of whether or not the stationary final state is a thermal state that can be described by the temperature corresponding to the total energy after the parameter quench or ramp, in fact requires improved approximations and larger reference systems.

Acknowledgments
We would like to thank K. Balzer for providing computer code for setting up the many-body basis and for CPT reference calculations as well as for helpful discussions. We would like to thank D. Golež, C. Gramsch, H. Strand for stimulating discussions. Support of this work by the Deutsche Forschungsgemeinschaft within the Sonderforschungsbereich 925 (projects B5 and B4) is gratefully acknowledged.

References

[1] Bloch I, Dalibard J and Zwerger W 2008 *Rev. Mod. Phys.* **80** 885
[2] Lewenstein M, Sanpera A, Ahufinger V, Damski B, Sen A and Sen U 2007 *Adv. Phys.* **56** 243
[3] Dutta O, Gajda M, Hauke P, Lewenstein M, Lühmann D S, Malomed B A, Sowinski T and Zakrzewski J 2015 *Rep. Prog. Phys.* **78** 066001
[4] Polkovnikov A, Sengupta K, Silva A and Vengalattore M 2011 *Rev. Mod. Phys.* **83** 863
[5] Eisert J, Friesdorf M and Gogolin C 2015 *Nat. Phys.* **11** 124
[6] Gutzwiller M C 1963 *Phys. Rev. Lett.* **10** 159
[7] Hubbard J 1963 *Proc. R. Soc. Lond. A* **276** 238
[8] Kanamori J 1963 *Prog. Theor. Phys.* **30** 275
[9] Fisher M P A, Weichman P B, Grinstein G and Fisher D S 1989 *Phys. Rev. B* **40** 546
[10] Jaksch D, Bruder C, Cirac J I, Gardiner C W and Zoller P 1998 *Phys. Rev. Lett.* **81** 3108
[11] Greiner M, Mandel O, Esslinger T, Hänsch T W and Bloch I 2002 *Nature* **415** 39
[12] Krutitsky K V 2015 *Preprint* `arXiv:1501.03125 [cond-mat.quant-gas]`
[13] Köhl M, Moritz H, Stöferle T, Günter K and Esslinger T 2005 *Phys. Rev. Lett.* **94** 080403
[14] Giorgini S, Pitaevskii L P and Stringari S 2008 *Rev. Mod. Phys.* **80** 1215
[15] Esslinger T 2010 *Annu. Rev. Condens. Matter Phys.* **1** 129
[16] Kubo R 1957 *J. Phys. Soc. Jap.* **12** 570; Matsubara T 1955 *Prog. Theor. Phys.* **14** 351; Schwinger J 1961 *J. Math. Phys.* **2** 407
[17] Thygesen K S and Rubio A 2007 *J. Chem. Phys.* **126** 091101
[18] Schmidt P and Monien H 2002 *Preprint* `arXiv:cond-mat/0202046 [cond-mat.str-el]`
[19] Freericks J K, Turkowski V M and Zlatić V 2006 *Phys. Rev. Lett.* **97** 266408
[20] Aoki H, Tsuji N, Eckstein M, Kollar M, Oka T and Werner P 2014 *Rev. Mod. Phys.* **86** 779
[21] Muõz E, Bolech C J and Kirchner S 2013 *Phys. Rev. Lett.* **110** 016601
[22] Jung C, Lieder A, Brener S, Hafermann H, Baxevanis B, Chudnovskiy A, Rubtsov A N, Katsnelson M I and Lichtenstein A I 2012 *Ann. Phys.* **524** 49
[23] Knap M, von der Linden W and Arrigoni E 2011 *Phys. Rev. B* **84** 115145

[24] Balzer K and Eckstein M 2014 *Phys. Rev. B* **89** 035148
[25] Lipavský P, Špička V and Velický B 1986 *Phys. Rev. B* **34** 6933
[26] Keldysh L 1965 *Sov. Phys. JETP* **20** 1018
[27] Maier T, Jarrell M, Pruschke T and Hettler M H 2005 *Rev. Mod. Phys.* **77** 1027
[28] Tsuji N, Barmettler P, Aoki H and Werner P 2014 *Phys. Rev. B* **90** 075117
[29] Eckstein M and Werner P 2014 *Preprint* `arXiv:1410.3956 [cond-mat.str-el]`
[30] Gros C and Valentí R 1993 *Phys. Rev. B* **48** 418
[31] Sénéchal D, Perez D and Pioro-Ladrière M 2000 *Phys. Rev. Lett.* **84** 522
[32] Balzer M and Potthoff M 2011 *Phys. Rev. B* **83** 195132
[33] Gramsch C and Potthoff M 2015 *Preprint* `arXiv:1509.05313 [cond-mat.str-el]`
[34] Dahnken C, Aichhorn M, Hanke W, Arrigoni E and Potthoff M 2004 *Phys. Rev. B* **70** 245110
[35] Potthoff M 2014 Making use of self-energy functionals: The variational cluster approximation *DMFT at 25: Infinite Dimensions* (*Modeling and Simulation* vol 4) ed Eva Pavarini Erik Koch D V and Lichtenstein A (Forschungszentrum Jülich)
[36] Potthoff M 2003 *Eur. Phys. J. B* **32** 429
[37] Hofmann F, Eckstein M, Arrigoni E and Potthoff M 2013 *Phys. Rev. B* **88** 165124
[38] Greif D, Jotzu G, Messer M, Desbuquois R and Esslinger T 2015 *Preprint* `arXiv:1509.00854 [cond-mat.quant-gas]`
[39] Atala M, Aidelsburger M, Barreiro J T, Abanin D, Kitagawa T, Demler E and Bloch I 2013 *Nat. Phys.* **9** 795
[40] Danielewicz P 1984 *Ann. Phys.* **152** 239; van Leeuwen R, Dahlen N E, Stefanucci G, Almbladh C O and von Barth U 2006 Introduction to the Keldysh Formalism *Time-Dependent Density Functional Theory* Lecture Notes in Physics vol 706 ed Marques M A L, Ullrich C A, Nogueira F, Rubio A, Burke K and Gross E K U (Berlin Heidelberg: Springer), p 33; Rammer J 2007 *Quantum field theory of nonequilibrium states* (Cambridge University Press); Kamenev A 2011 *Field theory of non-equilibrium systems* (Cambridge, New York: Cambridge University Press)
[41] Wagner M 1991 *Phys. Rev. B* **44** 6104
[42] Potthoff M 2006 *Condens. Mat. Phys.* **9** 557
[43] Luttinger J M and Ward J C 1960 *Phys. Rev.* **118** 1417
[44] Georges A, Kotliar G, Krauth W and Rozenberg M J 1996 *Rev. Mod. Phys.* **68** 13
[45] Baym G and Kadanoff L P 1961 *Phys. Rev.* **124** 287
[46] Baym G 1962 *Phys. Rev.* **127** 1391
[47] Balzer M and Potthoff M 2010 *Phys. Rev. B* **82** 174441
[48] Eckstein M 2009 *Nonequilibrium dynamical mean-field theory* Ph.D. thesis Universität Augsburg
[49] Eckstein M, Kollar M and Werner P 2010 *Phys. Rev. B* **81** 115131
[50] Brunner H and Houwen P J v d 1986 *The numerical solution of Volterra equations* (Amsterdam: North-Holland)
[51] Press W H, Teukolsky S A, Vetterling W T and Flannery B P 2007 *Numerical Recipes. The Art of Scientific Computing* 3rd ed (Cambridge University Press)
[52] For a real-time propagation also the time-evolution operator can be implemented in a high-order scheme, e.g., by making use of techniques suggested by Alvermann and Fehske [61]. This either makes it necessary to know the respective Hamilton operator explicitly at *unequally* spaced times within the propagation time interval, as needed for Gaussian quadrature, or to extrapolate beyond the time interval to past times for the application of "extended" open Newton-Cotes formulas [51]. The first case is not feasible here, since our algorithm works on an equidistant time grid. The second, however, turned out to be highly unstable for any integration scheme of higher order than the trapzoidal rule. This might be due to Runge's phenomenon [62, 63]. Another explanation could be the following: changing the value of the considered parameter at the last time affects the shape of the fitted function for *all* nodes, and hence implicitly induces some acausal propagation which spoils a stable propagation.
[53] Kelley C T 1987 *Solving nonlinear equations with Newtons method* Fundamentals of algorithms (Society for Industrial and Applied Mathematics)
[54] Broyden C G 1965 *Math. Comp.* **19** 577
[55] Sylvester J 1884 *C. R. Acad. Sc. Paris* **99** 6771, 115116
[56] Rosenblum M 1956 *Duke Math. J.* **23** 263
[57] Bhatia R and Rosenthal P 1997 *Bull. London Math. Soc.* **29** 1
[58] Su W P, Schrieffer J R and Heeger A J 1979 *Phys. Rev. Lett.* **42** 1698
[59] Rice M J and Mele E J 1982 *Phys. Rev. Lett.* **49** 1455
[60] Balzer M, Hanke W and Potthoff M 2008 *Phys. Rev. B* **77** 045133
[61] Alvermann A and Fehske H 2011 *J. Comput. Phys.* **230** 5930

Journal of Physics: Conference Series **696** (2016) 012002 doi:10.1088/1742-6596/696/1/012002

[62] Runge C 1901 *Zeitschrift für Mathematik und Physik* **46** 224
[63] Epperson J F 1987 *Am. Math. Monthly* **94** 329

Progress in Non-equilibrium Green's Functions (PNGF VI) IOP Publishing
Journal of Physics: Conference Series **696** (2016) 012003 doi:10.1088/1742-6596/696/1/012003

Quasiparticle excitations in steady state transport across a correlated layer

Antonius Dorda, Irakli Titvinidze, and Enrico Arrigoni

Institute of Theoretical and Computational Physics, Graz University of Technology, 8010 Graz, Austria

E-mail: dorda@tugraz.at, arrigoni@tugraz.at

Abstract. In this work we investigate the spectral and transport properties of a single correlated layer attached to two metallic leads, with particular focus on the low-energy physics. A steady state current is driven across the layer by applying a bias voltage between the leads. Extending previous work we introduce a nonzero temperature in the leads, which enables us to study the influence of quasiparticle excitations on the transport characteristics in detail. Even though the system is clearly three dimensional we obtain current-voltage curves that closely resemble those of single quantum dots. Furthermore, a splitting of the quasiparticle excitation with bias voltage is observed in the spectral function.

1. Introduction

Correlated systems out of equilibrium and in particular electronic transport through quantum dots [1–4] and correlated heterostructures [5–10] have recently attracted increasing interest. Related model systems of paramount importance are the single impurity Anderson model (SIAM) [11] and the Hubbard model [12–14]. At the present time the equilibrium properties of these systems are understood to large extent [15–19]. The so-called dynamical mean field theory (DMFT) [18–21] was a key step in the theoretical description and understanding of Hubbard-like models and furthermore, established a link between correlated lattice systems which exhibit a Mott transition and the Kondo physics of a SIAM. Within this framework, the coherent quasiparticle excitations are described as a self-consistent Kondo effect [19]. The close relation between the SIAM and the Hubbard model poses the question whether analogous behavior is seen in the transport characteristics of the two systems. Exactly this question is the topic of the investigation presented here.

At the heart of DMFT lies the self-consistent solution of a quantum impurity model, the SIAM. An accurate description of the nonequilibrium physics of the SIAM and the related Kondo model is challenging by itself and currently intensively studied with different methods. To mention just a few, central aspects could be established with the noncrossing approximation [22–24], perturbative renormalization group (RG) [25, 26], flow equations [27, 28], real-time RG [29, 30], time-dependent density matrix RG [31–34], numerical RG [35, 36] and Monte Carlo methods [37–39]. A method recently introduced by some of us, which is well-suited for an application within nonequilibrium DMFT, is the so-called auxiliary master equation approach (AMEA) [40, 41]. AMEA has proven to be an accurate method for the study of the nonequilibrium steady state physics of the SIAM [41, 42]. Special emphasis was

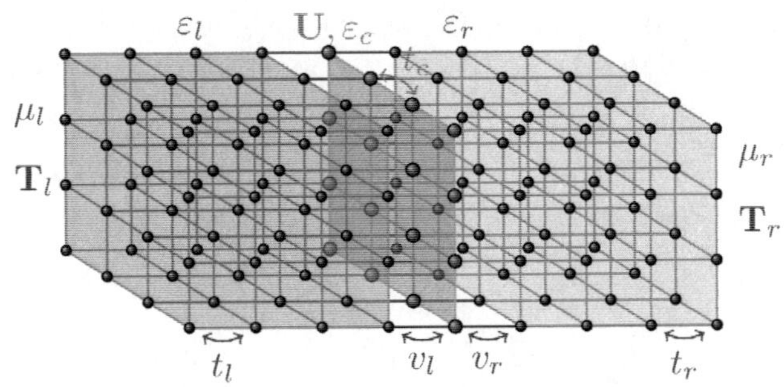

Figure 1. Sketch of the investigated heterostructure, consisting of a single correlated layer of infinite size (red) with local Hubbard interaction U, on-site energy ε_c and in-plane hopping amplitude t_c, sandwiched between two semi-infinite metallic leads (green). The hopping in the leads l and r is isotropic with amplitudes t_l and t_r, respectively, with v_l and v_r denoting the hybridizations between the respective lead and the layer. An applied bias voltage ϕ shifts the on-site energies ε_l and ε_r, as well as the chemical potentials μ_l and μ_r anti-symmetrically, which ensures together with $\varepsilon_c = -U/2$ particle-hole symmetry. All hoppings are for nearest neighbor terms only, and the temperature of the leads is labeled by $T = T_l = T_r$. We take t_c as unit of energy and consider the case with $U = 10\,t_c$, $v_l = v_r = t_c$ and $t_l = t_r = 2\,t_c$. We discuss results for different values of T and ϕ.

laid on investigating the evolution of the Kondo resonance upon driving the impurity model out of equilibrium by applying a bias voltage, and we found a transition from a single peak structure to a linear splitting of the Kondo resonance with bias. Furthermore, the current-voltage characteristics obtained for different temperatures showed clear signatures of the Kondo effect [42].

In the last years, rapid progress was made in the treatment of correlated lattice models out of equilibrium within DMFT, in explicitly time-dependent [43–46] and periodic or steady state situations [47–50]. In the study presented here we consider the special case of transport through a correlated heterostructure, consisting of a single correlated layer attached to two metallic leads at different chemical potentials. A similar setup was already treated in earlier studies with AMEA [40, 51], however, the influence of temperature was not investigated. Here we consider the transport and spectral properties of the system starting from a lowest temperature, which can still be well-resolved with the employed impurity solver and results in a strong quasiparticle excitation, and successively extending to larger values of T up to the quasigapped regime, analogous to a Mott insulator. Besides the bias-dependent spectral function, the experimentally well-accessible current-voltage characteristics is presented.

The work is organized as follows: In Sec. 2 the investigated model is defined, in Sec. 3 we briefly introduce the nonequilibrium DMFT approach together with AMEA, and in Sec. 4 the obtained results are presented and discussed. Concluding remarks are given in Sec. 5.

2. Model

The model system considered in this work is schematically depicted in Fig. 1. The corresponding Hamiltonian is given by

$$H = H_c + \sum_{\alpha=l,r} H_\alpha + H_{\text{coup}}\,, \tag{1}$$

consisting of a part for the central system H_c, a part for each decoupled lead $H_{l/r}$ and a coupling between the leads and the correlated layer H_{coup}. In the detail the Hamiltonian reads

$$H_c = -t_c \sum_{\langle ij\rangle,\sigma} c^\dagger_{i\sigma} c_{j\sigma} + U \sum_i n_{i\uparrow} n_{i_\downarrow} + \varepsilon_c \sum_{i,\sigma} n_{i\sigma}\,, \tag{2}$$

$$H_\alpha = -t_\alpha \sum_{\langle ij\rangle\sigma} c^\dagger_{\alpha i\sigma} c_{\alpha j\sigma} + \varepsilon_\alpha \sum_{i\sigma} c^\dagger_{\alpha i\sigma} c_{\alpha i\sigma}\,, \tag{3}$$

$$H_{\text{coup}} = \sum_{\langle ij\rangle\alpha\sigma} v_\alpha \left(c^\dagger_{i\sigma} c_{\alpha j\sigma} + h.c. \right)\,, \tag{4}$$

where $\langle i,j\rangle$ indicates neighboring sites, $c^\dagger_{i,\sigma}$ creates an electron at the i-th site of the correlated layer with spin $\sigma = \uparrow, \downarrow$, and $n_{i\sigma} = c^\dagger_{i\sigma} c_{i\sigma}$ denote the corresponding occupation number operators. The analogous fermionic creation/annihilation operators of lead α are labeled by $c^\dagger_{\alpha i\sigma}/c_{\alpha i\sigma}$. For the particular parameters see Fig. 1.

3. Method

In the following we outline the method only briefly and for details we refer to Ref. [40, 41, 51].

Nonequilibrium dynamics are conveniently formulated in terms of Keldysh Green's functions [52–57], whereby for the steady state limit it suffices to consider 2×2 objects on the Keldysh contour

$$\underline{G} = \begin{pmatrix} G^R & G^K \\ 0 & G^A \end{pmatrix}\,, \tag{5}$$

which we denote by an underscore. Here, the retarded G^R and the Keldysh component G^K are independent functions in a generic nonequilibrium situation, and the advanced part is given by $G^A = (G^R)^\dagger$. The spectral function is defined as usual: $A = i/2\pi\,(G^R - G^A)$.

Since the model outlined in Sec. 2 is translationally invariant in the in-plane direction, it is convenient to introduce the corresponding momentum variable $\mathbf{k}_{||}$. Furthermore, time translational invariance applies in the steady state limit and the governing equations can be formulated in the frequency domain ω. With this the Green's function of the correlated layer is given in terms of Dyson's equation by

$$\underline{G}^{-1}(\omega, \mathbf{k}_{||}) = \underline{g}_0^{-1}(\omega, \mathbf{k}_{||}) - \sum_{\alpha=l,r} v_\alpha^2\, \underline{g}_\alpha(\omega, \mathbf{k}_{||}) - \underline{\Sigma}(\omega, \mathbf{k}_{||})\,. \tag{6}$$

Here, the decoupled non-interacting Green's function of the layer is denoted by $\underline{g}_0(\omega, \mathbf{k}_{||})$, and those of the leads by $\underline{g}_\alpha(\omega, \mathbf{k}_{||})$. The non-interacting Green's functions are known exactly but the determination of the self-energy of the correlated layer $\underline{\Sigma}(\omega, \mathbf{k}_{||})$ is challenging and one has to resort to approximations. In particular we neglect spatial correlations and restrict ourselves to a local self-energy $\underline{\Sigma}(\omega, \mathbf{k}_{||}) = \underline{\Sigma}(\omega)$, as usually done in the context of DMFT [18–21, 43, 47]. Within DMFT, the local quantity $\underline{\Sigma}(\omega)$ is determined by mapping the lattice problem onto an equivalent quantum impurity model, with the same local parameters U and ε_c. However, the bath degrees of freedom of the impurity model depend on $\underline{\Sigma}(\omega)$, such that a self-consistent solution is needed, which is commonly obtained in an iterative manner. The bath for the impurity model is fully specified by the hybridization function

$$\underline{\Delta}_{\text{ph}}(\omega) = \underline{g}_0^{-1}(\omega) - \underline{G}_{\text{loc}}^{-1}(\omega) - \underline{\Sigma}(\omega)\,, \tag{7}$$

where $\underline{g}_0^{-1}(\omega)$ is the non-interacting Green's function of the disconnected impurity and the local Green's function is obtained by

$$\underline{G}_{\mathrm{loc}}(\omega) = \int\limits_{\mathrm{BZ}} \frac{d\mathbf{k}_{||}}{(2\pi)^2} \underline{G}(\omega, \mathbf{k}_{||}) \,. \tag{8}$$

In order to solve the nonequilibrium impurity problem we resort to AMEA, cf. [40, 41], which maps the original impurity problem onto an auxiliary one, with a finite number of bath sites N_B and additional Markovian environments. The resulting open quantum system is described by a Lindblad equation and is small enough to be solved accurately by numerical techniques. In contrast to other approaches, the parameters of the Lindblad equation are not obtained perturbatively but through an optimization procedure. In particular, we consider the hybridization function of the auxiliary system $\underline{\Delta}_{\mathrm{aux}}(\omega)$ and vary the auxiliary bath parameters in order to minimize the difference to the physical hybridization function $\underline{\Delta}_{\mathrm{ph}}(\omega)$, Eq. (7). In the limit of large N_B the approach becomes exact but even for small values of N_B we obtain $\underline{\Delta}_{\mathrm{aux}}(\omega) \approx \underline{\Delta}_{\mathrm{ph}}(\omega)$ to very good approximation. Typically, an exponential convergence with increasing N_B is achieved. After the mapping procedure, which can be done in a $U = 0$ calculation, the interacting impurity problem is solved. For this we introduced two different strategies in previous work: On the one hand, an implementation of AMEA which makes use of Krylov space methods, cf. Ref. [41], and on the other hand, a matrix product states based solution, cf. Ref. [42]. The latter allows for a highly accurate solution of the impurity problem but requires a rather large amount of CPU time. The former is not as accurate at low temperatures but faster in many cases and is used in the present work. The Krylov space solver enables us to consider up to $N_B = 6$ bath sites, which suffices to treat cases with strong Kondo or quasiparticle excitations reliably, cf. Ref. [41, 51]. Once the many-body problem of the auxiliary system is solved one obtains an approximation for the self-energy

$$\underline{\Sigma}(\omega) = \underline{G}_{\mathrm{aux},0}^{-1}(\omega) - \underline{G}_{\mathrm{aux}}^{-1}(\omega)\,, \tag{9}$$

from the knowledge of the non-interacting and the interacting auxiliary Green's functions. By inserting $\underline{\Sigma}(\omega)$ from Eq. (9) into Eqs. (6),(7),(8) we close the DMFT cycle and iterate until a self-consistent point is reached.

4. Results

In the following we present results for the transport and spectral properties of a correlated layer in a nonequilibrium steady state situation. The particular model is defined in Sec. 2 and depicted in Fig. 1. We place special emphasis on the low-energy properties and consider cases with rather low bias voltages ϕ. To investigate the role of resonant quasiparticle excitations, different temperatures $T = T_l = T_r$ are introduced in the leads. In contrast to previous work [51], a nonzero temperature is considered in the leads and additionally, the hopping parameters are chosen such that each correlated site has an equal hopping amplitude to neighboring sites ($v_l = v_r = t_c$). By this we expect to be in a regime in which a competition between the physics of an isolated 2D layer and the one of single quantum dots occurs.

In Fig. 2 the current-voltage characteristics $j(\phi)$ together with the differential conductance $\partial j/\partial\phi$ are displayed for various T. At low bias voltages $\phi \lesssim 2\,t_c$ the temperature has large influence on $j(\phi)$ and $\partial j/\partial\phi$, whereas for larger voltages $\phi \gtrsim 2.5\,t_c$ we find quantitatively similar current values for all of the temperatures. This is in close analogy to what was observed in Ref. [42] for the nonequilibrium properties of a quantum dot. In $\partial j/\partial\phi$ we obtain for $\phi \lesssim 1\,t_c$ a strong dependence on T. Again, the behavior is similar to what is known from Kondo systems [3, 4, 29, 30]. However, the accuracy of the present calculations does not permit

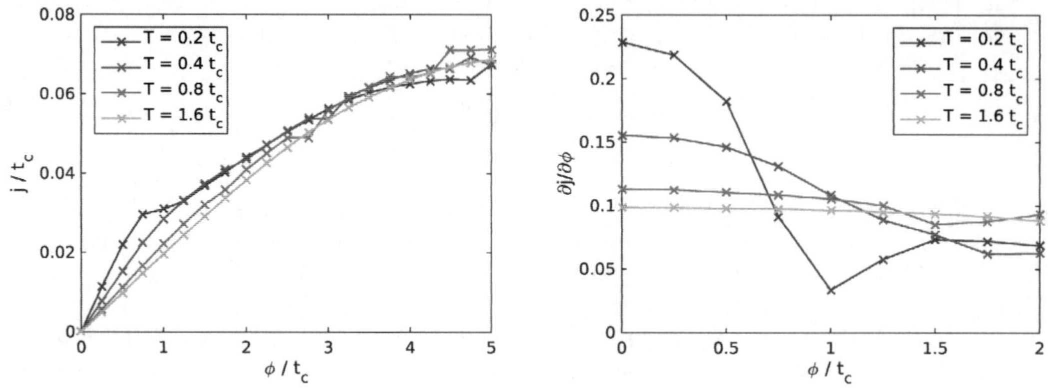

Figure 2. Current $j(\phi)$ and differential conductance $\partial j/\partial\phi$ as a function of bias voltage ϕ and for different temperatures T. For the particular model parameters see Fig. 1. Calculations for $T = 1.6\,t_c$ were performed with $N_B = 4$ and all others with $N_B = 6$.

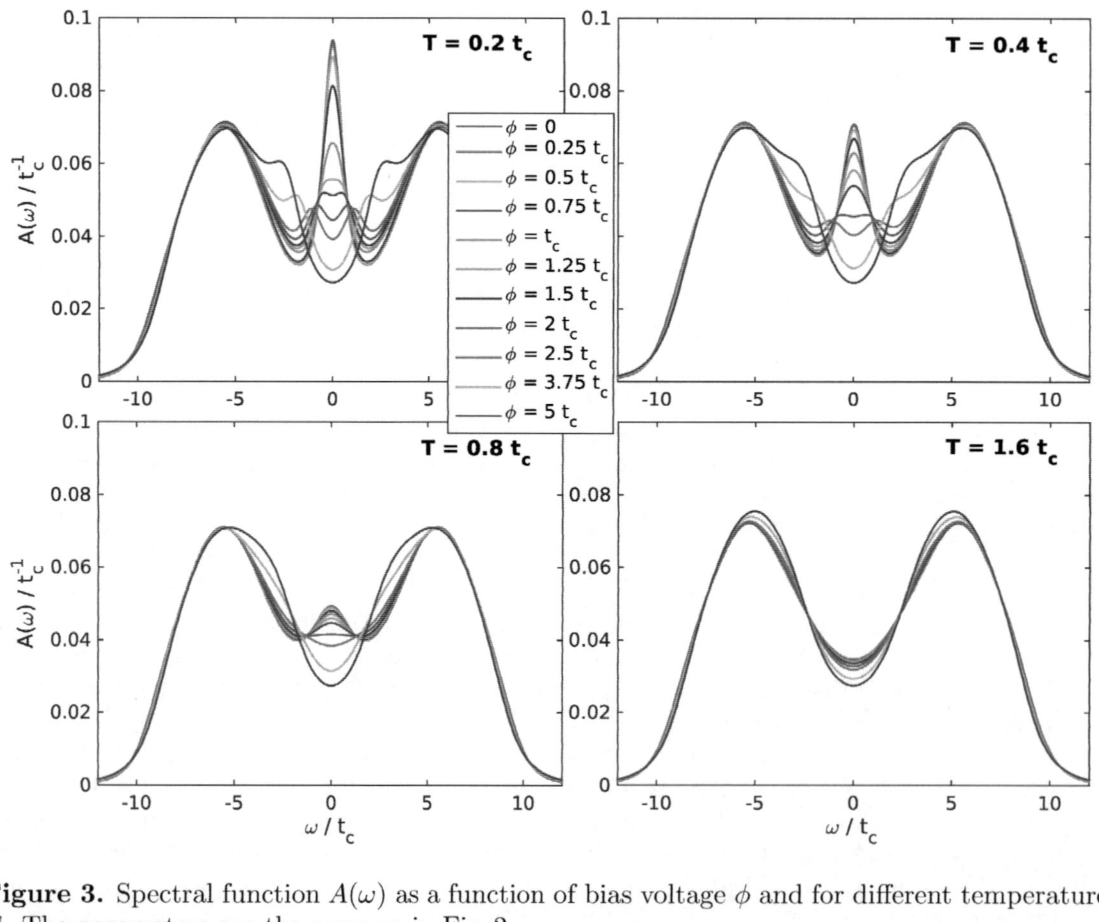

Figure 3. Spectral function $A(\omega)$ as a function of bias voltage ϕ and for different temperatures T. The parameters are the same as in Fig. 2

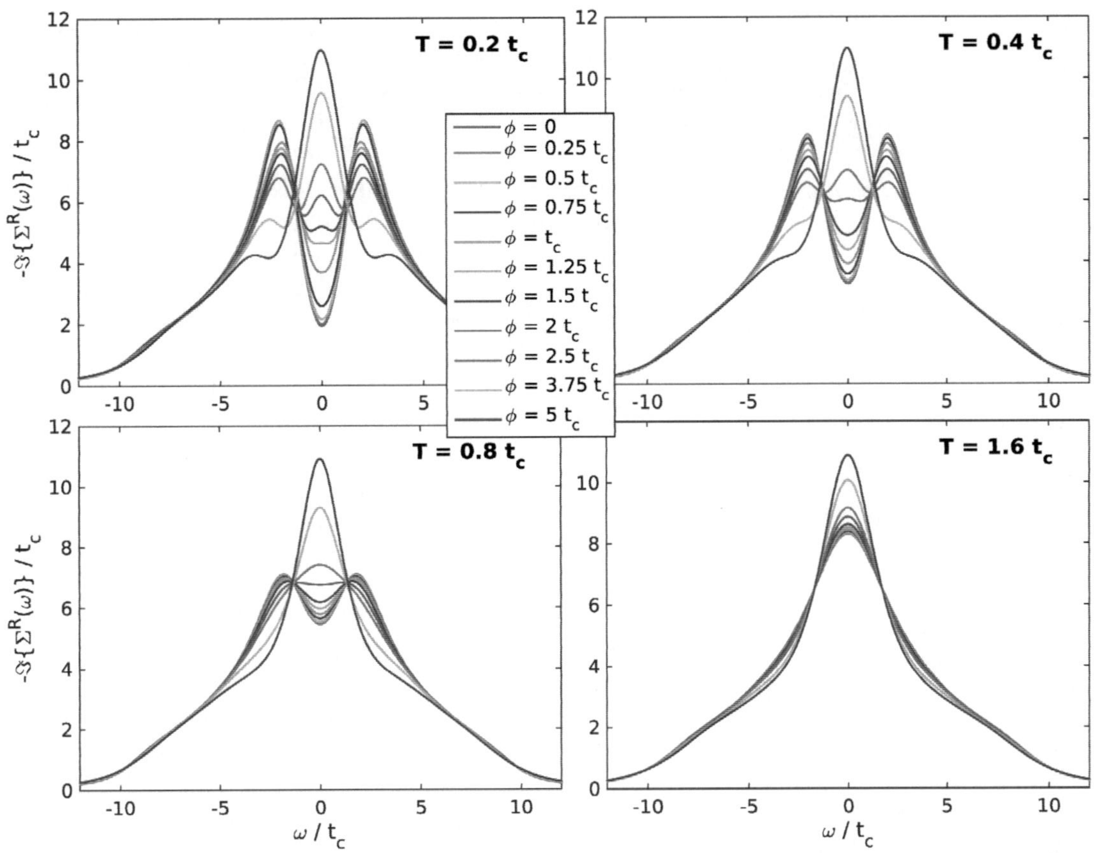

Figure 4. Retarded self-energy $\Sigma^R(\omega)$ as a function of bias voltage ϕ and for different temperatures T. The parameters are the same as in Fig. 2

us to investigate the scaling with ϕ in detail, in particular if a logarithmic dependence as in quantum dots is present. Still, one observes that the experimentally well-accessible quantity $\partial j/\partial\phi$ exhibits a strong temperature dependence and clear signatures of a Kondo-like behavior are visible for low T. From a technical point of view one should note that slight kinks or jumps in $j(\phi)$ are present, at different values of ϕ for the various T. These artefacts originate in the mapping procedure and appear at values $\phi = \phi_c$ at which more than one parameter set for the auxiliary system minimizes the difference to the physical hybridization function. Usually, one of these minima is better for $\phi < \phi_c$ while the other one for $\phi > \phi_c$. Therefore, such a crossing of minima leads to an abrupt change in parameters of the auxiliary system and may result in a slight shift of spectral weight, cf. Ref. [41]. In general, this effect is smallest in situations in which the difference between $\underline{\Delta}_{\rm aux}(\omega)$ and $\underline{\Delta}_{\rm ph}(\omega)$ is small in any case and thus, is reduced upon increasing the number of bath sites N_B.

A more detailed picture of the state of the system is obtained by investigating the spectral function, see Fig. 3. In the equilibrium case, $\phi = 0$, we find a strong quasiparticle excitation at $\omega = 0$ for $T = 0.2\,t_c$, which is attenuated with increasing temperature ($T = 0.4\,t_c$ and $T = 0.8\,t_c$) and is completely suppressed for $T = 1.6\,t_c$.[58] Especially interesting are the low temperature situations $T = 0.2\,t_c$ and $T = 0.4\,t_c$ in which a strong zero frequency excitation is visible in equilibrium, and for which we observe with increasing ϕ at first a reduction of the peak height before a splitting sets in. Similar to a quantum dot system [41, 42], we find two resonant excitations at the Fermi-energies of the two leads and thus a linear splitting with ϕ. For the case of $T = 0.2\,t_c$ the excitations are still clearly visible at rather large voltages up to $\phi \approx 5\,t_c$. The results for $T = 0.8\,t_c$ and $T = 1.6\,t_c$ reveal dissimilar spectral functions at low bias voltages, but surprisingly, the obtained current values in Fig. 2 are comparable. The reason for this is that the high temperatures average out details in $A(\omega)$ to large extent.

The presence of resonant excitations is also clearly visible in the retarded part of the self-energy, displayed in Fig. 4. For $\phi = 0$ we find for temperatures up to $T = 0.8\,t_c$ a local minimum in $-\Im\{\Sigma^R(\omega)\}$ at $\omega = 0$, indicating a quasiparticle excitation. But only in the cases $T = 0.2\,t_c$ and $T = 0.4\,t_c$ the temperature induced decoherence is weak enough to obtain a splitting of the single minimum when increasing the bias voltage. In contrast, the self-energy for $T = 1.6\,t_c$ is rather featureless and only weakly dependent on the bias voltage.

5. Conclusions

In this work we investigated the steady state properties of a correlated layer sandwiched between two metallic leads at different chemical potentials, induced by an externally applied bias voltage. For this we made use of a nonequilibrium DMFT approach together with AMEA as impurity solver. In addition to previous work [51], we studied the influence of temperature on the transport characteristics and on the bias-dependent spectral function, with focus on the low-bias regime. The parameters of the system were chosen such that a certain direction was not preferred in advance. In particular, all of the hopping amplitudes of a correlated site to its neighbors were of equal size. From investigating the spectral function and the differential conductance as a function of bias voltage and for various temperatures, we found that the considered system bore close analogy to the case of a single quantum dot. A result like this could be expected when considering the limit in which the hopping parallel to the 2D layer is much smaller than the longitudinal one, regarding the layer and the leads. But, since the hoppings to correlated sites were isotropic the result is not intuitive nor trivial.

Acknowledgements

We acknowledge valuable discussions with Martin Nuss, Markus Aichhorn, Michael Knap, and Wolfgang von der Linden. This work was supported by the Austrian Science Fund (FWF): P24081, P26508, as well as SFB-ViCoM project F04103, and NaWi Graz. The calculations were partly performed on the D-cluster Graz and on the VSC-3 cluster Vienna.

References

[1] Goldhaber-Gordon D, Shtrikman H, Mahalu D, Abusch-Magder D, Meirav U and Kastner M A 1998 *Nature (London)* **391** 156
[2] Cronenwett S M, Oosterkamp T H and Kouwenhoven L P 1998 *Science* **281** 540–544
[3] Kretinin A V, Shtrikman H and Mahalu D 2012 *Phys. Rev. B* **85**(20) 201301
[4] Zhang Y, Kahle S, Herden T, Stroh C, Mayor M, Schlickum U, Ternes M, Wahl P and Kern K 2013 *Nat Commun* **4**
[5] Ahn C H, Gariglio S, Paruch P, Tybell T, Antognazza L and Triscone J M 1999 *Science* **284** 1152–1155
[6] Izumi M, Ogimoto Y, Konishi Y, Manako T, Kawasaki M and Tokura Y 2001 *Materials Science and Engineering: B* **84** 53 – 57
[7] Ohtomo A, Muller D A, Grazul J L and Hwang H Y 2002 *Nature* **419** 378–380
[8] Ohtomo A and Hwang H Y 2004 *Nature* **427** 423–426
[9] Gariglio S, Ahn C H, Matthey D and Triscone J M 2002 *Phys. Rev. Lett.* **88**(6) 067002
[10] Zhu Q X, Wang W, Zhao X Q, Li X M, Wang Y, Luo H S, Chan H L W and Zheng R K 2012 *Journal of Applied Physics* **111** 103702
[11] Anderson P W 1961 *Phys. Rev.* **124**(1) 41–53
[12] Hubbard J 1963 *Proc. R. Soc. London* **276** 238
[13] Gutzwiller M C 1963 *Phys. Rev. Lett.* **10** 159
[14] Kanamori J 1963 *Prog. Theor. Phys.* **30** 275
[15] Hewson A C 1993 *The Kondo Problem to Heavy Fermions* (Cambridge University Press)

[16] Bulla R, Costi T A and Pruschke T 2008 *Rev. Mod. Phys.* **80** 395–450
[17] LeBlanc J P F, Antipov A E, Becca F, Bulik I W, Chan G K L, Chung C M, Deng Y, Ferrero M, Henderson T M, Jimnez-Hoyos C A, Kozik E, Liu X W, Millis A J, Prokof'ev N V, Qin M, Scuseria G E, Shi H, Svistunov B V, Tocchio L F, Tupitsyn I S, White S R, Zhang S, Zheng B X, Zhu Z and Gull E 2015 ArXiv:1505.02290
[18] Vollhardt D 2012 *Annalen der Physik* **524** 1–19
[19] Georges A 2004 *AIP Conference Proceedings* **715**
[20] Georges A, Kotliar G, Krauth W and Rozenberg M J 1996 *Rev. Mod. Phys.* **68**(1) 13–125
[21] Metzner W and Vollhardt D 1989 *Phys. Rev. Lett.* **62** 324
[22] Wingreen N S and Meir Y 1994 *Phys. Rev. B* **49**(16) 11040–11052
[23] Lebanon E and Schiller A 2001 *Phys. Rev. B* **65**(3) 035308
[24] Rosch A, Kroha J and Wölfle P 2001 *Phys. Rev. Lett.* **87**(15) 156802
[25] Rosch A, Paaske J, Kroha J and Wölfle P 2003 *Phys. Rev. Lett.* **90** 076804
[26] Shah N and Rosch A 2006 *Phys. Rev. B* **73**(8) 081309
[27] Kehrein S 2005 *Phys. Rev. Lett.* **95** 056602
[28] Fritsch P and Kehrein S 2010 *Phys. Rev. B* **81**(3) 035113
[29] Pletyukhov M and Schoeller H 2012 *Phys. Rev. Lett.* **108**(26) 260601
[30] Reininghaus F, Pletyukhov M and Schoeller H 2014 *Phys. Rev. B* **90** 085121
[31] Heidrich-Meisner F, Feiguin A E and Dagotto E 2009 *Phys. Rev. B* **79** 235336
[32] Holzner A, McCulloch I P, Schollwöck U, von Delft J and Heidrich-Meisner F 2009 *Phys. Rev. B* **80**(20) 205114
[33] Nuss M, Ganahl M, Evertz H G, Arrigoni E and von der Linden W 2013 *Phys. Rev. B* **88** 045132
[34] Nuss M, Ganahl M, Arrigoni E, von der Linden W and Evertz H G 2015 *Phys. Rev. B* **91** 085127
[35] Anders F B 2008 *Phys. Rev. Lett.* **101** 066804
[36] Schmitt S and Anders F B 2011 *Phys. Rev. Lett.* **107**(5) 056801
[37] Werner P, Oka T, Eckstein M and Millis A J 2010 *Phys. Rev. B* **81** 035108
[38] Han J E and Heary R J 2007 *Phys. Rev. Lett.* **99** 236808
[39] Cohen G, Gull E, Reichman D R and Millis A J 2014 *Phys. Rev. Lett.* **112**(14) 146802
[40] Arrigoni E, Knap M and von der Linden W 2013 *Phys. Rev. Lett.* **110** 086403
[41] Dorda A, Nuss M, von der Linden W and Arrigoni E 2014 *Phys. Rev. B* **89** 165105
[42] Dorda A, Ganahl M, Evertz H G, von der Linden W and Arrigoni E 2015 *Phys. Rev. B* **92**(12) 125145
[43] Freericks J K, Turkowski V M and Zlatić V 2006 *Phys. Rev. Lett.* **97** 266408
[44] Eckstein M, Kollar M and Werner P 2009 *Phys. Rev. Lett.* **103** 056403
[45] Aoki H, Tsuji N, Eckstein M, Kollar M, Oka T and Werner P 2014 *Rev. Mod. Phys.* **86** 779–837
[46] Balzer K, Wolf F A, McCulloch I P, Werner P and Eckstein M 2015 *Phys. Rev. X* **5**(3) 031039
[47] Schmidt P and Monien H ArXiv:cond-mat/0202046
[48] Joura A V, Freericks J K and Pruschke T 2008 *Phys. Rev. Lett.* **101** 196401
[49] Okamoto S 2008 *Phys. Rev. Lett.* **101** 116807
[50] Aron C, Kotliar G and Weber C 2012 *Phys. Rev. Lett.* **108** 086401
[51] Titvinidze I, Dorda A, von der Linden W and Arrigoni E 2015 ArXiv:1508.02953
[52] Kadanoff L P and Baym G 1962 *Quantum Statistical Mechanics: Green's Function Methods in Equilibrium and Nonequilibrium Problems* (Redwood City, CA: Addison-Wesley)
[53] Schwinger J 1961 *J. Math. Phys.* **2** 407
[54] Keldysh L V 1965 *Sov. Phys. JETP* **20** 1018
[55] Haug H and Jauho A P 1998 *Quantum Kinetics in Transport and Optics of Semiconductors* (Heidelberg: Springer)
[56] Rammer J and Smith H 1986 *Rev. Mod. Phys.* **58**(2) 323–359
[57] van Leeuwen R, Dahlen N E, Stefanucci G, Almbladh C O and von Barth U 2006 *Introduction to the Keldysh Formalism* (*Lect. Notes. Phys.* vol 706) (Berlin: Springer-Verlag) chap 3, pp 33–59
[58] One should note that the $T = 0$ Fermi liquid limit is given by $A(\omega = 0) \approx 0.1667 t_c^{-1}$. The accurate consideration of temperatures below $T = 0.2\, t_c$ would require more bath sites N_B and thus more advanced manybody solution strategies, cf. Ref. [42].

Journal of Physics: Conference Series **696** (2016) 012004 doi:10.1088/1742-6596/696/1/012004

Time-resolved photoabsorption in finite systems: A first-principles NEGF approach

E. Perfetto[1], A.-M. Uimonen[2,3], R. van Leeuwen[2,4] and G. Stefanucci[1,4,5]

[1] Dipartimento di Fisica, Università di Roma Tor Vergata, Via della Ricerca Scientifica 1, I-00133 Rome, Italy

[2] Department of Physics, Nanoscience Center, FIN 40014, University of Jyväskylä, Jyväskylä, Finland

[3] Clarendon Laboratory, University of Oxford, Parks Road, Oxford OX1 3PU, United Kingdom

[4] European Theoretical Spectroscopy Facility (ETSF)

[5] Laboratori Nazionali di Frascati, Istituto Nazionale di Fisica Nucleare, Via E. Fermi 40, 00044 Frascati, Italy

E-mail: `gianluca.stefanucci@roma2.infn.it`

Abstract. We describe a first-principles NonEquilibrium Green's Function (NEGF) approach to time-resolved photoabsortion spectroscopy in atomic and nanoscale systems. The method is used to highlight a recently discovered dynamical correlation effect in the spectrum of a Krypton gas subject to a strong ionizing pump pulse. We propose a minimal model that captures the effect, and study the performance of time-local approximations versus time-nonlocal ones. In particular we implement the time-local Hartree-Fock and Markovian second Born (2B) approximation as well as the exact adiabatic approximation within the Time-Dependent Density Functional Theory framework. For the time-nonlocal approximation we instead use the 2B one. We provide enough convincing evidence for the fact that a proper description of the spectrum of an evolving admixture of ionizing atoms requires the simultaneous occurrence of *correlation* and *memory* effects.

1. Introduction

Time-resolved photoabsorption spectroscopy is a cutting edge experimental technique to probe quantum systems in nonstationary states [1–5]. The sample is driven out of equilibrium by a strong pump pulse (whose intensity can exceed 10^{15} W/cm^2) and subsequently probed with a weak ultrashort (down to hundreds of as) pulse. By measuring the energy per unit frequency carried by the transmitted probe as a function of the delay between the pump and probe pulses one obtains the time-dependent photoabsorption spectrum. From this spectrum it is possible to read out information like, e.g., charge/exciton dynamics [6–19], dynamical screening effects [20–23] and several other nonequilibrium properties [24–33].

For optically thin samples (atoms, molecules and more generally nanostructures) the transmitted probe pulse can be expressed in terms of the time-dependent *probe-induced* change of the dipole moment [34–41]. The latter is defined as the difference between the time-dependent dipole moment originating from the pump+probe fields and the time-dependent dipole moment originating from the pump only. Both quantities can be calculated from the time-dependent quantum average of the dipole operator over the many-electron state of the system evolving according to the Schrödinger equation with the appropriate electric field. Due to electron correlations these averages are difficult to compute using the Configuration Interaction (CI) method since the number of configurations scales exponentially with the number of electrons. CI is often implemented in the so called Single Active Electron (SAE) approximation [42, 43] which, by construction, neglects double or higher excitations. An alternative approach, statistical in nature, to tackle the problem is provided by Time-Dependent Density Functional Theory (TDDFT) [44]. In TDDFT the interacting many-body system is mapped onto a fictitious noninteracting system having the same density, and hence the same dipole moment, as the interacting one [45, 46]. Therefore, TDDFT is ideal to study much larger systems than those accessible by CI. However, most TDDFT calculations are performed using the Adiabatic Local Density Approximation (ALDA) for the potential of the fictitious system but, unfortunately, the ALDA potential suffers from the same drawbacks of the SAE approximation in CI [47, 48]. Other critical drawbacks of the ALDA potential in the context of photoabsorption are that (i) the renormalization of the molecular levels due to the nearby presence of a highly polarizable structure or medium is poorly described [49–54] and (ii) single-particle excitation with a long-range charge-transfer character are missed [55–57].

In Ref. [41] we investigated the possibility of using a statistical approach based on NonEquilibrium Green's Functions (NEGF) [58–62]. The NEGF approach is computationally more expensive than TDDFT but still more convenient than CI. From the theoretical point of view the advantage of NEGF over TDDFT is that the relevant electron-electron scattering processes can be incorporated through a proper selection of Feynman diagrams for the self-energy Σ from which the Green's function $G = G_0 + G_0 \Sigma G$ can be self-consistently calculated, G_0 being the noninteracting Green's function. The aformentioned drawbacks of the ALDA potential in TDDFT are absent in NEGF already with rather simple self-energies. It is also worth observing that the self-energy $\Sigma[G]$ is a functional of the Green's function and hence the NEGF approach is a *nonperturbative* approach as it amounts to sum a subset of Feynman diagrams to infinite order in the interaction strength.

Actually, the most convenient method to calculate the nonequilibrium G is not through the Dyson equation $G = G_0 + G_0 \Sigma G$ but through the so called Kadanoff-Baym Equations (KBE) [58–62]. The KBE are nonlinear integro-differential equations for the lesser and greater components of G, $G^<$ and $G^>$ respectively. For a given self-energy the computational cost for the solution of the KBE scales like N_t^3 [63–71], where N_t is the number of time steps. This poses severe limitations to the maximum propagation time, essentially preventing the use of NEGF for an acceptable frequency resolution of the time-resolved photoabsorption spectra. The scaling of the computational cost can be reduced from N_t^3 to N_t^2 using the Generalized Kadanoff-Baym Ansatz (GKBA) [72]. The GKBA is an ansatz for $G^{\lessgtr}(t, t')$ in terms of the one-particle density matrix $\rho(t) = -iG^<(t, t)$, and allows for collapsing the KBE into a single nonlinear integro-differential equation for the one-time quantity $\rho(t)$. The ansatz is exact for the Hartree-Fock (HF) self-energy and an approximation for correlated self-energies (beyond HF). Nevertheless, this approximation turns out to be extremely accurate in systems with well-defined quasiparticles (like atoms and molecules) and more generally whenever the collision time is much smaller than the quasiparticle lifetime [72]. We wish to emphasize here that within the GKBA the NEGF formalism is converted into a theory for the one-particle density matrix which could provide useful hints for the development of TDDFT functionals beyond the ALDA.

In this work we briefly describe the NEGF+GKBA approach to time-resolved photoabsorption spectroscopy and apply it to model Hamiltonians with few bound states and a continuum of states. Our purpose is to shed light on a recently discovered dynamical correlation effect in a gas of Krypton atoms [28, 41]. In the experiment the gas is perturbed by a strong near infrared pump pulse responsible for the ionization of some of the atoms. During the action of the pump the neutral gas evolves into an admixture of Kr atoms and Kr^{n+} ions with $n = 1, 2, 3, \ldots$. By probing the admixture with a weak, extreme ultraviolet attosecond pulse having a delay τ with respect to the pump pulse one can monitor the evolution of the photoabsorption spectrum as a function of τ. The development of absorption peaks in different spectral regions are fingerprints of the presence of Kr ions with different charge. Our calculations have shown that the HF self-energy as well as other time-local self-energies, e.g., the Markovian second-Born (M2B), capture at most the absorption of Kr^{1+} [41]. Thus, time-local approximations to the self-energy do not describe the absorption of multiply ionized Kr atoms. Instead the full (nonlocal in time) second-Born (2B) approximation correctly captures the absorption of Kr^{1+} and Kr^{2+} as well as the retardation in the onset of the absorption peak of Kr^{2+} with respect to the onset of the absorption peak of Kr^{1+}. In this work we introduce a simple model Hamiltonian to clarify the role of time-nonlocal correlations in the formation of an admixture of multiply ionized atoms. The simplicity of the model also allows for obtaining an analytic expression of the exact exchange-correlation potential of static DFT, and hence to perform exact adiabatic TDDFT simulations. We show that similarly to the HF and the M2B approximations also the exact adiabatic approximation of TDDFT fails in reproducing the absorption peaks of multiply ionized atoms. This provides a conclusive evidence for the fact that a proper description of the time-resolved spectrum of an evolving admixture of ions requires the simultaneous inclusion of *correlations* and *memory* effects.

2. Time-resolved photoabsorption spectrum

We consider a finite system like an atom or a molecule driven out of equilibrium by a pump pulse. Choosing the pump frequency appropriately electrons can be expelled from the system, thus changing its absorption properties. Our goal is to study the photoabsorption spectrum of the system during the ionization process. For this purpose we assign a single-particle basis consisting of a certain number of orthonormal bound wavefunctions $\{\varphi_i(\mathbf{r})\}$ and of a continuum of extended wavefunctions $\{\varphi_k(\mathbf{r})\}$. Let $\hat{c}_{\alpha\sigma}$ ($\hat{c}^\dagger_{\alpha\sigma}$) be the annihilation (creation) operator for an electron with spin $\sigma = \uparrow, \downarrow$ in $\varphi_\alpha(\mathbf{r})$, where α can be either a bound index or a continuum index. For ionization processes from the valence shells Auger recombination are unlikely to occur. Furthermore, as we are not interested in the energy of the photoelectrons, we ignore the Coulomb repulsion between bound and extended electrons. The equilibrium Hamiltonian then reads

$$\hat{H}_{\rm eq} = \sum_{k,\sigma} \epsilon_k \hat{c}^\dagger_{k\sigma}\hat{c}_{k\sigma} + \sum_{ij,\sigma} h_{ij}\hat{c}^\dagger_{i\sigma}\hat{c}_{j\sigma} + \frac{1}{2}\sum_{\substack{ijmn\\ \sigma\sigma'}} v_{ijmn}\hat{c}^\dagger_{i\sigma}\hat{c}^\dagger_{j\sigma'}\hat{c}_{m\sigma'}\hat{c}_{n\sigma}. \tag{1}$$

Without any loss of generality we have chosen the set $\{\varphi_k(\mathbf{r})\}$ to be the set of continuum eigenstates with energy ϵ_k of the single particle Hamiltonian $h(\mathbf{r}) = -\frac{\nabla^2}{2} + V_{\rm n}(\mathbf{r})$, $V_{\rm n}(\mathbf{r})$ being the nuclear potential. The one- and two-electron integrals appearing in Eq. (1) are defined according to $h_{ij} \equiv \int d\mathbf{r}\, \varphi^*_i(\mathbf{r}) h(\mathbf{r}) \varphi_j(\mathbf{r})$ and $v_{ijmn} \equiv \int d\mathbf{r}\, d\mathbf{r}'\, \varphi^*_i(\mathbf{r})\varphi^*_j(\mathbf{r}')\varphi_m(\mathbf{r}')\varphi_n(\mathbf{r})/|\mathbf{r}-\mathbf{r}'|$. The interaction between light and matter is treated in the dipole approximation. Let $\boldsymbol{\mathcal{E}}(t)$ be the electric field and $\mathbf{d}_{\alpha\beta} \equiv \int d\mathbf{r}\, \varphi^*_\alpha(\mathbf{r})\mathbf{r}\varphi_\beta(\mathbf{r})$ the matrix elements of the dipole vector. The Hamiltonian of the light-matter interaction reads

$$\hat{H}_{\rm LM}(t) = \sum_{\alpha\beta,\sigma} \left[\boldsymbol{\mathcal{E}}(t)\cdot\mathbf{d}_{\alpha\beta}\right] \hat{c}^\dagger_{\alpha\sigma}\hat{c}_{\beta\sigma} \tag{2}$$

where the sum runs over all bound and continuum states. The time-evolution of any many-body state $|\Psi(t)\rangle$ is therefore governed by the Schrödinger equation $i\frac{d}{dt}|\Psi(t)\rangle = \hat{H}(t)|\Psi(t)\rangle$ where

$$\hat{H}(t) = \hat{H}_{\rm eq} + \hat{H}_{\rm LM}(t) \tag{3}$$

is the total Hamiltonian.

Let $\mathbf{E}(t)$ be the pump field and $\mathbf{e}(t)$ the probe field. The pump field is typically a few-cycle pulse with frequency centered around the transition from a valence state to a low-energy continuum state. Thus for $\boldsymbol{\mathcal{E}} = \mathbf{E}$ we can ignore the terms in the sum of Eq. (2) where both indices α, β are continuum indices. On the contrary the main effect of the probe field is to promote transitions in the bound sector. Therefore we can restrict the sum in Eq. (2) to bound indices $\alpha, \beta = i, j$ for $\boldsymbol{\mathcal{E}} = \mathbf{e}$. It follows that for $\boldsymbol{\mathcal{E}} = \mathbf{E}$ we have

$$\hat{H}_{\rm LM}(t) = \hat{H}_P(t) \equiv \sum_{ij,\sigma} [\mathbf{E}(t) \cdot \mathbf{d}_{ij}] \, \hat{c}^\dagger_{i\sigma} \hat{c}_{j\sigma} + \sum_{ik,\sigma} \Big([\mathbf{E}(t) \cdot \mathbf{d}_{ik}] \, \hat{c}^\dagger_{i\sigma} \hat{c}_{k\sigma} + {\rm H.c.} \Big), \tag{4}$$

whereas for $\boldsymbol{\mathcal{E}} = \mathbf{E} + \mathbf{e}$ we have

$$\hat{H}_{\rm LM}(t) = \hat{H}_{P\&p}(t) \equiv \hat{H}_{\rm P}(t) + \sum_{ij,\sigma} [\mathbf{e}(t) \cdot \mathbf{d}_{ij}] \, \hat{c}^\dagger_{i\sigma} \hat{c}_{j\sigma}. \tag{5}$$

These approximations are based on physical considerations and simplify the presentation. However, they are not necessary for the theory developed in the next section.

We are interested in calculating the time-resolved photoabsorption spectrum $\mathfrak{S}(\omega)$ for frequencies corresponding to transitions between bound many-particle states. Using the convention that quantities with a tilde denote the Fourier transform of the corresponding time-dependent quantities, the spectrum for the frequencies of interest is given by

$$\mathfrak{S}(\omega) = -2{\rm Im} \left[\omega \, \tilde{\mathbf{e}}^*(\omega) \cdot \tilde{\mathbf{d}}_p(\omega) \right] \tag{6}$$

where $\mathbf{d}_p$ is the *probe-induced* change of the time-dependent average of the dipole operator $\hat{\mathbf{d}} = \sum_{ij,\sigma} \mathbf{d}_{ij} \hat{c}^\dagger_{i\sigma} \hat{c}_{j\sigma}$. In order to calculate $\mathbf{d}_p$ one has to calculate the time-dependent average $\mathbf{d}_{P\&p}(t)$ of the dipole operator with $\hat{H}_{\rm LM} = \hat{H}_{P\&p}$, the time-dependent average $\mathbf{d}_P(t)$ of the dipole operator with $\hat{H}_{\rm LM} = \hat{H}_P$ and then subtract the two

$$\mathbf{d}_p(t) \equiv \mathbf{d}_{P\&p}(t) - \mathbf{d}_P(t). \tag{7}$$

We observe that $\tilde{\mathbf{d}}_p(\omega)$ is *not* proportional to $\tilde{\mathbf{e}}(\omega)$ since the system is out of equilibrium. The spectrum $\mathfrak{S}(\omega)$ depends on the entire function $\mathbf{e}(t)$ and not only on its frequency component $\tilde{\mathbf{e}}(\omega)$. In particular $\mathfrak{S}(\omega)$ depends on the delay between the pump and probe pulses.

3. NEGF+GKBA approach

In this section we describe the NEGF+GKBA approach to calculate $\mathbf{d}_{P\&p}$. The same approach can be used for $\mathbf{d}_P$ by setting $\mathbf{e} = 0$. The average $\mathbf{d}_{P\&p}(t) = -2i \sum_{ij} \mathbf{d}_{ij} G^<_{ji}(t,t)$ where $G^<(t,t)$ is the equal-time lesser component of the Green's function $G(z,z')$ with arguments on the Keldysh contour. The latter is a matrix function with indices in both the bound and extended sector and it satisfies the equation of motion

$$\Big[i\frac{d}{dz} - h_{\rm HF}(z)\Big] G(z,z') = \delta(z,z') + \int d\bar{z} \, \Sigma(z,\bar{z}) G(\bar{z},z'). \tag{8}$$

In accordance with the considerations of the previous section the HF hamiltonian in Eq. (8) reads

$$h_{{\rm HF},\alpha\beta}(z) = \begin{cases} \delta_{kk'} \epsilon_k & (\alpha,\beta) = (k,k') \\ \mathbf{E}(z) \cdot \mathbf{d}_{ik} & (\alpha,\beta) = (i,k) \\ \mathbf{E}(z) \cdot \mathbf{d}_{ki} & (\alpha,\beta) = (k,i) \\ h_{ij} + [\mathbf{E}(z) + \mathbf{e}(z)] \cdot \mathbf{d}_{ij} + \sum_{mn} w_{imnj} \rho_{nm}(z) & (\alpha,\beta) = (i,j) \end{cases} \tag{9}$$

Journal of Physics: Conference Series **696** (2016) 012004 doi:10.1088/1742-6596/696/1/012004

with $w_{imnj} \equiv 2v_{imnj} - v_{imjn}$ and $\rho_{nm}(z) = -iG_{nm}(z, z^+) = -iG^<_{nm}(t,t)$ independently of the branch of the contour where z lies. The correlation self-energy Σ is a functional of G and has nonvanishing matrix elements only in the bound sector. As for the calculation of $\mathbf{d}_{P\&p}$ we only need G with bound indices we project Eq. (8) onto the bound sector and find

$$\sum_m \left[i\frac{d}{dz}\delta_{im} - h_{\mathrm{HF},im}(z)\right]G_{mj}(z,z') - \sum_k \left[\mathbf{E}(z)\cdot\mathbf{d}_{ik}\right]G_{kj}(z,z') = \delta(z,z')\delta_{ij} + \sum_m \int d\bar{z}\,\Sigma_{im}(z,\bar{z})G_{mj}(\bar{z},z'). \tag{10}$$

Similarly we can obtain an equation for G_{kj}. Taking into account Eq. (9) we find

$$\left[i\frac{d}{dz} - \epsilon_k\right]G_{kj}(z,z') - \sum_m \left[\mathbf{E}(z)\cdot\mathbf{d}_{km}\right]G_{mj}(z,z') = \delta(z,z'). \tag{11}$$

Next we define the noninteracting Green's function g_k as the solution of

$$\left[i\frac{d}{dz} - \epsilon_k\right]g_k(z,z') = \delta(z,z') \tag{12}$$

with Kubo-Martin-Schwinger boundary conditions, and rewrite Eq. (11) in integral form

$$G_{kj}(z,z') = \int d\bar{z}\,g_k(z,\bar{z})\sum_m \left[\mathbf{E}(\bar{z})\cdot\mathbf{d}_{km}\right]G_{mj}(\bar{z},z'). \tag{13}$$

Substitution of this result into Eq. (10) leads to

$$\sum_m \left[i\frac{d}{dz}\delta_{im} - h_{\mathrm{HF},im}(z)\right]G_{mj}(z,z') = \delta(z,z')\delta_{ij} + \sum_m \int d\bar{z}\left[\Sigma_{\mathrm{ion}}(z,\bar{z}) + \Sigma(z,\bar{z})\right]_{im}G_{mj}(\bar{z},z'), \tag{14}$$

where we defined the ionization self-energy according to

$$\Sigma_{\mathrm{ion},ij}(z,z') = \sum_k \left[\mathbf{E}(z)\cdot\mathbf{d}_{ik}\right]g_k(z,z')\left[\mathbf{E}(z')\cdot\mathbf{d}_{kj}\right]. \tag{15}$$

Although the ionization self-energy resembles the embedding self-energy in quantum transport [73–75] its physical origin is different, see below.

Equation (14) is a closed equation for the Green's function in the bound sector. Subtracting from it the adjoint of Eq. (14) and setting $z = z^+$ we obtain an equation for the one-particle density matrix $\rho_{ij}(t)$ with indices in the bound sector

$$-i\frac{d}{dt}\rho(t) + \left[h_{\mathrm{HF}}(t), \rho(t)\right] = i\left[I(t) + I_{\mathrm{ion}}(t)\right] - \mathrm{H.c.}. \tag{16}$$

where the collision integral $I(t)$ reads

$$I(t) = \int_{-\infty}^{t} dt'\left[\Sigma^<(t,t')G^>(t',t) + \Sigma^>(t,t')G^<(t',t)\right], \tag{17}$$

and the ionization integral $I_{\mathrm{ion}}(t)$ is defined as $I(t)$ with Σ_{ion} in place of Σ. Due to the presence of the off-diagonal in time $G^{\lessgtr}$ in the collision and ionization integrals Eq. (16) is not a closed equation for ρ. The GKBA is an approximation that relates $G^{\lessgtr}(t,t')$ to $\rho(t)$ and $\rho(t')$ [72],

Journal of Physics: Conference Series **696** (2016) 012004 doi:10.1088/1742-6596/696/1/012004

thereby transforming Eq. (16) into a closed integro-differential equation for ρ. According to the GKBA

$$G^{<}(t,t') = -G^{\rm R}(t,t')\rho(t') + \rho(t)G^{\rm A}(t,t'), \quad (18a)$$
$$G^{>}(t,t') = +G^{\rm R}(t,t')\bar{\rho}(t') - \bar{\rho}(t)G^{\rm A}(t,t'), \quad (18b)$$

with $\bar{\rho}_{ij} = \delta_{ij} - \rho_{ij}$. It is worth noting that the GKBA results are extremely sensitive to the choice of the quasiparticle propagator $G^{\rm R}(t,t') = [G^{\rm A}(t',t)]^{\dagger}$. For $\Sigma = 0$ (no correlations) the Green's functions $G^{\lessgtr}$ that solve the equation of motion (10) can be written as in Eqs. (18) provided that

$$G^{\rm R}(t,t') = -i\theta(t-t')\,T\left[e^{-i\int_{t'}^{t}d\bar{t}\,h_{\rm HF}(\bar{t})}\right] \quad (19)$$

is the HF propagator. The numerical simulations presented in Ref. [41] show that for atomic systems the HF propagator yields accurate results in the correlated case too. Similar accuracies were recently reported for Hubbard clusters [76] and for the two-dimensional Anderson–Hubbard model [77]. A nice property of the HF propagator is that it guarantees the satisfaction of all basic conservation laws provided that the self-energies is conserving [78]. The results presented in this work have been obtained using Eq. (19). Nevertheless, it should be emphasized that the performance of the HF propagator is not always this good for all systems. In open systems, e.g., in molecular junctions attached to metallic electrodes, it is pivotal to include relaxation effects due to electron collisions and to the presence of the electrodes (electronic bath) [79].

Before closing this section let us discuss and simplify the ionization self-energy. From Eq. (12) we have

$$g_{k}^{\lessgtr}(t,t') = -if^{\lessgtr}(\epsilon_k)e^{-i\epsilon_k(t-t')} \quad (20)$$

with $f^{<}(\epsilon_k)$ the Fermi function and $f^{>} = 1 - f^{<}$. As the energy of the extended states is larger than the chemical potential we have $f^{<}(\epsilon_k) = 0$ and $f^{>}(\epsilon_k) = 1$. This implies that the lesser ionization self-energy vanishes. On the other hand the greater component reads

$$\Sigma^{>}_{{\rm ion},ij}(t,t') = -i\sum_{k}\left[\mathbf{E}(t)\cdot\mathbf{d}_{ik}\right]e^{-i\epsilon_k(t-t')}\left[\mathbf{E}(t')\cdot\mathbf{d}_{kj}\right]$$
$$= \mathbf{E}(t)\overleftrightarrow{\sigma}_{ij}(t-t')\mathbf{E}(t'), \quad (21)$$

where in the last equality we defined the dyad $\overleftrightarrow{\sigma}_{ij}(t) = -i\sum_k \mathbf{d}_{ik}e^{-i\epsilon_k t}\mathbf{d}_{kj}$. In Fourier space

$$\overleftrightarrow{\sigma}_{ij}(\omega) = -2\pi i\sum_{k}\mathbf{d}_{ik}\,\delta(\omega-\epsilon_k)\,\mathbf{d}_{kj} \approx 2i\sum_{k}\mathbf{d}_{ik}\,{\rm Im}\left[\frac{1}{\omega-\epsilon_k+i\eta}\right]\mathbf{d}_{kj}, \quad (22)$$

where η is a positive constant of the order of the level spacing of the continuum states. We already mentioned that the pump pulse is a few cycles pulse centered around some ionization frequency ω_P. Therefore, $\Sigma^{>}_{\rm ion}$ is dominated by those terms in $\overleftrightarrow{\sigma}(t-t')$ that oscillate at frequency $\epsilon_k \simeq \omega_P$. Then, for simplicity, we implement a time-local approximation $\overleftrightarrow{\sigma}_{ij}(\omega) \approx \overleftrightarrow{\sigma}_{ij}(\omega_P)$ which implies $\overleftrightarrow{\sigma}_{ij}(t-t') = \overleftrightarrow{\sigma}_{ij}(\omega_P)\delta(t-t')$. Substituting this result into Eq. (21) yields

$$\Sigma^{>}_{\rm ion}(t,t') = -i\delta(t-t')\Gamma(t), \quad (23)$$

where

$$\Gamma_{ij}(t) = i\mathbf{E}(t)\overleftrightarrow{\sigma}_{ij}(\omega_P)\mathbf{E}(t) \quad (24)$$

is a self-adjoint positive-definite matrix for all times t. Inserting this result into the ionization integral the equation of motion (16) for ρ becomes

$$-i\frac{d}{dt}\rho(t) + [h_{\rm HF}(t),\rho(t)] - i\left\{\Gamma(t),\rho(t)\right\} = iI(t) - {\rm H.c.} \quad (25)$$

Journal of Physics: Conference Series **696** (2016) 012004 doi:10.1088/1742-6596/696/1/012004

where the curly brackets signify an anticommutator. From Eq. (25) we clearly see that the ionization self-energy is responsible for taking away particles from the system. We also see that the draining of particles is effective only during the pump since $\Gamma(t)$ vanishes for $\mathbf{E}(t) = 0$. This corresponds to the intuitive picture that no ionization occurs after the pump pulse has passed through the system.

4. Numerical Results

In this Section we present numerical results on a simple model system in order to illustrate the performance of various approximations in describing the strong-field multiple ionization and the subsequent formation of a mixture of different ions. The model consists of two levels, one describing a $3d$ orbital (level 1) and the other describing a $4p$ orbital (level 2) of a Kr atom. The pump couples level 1 and 2 via the dipole matrix $\mathbf{d}_{ij} = \hat{\boldsymbol{x}} d_0(1-\delta_{ij})$, and at the same time couples level 2 to the continuum states, thus generating a ionization dyad $\overleftrightarrow{\sigma}_{ij}(\omega_P) = -i\sigma_0\delta_{i2}\delta_{j2}$. The remaining quantities of the model are $h_{ij} = \varepsilon_i\delta_{ij}$ and $v_{imnj} = U_{im}\delta_{ij}\delta_{mn}$, with $i, j = 1, 2$. The numerical values of the model parameters have been chosen to closely reproduce energy- and time-scales of the first-principle simulations of Ref. [41]. These values are all reported in the caption of Fig. 1.

Equation (25) has been solved within two different approximations, namely HF and 2B. In the first case the collision integral $I(t)$ is identically zero, while within 2B we have [61]

$$I^{(2B)}_{ik}(t) = \sum_{\substack{nm\\pq\\sr}}\sum_{j} v_{irpn}w_{mqsj}\int_{-\infty}^{t} dt' \Big[G^{<}_{nm}(t,t')G^{<}_{pq}(t,t')G^{>}_{sr}(t',t)G^{>}_{jk}(t',t) + G^{>}_{nm}(t,t')G^{>}_{pq}(t,t')G^{<}_{sr}(t',t)G^{<}_{jk}(t',t)\Big]. \quad (26)$$

The real-time propagation has been performed with a time step $\Delta = 0.1$ a.u. and the spectra have been obtained by a discrete Fourier transform of the probe-induced dipole with $N_t = 10^4$ time-steps.

The atom, initially in its ground state with both levels filled, is ionized by a few-cycle pump $\mathbf{E}(t) = \boldsymbol{\eta}_P E(t)$ where $E(t) = E_0 \sin^2(\pi t/\Delta_P)\sin(\omega_P t)$. The expelled charge as a function of time is diplayed in Fig. 1. During the ionization process the occupation of level 1 remains roughly constant, meaning that the atom is loosing charge mainly from level 2. We also notice that the time-evolution of the expelled charge calculated within HF and 2B is essentially the same. This fact could suggest that the system is weakly correlated and that HF well captures the physics of the problem. However, the expelled charge is just one possible observable. Below we show that the HF and 2B photoabsorption spectra exhibit qualitative different features (although the local density in these two approximations agrees with high accuracy).

After a time τ the system is probed with an ultrashort pulse $\mathbf{e}(t) = \boldsymbol{\eta}_p e(t)$, with $e(t) = e_0\sin^2(\pi(t-\tilde{\tau})/\Delta_p)\sin(\omega_p(t-\tilde{\tau}))$ for $\tilde{\tau} < t < \tilde{\tau} + \Delta_p$. Here $\tau = \tilde{\tau} - (\Delta_P - \Delta_p)/2$ is the time distance between the maxima of the pump and probe pulses. The frequency ω_p of the probe is chosen to promote the intra-atomic excitation between the levels 1 and 2.

The transient absorption spectrum within the HF approximation is displayed in the top-left panel of Fig. 2. Only one absorption peak appears, corresponding to the absorption of a singly ionized atom. We verified that the only effect of increasing the intensity of the pump is a shift in the position of the peak. We also varied the frequency and duration of the pump but we found no sign of absorption peaks due multiply ionized atoms. A single-peak absorption spectrum is also observed within the M2B approximation [41] (not shown), thus suggesting that the solution of Eq. (25) without the inclusion of memory effects in the self-energy cannot describe multiple ionization. However, the occurrence of multiple ionization (and subsequent absorption) for a strong enough pump is physically expected [28].

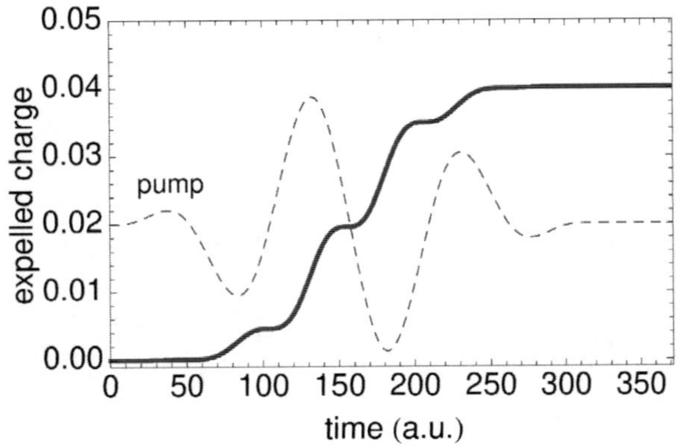

Figure 1: Total expelled charge (solid line) as a function of time. The pump profile $E(t)$ (in arbitrary units) is displayed as a dashed line. The model parameters (in atomic units) are: $\varepsilon_1 = -6.82$, $\varepsilon_2 = -2.25$, $U_{11} = 1.8$, $U_{22} = 0.5$, $U_{12} = U_{21} = 0.6$, $d_0 = 0.1$, $\sigma_0 = 1.8$, $E_0 = 0.14$, $\omega_P = 0.061$, $\Delta_P = 314$, $\boldsymbol{\eta}_P = \hat{\boldsymbol{x}}$. The expelled charge does not vary appreciably within the different approximations used in this work, see main text.

We then performed a full 2B simulation, by including time-nonlocal correlation effects according to Eq. (26), and calculated the time-resolved spectrum, see the top-right panel of Fig. 2. We clearly see the development of a second peak at higher energy, corresponding to the absorption of a doubly ionized atom. Furthermore the onset of the high-energy peak as function of τ is *retarded* with respect to the onset of the absorption peak at lower energy. This is consistent with the fact that expelling two electron requires more time than expelling one electron, and is also in qualitative agreement with both the experimental data [28] and our previous analysis [41].

We argue that the difference between time-local approximations, like HF and M2B, and the 2B approximation in describing the double ionization process is due to the absence of multiple excitations in the formers. Multiple excitation are a direct consequence of the simultaneous presence of *correlation* and *memory* effects in the self-energy. We therefore conjecture that correlations alone are not sufficient to describe the absorption of multiply ionized atoms as obtained from a real-time propagation. To reinforce this conjecture we have performed a TDDFT calculation using the exact-adiabatic (EX-ADIA) exchange-correlation potential of the model system. The EX-ADIA approximation produces the exact level densities in equilibrium. However, being it an adiabatic approximation, the EX-ADIA ignores memory effects entirely. Within this approximation the time propagation is performed by setting the collision integral $I(t) = 0$ and by replacing in Eq. (25) the HF Hamiltonian $h_{\rm HF}(t)$ with the Kohn-Sham Hamiltonian $h_{\rm KS}(t)$. The matrix element of the Kohn-Sham Hamiltonian are $h_{{\rm KS},ij}(t) = h_{ij} + [\mathbf{E}(t) + \mathbf{e}(t)] \cdot \mathbf{d}_{ij} + \delta_{ij} v_{\rm Hxc}[n_1, n_2]$, with $n_i = \rho_{ii}$ the density at level i. Since the occupation numbers commute with the equilibrium Hamiltonian, the exact Hartree+exchange-correlation potential $v_{\rm Hxc}$ of the model is a step-wise function of the total density $n_1 + n_2$ [80, 81], and is explicitly displayed in the bottom-left panel Fig. 2. In the bottom-right panel of Fig. 2 we show the corresponding EX-ADIA absorption spectrum. Notice that during and after the action of the pump the total number of electrons $n_1 + n_2$ remains close to four and hence the $v_{\rm Hxc}$ value stays on the highest energy plateau. This implies that $v_{\rm Hxc}$ is essentially a constant energy shift and consequently the spectrum is similar to the noninteracting spectrum. In particular the position of the peak is very close to the single-particle energy difference $\varepsilon_2 - \varepsilon_1 = 4.57$ a.u.. Even more important for our analysis is that in this case too only a single peak develops, no matter of the strenght of the pump, thus confirming our conjecture that time-local approximations do

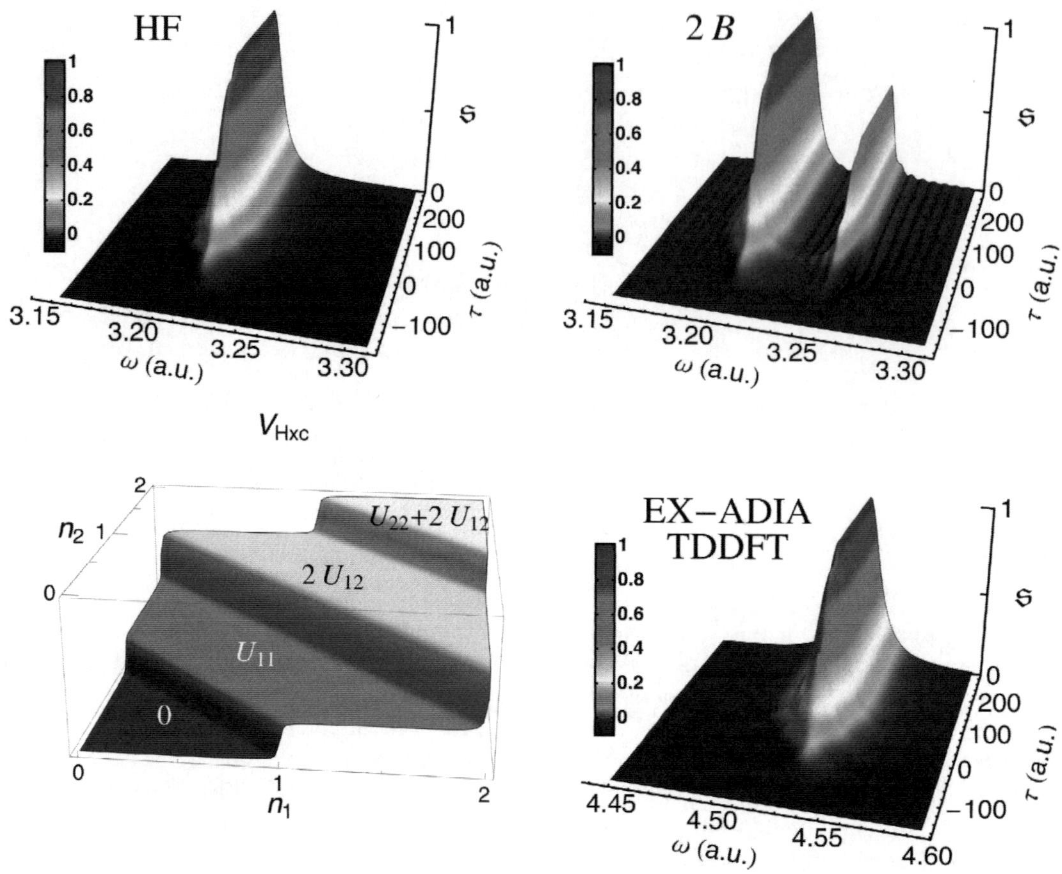

Figure 2: TPA spectrum (normalized to the maximum height) within HF (top-left), 2B (top-right) and EX-ADIA TDDFT (bottom-right). The probe parameters are (in atomic units) $e_0 = 1.7 \times 10^{-3}$, $\omega_p = 2.94$, $\Delta_p = 6.19$, $\boldsymbol{\eta}_p = \hat{\boldsymbol{x}}$. A broadening of 0.03 a.u. has been introduced in order to smoothen the finite-time Fourier transform of $\mathbf{d}_p(t)$. In the bottom-left panel we show the exact Hartree+exchange-correlation potential $v_{\rm Hxc}$ of the model as a function of the densities n_i of the two levels. The value of $v_{\rm Hxc}$ in the different plateaus is reported explicitly as a function of the repulsion parameters U_{ij}. The exact sharp discontinuities occurring at integer values of the total density $n_1 + n_2$ have been artificially broadened for graphical reasons.

not describe the absorption of multiply ionized atoms.

5. Conclusions

We discussed the recently proposed NEGF-based approach to calculate the time-resolved photoabsorption spectrum of finite systems [41]. The approach relies on the GKBA and amounts to solving an approximate equation of motion for the one-particle density matrix of the system. The external driving fields are treated to all orders, while correlation effects are included systematically by means of non-equilbrium many-body perturbation theory. Strong-field ionization can be accurately described via an appropriate ionization self-energy $\Sigma_{\rm ion}$, that accounts for the coupling to a continuum of extended states. The inclusion of $\Sigma_{\rm ion}$ renders the real-time evolution not unitary and it is responsible for the loss of electrons during the ionization process.

We studied a two-level model coupled to a continuum to illustrate how different approximations perform in describing the time-resolved photoabsorption spectrum of finite systems under the action of a strong ionizing pump. The model stems from our analysis [41] of a recent experiment on Krypton [28] where the pump causes multiple ionization, as one can infer from the presence of different absorption peaks attributable to Kr^{+}, Kr^{2+} and Kr^{3+}. The proper description of the produced admixture of ions constitutes the central issue of the present work. We computed the spectrum using the time-local approximations HF and M2B for the collision integral as well as using the EX-ADIA approximation in TDDFT. In all cases we only observed the absorption from a single ion. Our numerical results show unambiguosly that 2B is the only scheme able to provide more than a single absorption peak. We conclude that the inclusion of time-nonlocal correlations is crucial for a correct description of the absorption spectra of an evolving admixture of ionizing atoms.

Acknowledgments

EP and GS acknowledge funding by MIUR FIRB Grant No. RBFR12SW0J. RvL thanks the Academy of Finland for support.

References

[1] F. Krausz and M. Ivanov, Rev. Mod. Phys. **81**, 163 (2009).

[2] R. Berera, R. van Grondelleand, and J. T. M. Kennis, Photosynth. Res. **101**, 105 (2009).

[3] G. Sansone, T. Pfeifer, K. Simeonidis, and A. I. Kuleff, Chem. Phys. Chem. **13**, 661 (2012).

[4] L. Gallmann, J. Herrmann, R. Locher, M. Sabbar, A. Ludwig, M. Lucchini, and U. Keller, Mol. Phys. **111**, 2243 (2013).

[5] A. I. Kuleff and L. S. Cederbaum, J. Phys. B: At. Mol. Opt. Phys. **47**, 124002 (2014).

[6] R. A. Kaindl, M. A. Carnahan, D. Hägele, R. Lövenich, and D. S. Chemla, Nature (London) **423**, 734 (2003).

[7] F. Wang, G. Dukovic, L. E. Brus, and T. F. Heinz, Science **308**, 838 (2005).

[8] S. W. Koch, M. Kira, G. Khitrova, and H. M. Gibbs, Nat. Mater. **5**, 523 (2006).

[9] J. Piris, T. E. Dykstra, A. A. Bakulin, P. H. M. van Loosdrecht, W. Knulst, M. T. Trinh, J. M. Schins, and L. D. A. Siebbeles, J. Phys. Chem. C **113**, 14500 (2009).

[10] S. R. Cowan, N. Banerji, W. L. Leong, and A. J. Heeger, Adv. Funct. Mater. **22**, 1116 (2012).

[11] V. May and O. Kühn, *Charge and Energy Transfer Dynamics in Molecular Systems* (Wiley, Weinheim, 2004).

[12] J. B. Asbury, E. Hao, Y. Wang, and T. Lian, J. Phys. Chem. B **104**, 11957 (2000).

[13] J. Schnadt, P. A. Brühwiler, L. Patthey, J. N. O'Shea, S. Södergren, M. Odelius, R. Ahuja, O. Karis, M. Bässler, P. Persson et al., Nature (London) **418**, 620 (2002).

[14] A . Föhlisch, P. Feulner, F. Hennies, A. Fink, D. Menzel, D. Sanchez-Portal, P. M. Echenique, and W. Wurth, Nature (London) **436**, 373 (2005).

[15] H. B. Gray and J. R. Winkler, Proc. Natl. Acad. Sci. USA **102**, 3534 (2005).

[16] F. Remacle and R. D. Levine, Proc. Natl. Acad. Sci. USA **103**, 6793 (2006).

[17] C. A. Rozzi, S. M. Falke, N. Spallanzani, A. Rubio, E. Molinari, D. Brida, M. Maiuri, G. Cerullo, H. Schramm, J. Christoffers et al., Nat. Commun. **4**, 1602 (2013).

Journal of Physics: Conference Series **696** (2016) 012004 doi:10.1088/1742-6596/696/1/012004

[18] S. M. Falke, C. A. Rozzi, D. Brida, M. Maiuri, M. Amato, E. Sommer, A. D. Sio, A. Rubio, G. Cerullo, E. Molinari et al., Science **344**, 1001 (2014).

[19] S. Pittalis, A. Delgado, J. Robin, L. Freimuth, J. Christoffers, C. Lienau, and C. A. Rozzi, Adv. Funct. Mater. **25**, 2047 (2015).

[20] R. Huber, F. Tauser, A. Brodschelm, M. Bichler, G. Abstreiter, and A. Leitenstorfer, Nature (London) **414**, 286 (2001).

[21] M. Hase, M. Kitajima, A. M. Constantinescu, and H. Petek, Nature (London) **426**, 51 (2003).

[22] A. Borisov, D. Sánchez-Portal, R. Muĩno, and P. Echenique, Chem. Phys. Lett. 387, **95** (2004).

[23] A. S. Moskalenko, Y. Pavlyukh, and J. Berakdar, Phys. Rev. A **86**, 013202 (2012).

[24] M. Drescher, M. Hentschel, R. Kienberger, M. Uiberacker, V. Yakovlev, A. Scrinzi, T. Westerwalbesloh, U. Kleineberg, U. Heinzmann, and F. Krausz, Nature (London) **419**, 803 (2002).

[25] H. Niikura, D. M. Villeneuve, and P. B. Corkum, Phys. Rev. Lett. **94**, 083003 (2005).

[26] O. Smirnova, Y. Mairesse, S. Patchkovskii, N. Dudovich, D. Villeneuve, P. Corkum, and M. Y. Ivanov, Nature (London) **460**, 972 (2009).

[27] O. Smirnovaa, S. Patchkovskiia, Y. Mairessea, N. Dudovich, and M. Y. Ivanov, Proc. Natl. Acad. Sci. USA **106**, 16556 (2009).

[28] E. Goulielmakis, Z. Loh, A. Wirth, R. Santra, N. Rohringer, V. S. Yakovlev, S. Zherebtsov, T. Pfeifer, A. M. Azzeer, M. F. Kling et al., Nature (London) **466**, 739 (2010).

[29] S. Pabst, L. Greenman, P. J. Ho, D. A. Mazziotti, and R. Santra, Phys. Rev. Lett. **106**, 053003 (2011).

[30] A. H. Zewail, J. Phys. Chem. A **104**, 5660 (2000).

[31] L. Chen, Annu. Rev. Phys. Chem. **56**, 221 (2005).

[32] G. Sansone, F. Kelkensberg, J. F. Pérez-Torres, F. Morales, M. F. Kling, W. Siu, O. Ghafur, P. Johnsson, M. Swoboda, E. Benedetti et al., Nature (London) **465**, 763 (2010).

[33] C. P. Lawrence and J. L. Skinner, Chem. Phys. Lett. **369**, 472 (2003).

[34] M. B. Gaarde, C. Buth, J. L. Tate, and K. J. Schafer, Phys. Rev. A **83**, 013419 (2011).

[35] R. Santra, V. S. Yakovlev, T. Pfeifer, and Z.-H. Loh, Phys. Rev. A **83**, 033405 (2011).

[36] W.-C. Chu and C. D. Lin, Phys. Rev. A **85**, 013409 (2012).

[37] J. C. Baggesen, E. Lindroth, and L. B. Madsen, Phys. Rev. A **85**, 013415 (2012).

[38] A. D. Dutoi, K. Gokhberg, and L. S. Cederbaum, Phys. Rev. A **88**, 013419 (2013).

[39] E. Perfetto and G. Stefanucci, Phys. Rev. A **91**, 033416 (2015).

[40] E. Perfetto, D. Sangalli, A. Marini, and G. Stefanucci, cond-mat/arXiv:1507.01786.

[41] E. Perfetto, A.-M. Uimonen, R. van Leeuwen and G. Stefanucci, Phys. Rev. A **92**, 033419 (2015).

[42] K. C. Kulander, T. N. Rescigno, Comp. Phys. Comm. **63**, 523 (1991).

[43] K. J. Schafer, in *Strong Field Laser Physics*, Eds. T. Brabec (Springer, 2008), pp 111-145.

[44] U. De Giovannini, G. Brunetto, A. Castro, J. Walkenhorst, and A. Rubio, Chem. Phys. Chem. **14**, 1363 (2013).

[45] E. Runge and E. K. U. Gross, Phys. Rev. Lett. **52**, 997 (1984).

[46] M. Ruggenthaler, and R. van Leeuwen, EPL **95**, 13001 (2011).

[47] N. T. Maitra, F. Zhang, R. J. Cave, and K. Burke, J. Chem. Phys. **120**, 5932 (2004).

[48] S. Kümmel and L. Kronik, Rev. Mod. Phys. **80**, 3 (2008).

[49] J. B. Neaton, M. S. Hybertsen, and S. G. Louie, Phys. Rev. Lett. **97**, 216405 (2006).

[50] K. S. Thygesen and A. Rubio, Phys. Rev. Lett. **102**, 046802 (2009).

[51] J. M. Garcia-Lastra, C. Rostgaard, A. Rubio, and K. S. Thygesen, Phys. Rev. B **80**, 245427 (2009).

[52] K. Kaasbjerg and K. Flensberg, Nano Lett. **8**, 3809 (2008).

[53] K. Kaasbjerg and K. Flensberg, Phys. Rev. B **84**, 115457 (2011).

[54] P. Myöhänen, R. Tuovinen, T. Korhonen, G. Stefanucci, and R. van Leeuwen, Phys. Rev. B **85**, 075105 (2012).

[55] O. Gritsenko and E. J. Baerends, J. Chem. Phys. **121**, 655 (2004).

[56] N. T. Maitra, J. Chem. Phys. **122**, 234104 (2005).

[57] N. T. Maitra and D. G. Tempel, J. Chem. Phys. **125**, 184111 (2006).

[58] L. P. Kadanoff and G. Baym, *Quantum Statistical Mechanics* (W. A. Benjamin, New York, 1962).

[59] P. Danielewicz, Ann. Phys. (N.Y.) **152**, 239 (1984).

[60] H. Haug and A.-P. Jauho, *Quantum Kinetics in Transport and Optics of Semiconductors* (Springer, Berlin, 2007).

[61] G. Stefanucci and R. van Leeuwen, *Nonequilibrium Many-Body Theory of Quantum Systems: A Modern Introduction* (Cambridge University Press, Cambridge, UK, 2013).

[62] K. Balzer and M. Bonitz, *Nonequilibrium Greens Functions Approach to Inhomogeneous Systems*, Lecture Notes in Physics, Vol. 867 (Springer-Verlag, Berlin, Heidelberg, 2013).

[63] W. Schäfer, J. Opt. Soc. Am. B **13**, 1291 (1996).

[64] R. Binder, H. S. Köhler, M. Bonitz, and N. Kwong, Phys. Rev. B **55**, 5110 (1997).

[65] N.-H. Kwong and M. Bonitz, Phys. Rev. Lett. **84**, 1768 (2000).

[66] N. E. Dahlen and R. van Leeuwen, Phys. Rev. Lett. **98**, 153004 (2007).

[67] P. Myöhänen, A. Stan, G. Stefanucci, and R. van Leeuwen, Europhys. Lett. **84**, 67001 (2008).

[68] P. Myöhänen, A. Stan, G. Stefanucci, and R. van Leeuwen, Phys. Rev. B **80**, 115107 (2009).

[69] M. P. von Friesen, C. Verdozzi, and C.-O. Almbladh, Phys. Rev. Lett. **103**, 176404 (2009).

[70] M. P. von Friesen, C. Verdozzi, and C.-O. Almbladh, Phys. Rev. B **82**, 155108 (2010).

[71] K. Balzer, S. Bauch, and M. Bonitz, Phys. Rev. A **81**, 022510 (2010).

[72] P. Lipavský and V. Špička and B. Velický, Phys. Rev. B **34**, 6933 (1986).

[73] Y. Meir and N. S. Wingreen, Phys. Rev. Lett. **68**, 2512 (1992).

[74] A.-P. Jauho, N. S. Wingreen, and Y. Meir, Phys. Rev. B **50**, 5528 (1994).

[75] G. Stefanucci and C.O. Almbladh, Phys. Rev. B **69**, 195318 (2004).

[76] N. Schlünzen, S. Hermanns, M. Bonitz, and C. Verdozzi, cond-mat/arXiv:1508.02947.

[77] Y. Bar Lev and D. R. Reichman, cond-mat/arXiv:1508.05391

[78] S. Hermanns, N. Schlünzen, and M. Bonitz, Phys. Rev. B **90**, 125111 (2014).

[79] S. Latini, E. Perfetto, A.-M. Uimonen, R. van Leeuwen, and G. Stefanucci, Phys. Rev. B **89**, 075306 (2014).

[80] S. Kurth and G. Stefanucci, Phys. Rev. Lett. **111**, 030601 (2013).

[81] G. Stefanucci and S. Kurth, Phys. Status Solidi **250**, 2378 (2013).

Journal of Physics: Conference Series **696** (2016) 012005 doi:10.1088/1742-6596/696/1/012005

Single and double photoemission and generalizations

Yaroslav Pavlyukh

E-mail: yaroslav.pavlyukh@physik.uni-halle.de
Institut für Physik, Martin-Luther-Universität Halle-Wittenberg, 06120 Halle, Germany

Abstract. A unified diagrammatic treatment of single and double electron photoemission currents is presented. The irreducible lesser density-density response function is the starting point of these derivations. Diagrams for higher order processes in which several electrons are observed in coincidence can likewise be obtained. For physically relevant situations, in which the photoemission cross-section can be written as the Fermi Golden rule, the diagrams from the nonequilibrium Green's function approach can be put in direct correspondence with the diagrams of the scattering theory.

1. Introduction

The electron photoemission is a process in which classical or quantized electromagnetic field interacts with a many-body target leaving the system in an unbound electronic state [1]. This subclass of optical absorption processes can further be categorized according to the number of emitted particles in one scattering event, i.e., single, double or, more generally, n-electron photoemission. The exact quantum state the ionized target was left in is typically not known. However, since detectors can determine the asymptotic scattering state of emitted electrons some information about the ionized target can be inferred by virtue of energy and momentum conservation laws. In particular, one can tell whether the target was left in the ground or an excited state. Thus, we can distinguish between the no-loss photoelectron current and the photoelectron signal incorporating losses. The processes leading to the emission of an electron with a reduced energy can have various origins. On the basis of the empirical three step model [2] we have a picture where a liberated electron propagates through the sample loosing its energy on the way. In this way the electron-photon interaction and the inelastic electron scattering are considered as two independent processes and for the latter one speaks about the *extrinsic* energy losses [3, 4]. Besides electronic excitations the phonon [5] and impurity [6] scatterings, and Auger processes [7, 8] are important.

On the other hand electron-electron interaction in the system causes the spectral function (or the density of states) to deviate from the non-interacting one in which all spectral weight is confined to a quasiparticle peak. A typical picture would be a quasiparticle peak followed by a sequence of satellite peaks. Their origin can be different as well [9, 10, 11]. For instance, in the photoemission from a deep core level it is the density oscillations in the system — plasmons — that form scattering channels responsible for occurrence of plasmonic satellites. S-model, proposed by Lundqvist [12] and solved by Langreth [13], captures exactly this effect and predicts equally spaced satellite peaks with oscillator strengths satisfying the Poisson distribution. If an electron originates from one of such peaks, it will come to a detector with a reduced energy. One speaks here about *intrinsic* energy losses because it is the ground state spectral density that is modified by plasmon scattering [14, 15].

Although first microscopic theories of photoemission [16, 17] have taken underlying electronic structure into account, they have not provided a description of inelastic energy losses experienced by the liberated electron. Such formulation was achieved with the help of nonequilibrium Green's function approach by Caroli *et al.* [18]. The starting point of their treatment is the zeroth order triangular diagram where two vertices are situated on the forward and backward tracks of the Keldysh contour [19] and describe the light-matter interaction. The third vertex lying at the common points of two tracks in the remote future describes the detector state. There are no interaction lines because the photoemission is possible even for non-interacting systems. Each electron-photon interaction vertex is associated with the minimal coupling part of the Hamiltonian and is linearly proportional to the field amplitude. The whole diagram, as expected, is quadratic in field strength and is linearly proportional to the intensity or the number of photons.

Progressing diagrammatically further [20] one realizes that dressing of this diagram leads not only to the renormalization of each electronic propagator involved, but also generates vertex functions [21]. All these ingredients are important in different physical scenarios, they account for the effects of optical field screening [22, 23, 24], intrinsic and extrinsic losses as well as their interference [25, 26, 27], and for the formation of the scattering states [28]. The goal of this work is to give a diagrammatic description of the multiparticle emission. Before proceeding with corresponding diagrammatic construction, a short account of the single photoemission will be given in Sec. II. Properties of double photoemission diagrams are discussed in Sec. III and generalization to n-particle emission is presented in Sec. IV. The formalism looks more natural if the irreducible density-density response function describing general photoabsorption process is used as a starting point. Such a point of view has already been mentioned by Feibelman and Eastman [29] in their derivation of the Fermi Golden rule (FGR) expression for photoemission starting with diagrammatic approach of Caroli *et al.* [18]. Supported by work on positive definite diagrammatic approximations [21, 30, 31] a one-to-one correspondence between the NEGF (provided they can be written in the Fermi Golden rule form) and Goldstone diagrams in scattering theory [32, 33, 34] can established and is outlined below.

In this work atomic units are used throughout.

2. Ingredients of SPE

The differential cross-section fully describes the photoemission experiment. It relates the electron current $J_{\mathbf{k}}$ to the incoming photon flux density F:

$$\frac{d\sigma}{d\mathbf{k}} = \frac{J_{\mathbf{k}}}{F}$$

The incident photon flux is obtained by dividing the incident intensity $I = \frac{\omega^2 A^2}{2\pi c}$ by the photon energy ω. The current is determined by the electronic structure of the system and by details of its coupling to the electromagnetic field, in velocity gauge we have $\hat{\Delta} = \frac{1}{c}\mathbf{A}\cdot\hat{\mathbf{p}}$, where $\hat{\mathbf{p}}$ is the momentum operator. Many-body perturbation theory yields diagrammatic expansion of the current [16, 18, 21, 20, 24, 35] in terms of Green's functions on the Keldysh contour [19]. It results in

$$J_{\mathbf{k}} = \lim_{\eta\to 0} 2\eta \int_{-\infty}^{0} d(tt') e^{\eta(t+t')} (\mathbf{\Delta}^* \mathbf{Z}^{(1)}(t,t')\mathbf{\Delta})_{\mathbf{k}}, \tag{1}$$

with the following ground state correlator

$$Z^{(1)}(t,t') = \langle \Psi_0 | c_b^\dagger(t) c_a(t) c_{\mathbf{k}}^\dagger(0) c_{\mathbf{k}}(0) c_c^\dagger(t') c_d(t') | \Psi_0 \rangle, \tag{2}$$

where the field operators are in the Heisenberg representation. Three parts can be seen: operators dependent on the times $t, t' \in (-\infty, 0]$ originate from the light-matter couplings $\hat{\Delta}^\dagger(t) = \sum_{ab} \Delta_{ba}^* c_b^\dagger(t) c_a(t)$ and $\hat{\Delta}(t') = \sum_{cd} \Delta_{cd} c_c^\dagger(t') c_d(t')$, whereas $c_{\mathbf{k}}^\dagger(0) c_{\mathbf{k}}(0)$ yields the current operator. This is the reason why $Z^{(1)}(t,t')$ is sometimes called the *three-current* correlator. The system interacts with electromagnetic field in the interval of time from remote past to the moment $t = 0$ where the observation takes place. This entails the specific time-order of the operators in the correlator and is

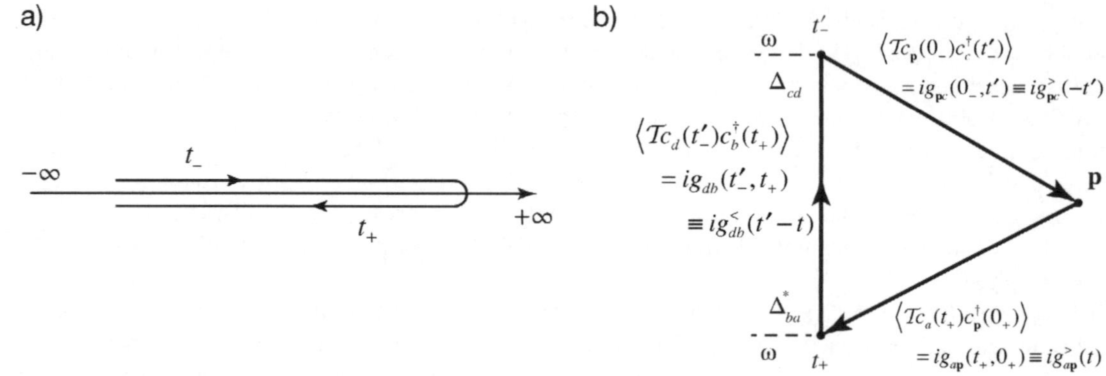

Figure 1. (a) The Keldysh time-loop contour C. The forward branch is denoted with a "$-$" label while the backward branch is denoted by a "+" label. (b) Lowest order diagram for single photoemission. $\hat{\Delta}$ is the operator for light-matter coupling. It is associated with vertices having time arguments t'_- and t_+.

the reason why the standard time-ordered formalism is insufficient for its computation (no Kubo formula for photoemission [17]). On the other hand the time-ordering in (2) can naturally be represented on the Keldysh contour, Fig. 1(a). The correlator can be perturbatively computed by applying the Wick's theorem.

Specification of the single-particle indices for the creation and annihilation operators in Eq. (2) is important. The unbound (continuum) states are denoted in bold-style to distinguish them from general indices denoted as a, b, c, ..., which also comprise bound states i, j, k, ... In zeroth order there is a single diagram resulting from the following contraction

$$\langle c_b^\dagger(t_+)c_a(t_+)c_{\mathbf{p}}^\dagger(0)c_{\mathbf{p}}(0)c_c^\dagger(t'_-)c_d(t'_-)\rangle .$$

Other contractions vanish because the target is bound in its ground state, i.e.

$$c_{\mathbf{k}}|\Psi_0\rangle = 0. \tag{3}$$

Introducing the *lesser* and *greater* Green's function (GF) components

$$G^<_{ab}(t_1,t_2) \equiv G^{-+}_{ab}(t_1,t_2) \;=\; i\langle c_b^\dagger(t_2)c_a(t_1)\rangle, \tag{4a}$$

$$G^>_{ab}(t_1,t_2) \equiv G^{+-}_{ab}(t_1,t_2) \;=\; -i\langle c_a(t_1)c_b^\dagger(t_2)\rangle. \tag{4b}$$

condition (3) can be formulated as the requirement that lesser GFs with one or more continuum indices vanish:

$$G^<_{\mathbf{k}a} = G^<_{a\mathbf{k}} = 0.$$

In these notations zeroth order (in Coulomb interaction v) correlator is given by the product of one lesser GF describing the propagation of an added hole and two greater GFs describing propagation of the emitted particle [see Fig. 1(b)]:

$$Z^{(1)}(t,t') = ig^>_{a\mathbf{p}}(t)g^<_{db}(t'-t)g^>_{\mathbf{p}c}(-t'). \tag{5}$$

The constituent GFs are mean-field ones, i. e., they are averages over the ground state of corresponding Hartree-Fock Hamiltonian (we assume that it is bound as well fulfilling a condition similar to (3)). The vertical electron propagator, i. e., the one that connects two electron-photon vertices, is of the *lesser* kind $G^< \equiv G^{-+}$ because its first and second time arguments belong to different tracks, "$-$" and "+"

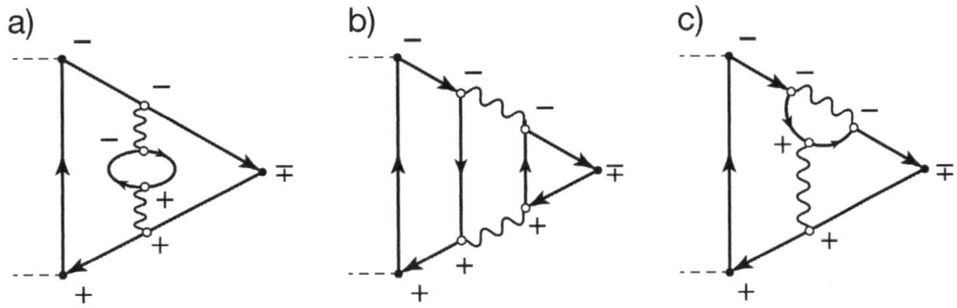

Figure 2. Lowest order diagrams with extrinsic losses. Wavy lines denote the bare Coulomb interaction. Pluses and minuses indicate to which branch of the Keldysh contour [Fig. 1(a)] corresponding points are belonging.

respectively. It contains the information on the spectral density of the occupied states. Replacing the bare with the full propagator means that mean-field spectral density of occupied states is replaced with the exact interacting one, which also incorporates the intrinsic energy losses. Computation of the correlated spectral densities is the main goal of electronic structure theory. The electronic lines flowing to or from the detector are of the *greater* type and, thus, are associated with unoccupied electronic states. Their dressing describes the formation of *scattering* states observed in the detector [36, 37, 28].

In addition to renormalization of fermionic lines, the evaluation of next orders of Eq. (2) leads to the appearance of vertex functions. As Almbladh explains [21, 22] they amount to modification of the bare light-matter coupling operator $\hat{\Delta}$ by the vector-coupling vertex function $\hat{\Lambda}(\epsilon + \omega, \epsilon)$. All other possible renormalizations of $Z^{(1)}(t, t')$ contain the renormalized interaction lines connecting points on the forward and backward branches of the contour and are associated with extrinsic energy losses. In the next section we consider a particular example of such processes involving emission of a secondary electron.

3. Double photoemission

There are several possibilities to introduce double photoemission. One way is to regard it as a single photoemission process accompanied by extrinsic energy losses whereby a liberated electron interacts with the target system knocking out a secondary electron. In atomic, molecular and optical (AMO) physics community [38, 34] this is known as the *knock out* mechanism. As an illustration let us consider second order diagrams at Fig. 2. They are discussed in details by Caroli *et al.* as examples of extrinsic losses (cf. Fig. 9 of Ref. [18]). There are no first order processes because the bare Coulomb interaction is local in time and, therefore, cannot connect points on different branches of Keldysh contour. In all three cases there is a fermionic line flowing from "−" to "+" vertices, i. e., given by the greater GF. This is a potential candidate for a secondary electron that can be detected in coincidence measurement. Intuitively corresponding DPE diagrams can be drawn by cutting this line and adding an intermediate point with well defined momentum $\mathbf{p}_2$. These and several more diagrams describing double photoemission can be obtained applying the Wick theorem to the following correlator [35]:

$$Z^{(2)}(t,t') = \langle \Psi_0 | c_b^\dagger(t) c_a(t) c_{\mathbf{k}_1}^\dagger(0) c_{\mathbf{k}_2}^\dagger(0) c_{\mathbf{k}_2}(0) c_{\mathbf{k}_1}(0) c_c^\dagger(t') c_d(t') | \Psi_0 \rangle. \qquad (6)$$

The derivation is simpler if this correlator is written in real space representation. For completeness all leading order (second order in bare Coulomb interaction) diagrams are shown on Fig. 3, where points labeled by α, β, γ denote integration over space, summation over spin, and integration over the time on "−"-branch of the Keldysh contour. Similarly, χ, υ, ζ correspond to the points on the "+"-branch. Specifically, $\alpha \equiv (\mathbf{x}_\alpha, t_\alpha) \equiv (\mathbf{r}_\alpha, \sigma_\alpha, t_\alpha)$, $\chi \equiv (\mathbf{x}_\chi, t_\chi) \equiv (\mathbf{r}_\chi, \sigma_\chi, t_\chi)$, etc. Let us recall explicit form of the

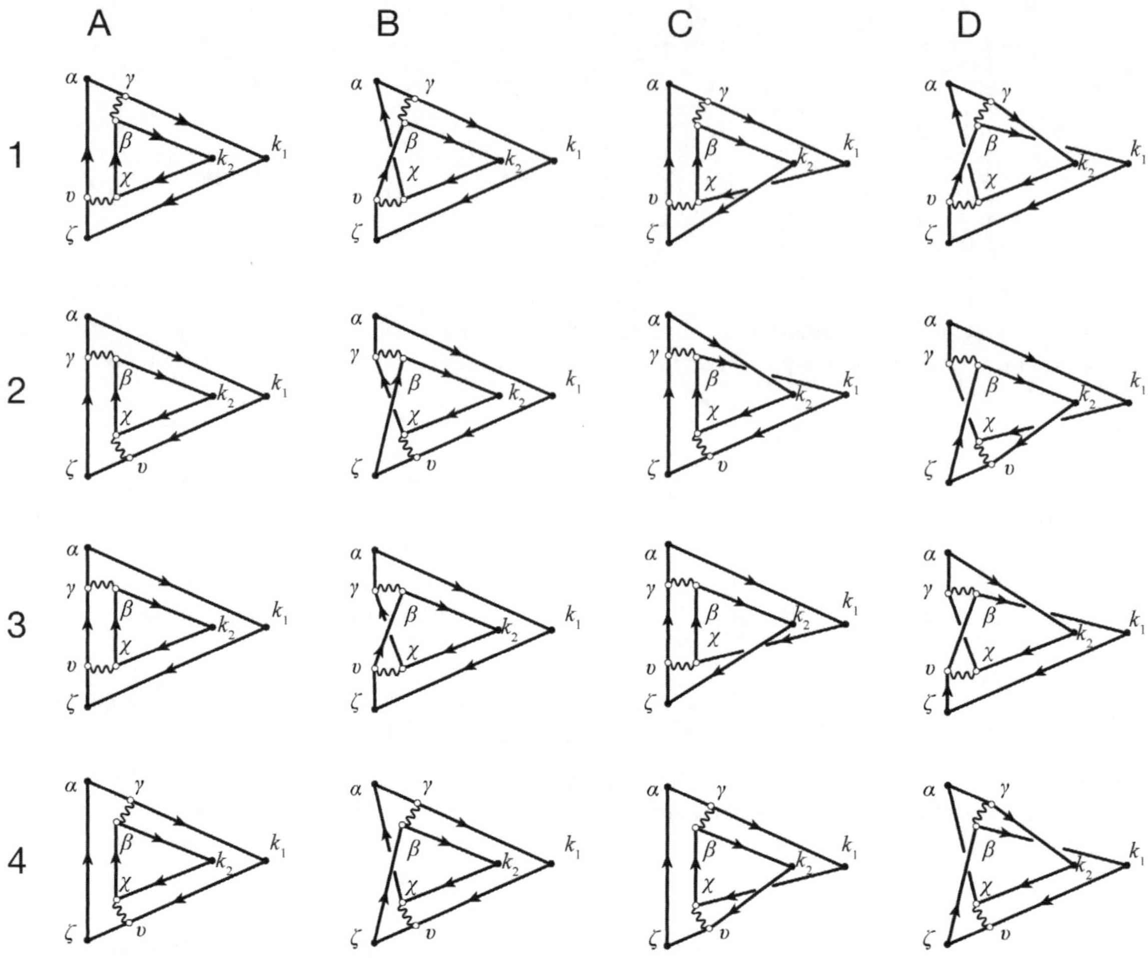

Figure 3. Leading order diagrams describing double photoemission. Notation 1A, 1B, etc., is used. Exchange of $\mathbf{k}_1$ and $\mathbf{k}_2$ generates topologically new ones. Corresponding columns will be denoted as A′, B′, and so on.

non-interacting greater GF:

$$g^{>}_{\mathbf{x}_\alpha \mathbf{x}_\zeta}(t,t') = -i \sum_a \bar{f}(\epsilon_a) e^{-i\epsilon_a(t-t')} \langle \mathbf{x}_\alpha | a \rangle \langle a | \mathbf{x}_\zeta \rangle, \tag{7}$$

where $f(\epsilon) = 1 - \bar{f}(\epsilon)$ is the zero temperature Fermi distribution function and the sum runs over all one-particle states. For times $t, t' \in (-\infty, 0]$ the product of two greater GFs can be expressed as:

$$\begin{aligned}\sum_{\mathbf{y}} g^{>}_{\mathbf{x}_\alpha \mathbf{y}}(t,0) g^{>}_{\mathbf{y}\mathbf{x}_\zeta}(0,t') &= (-i)^2 \sum_{ab} \bar{f}(\epsilon_a)\bar{f}(\epsilon_b) e^{-i\epsilon_a t + i\epsilon_b t'} \sum_{\mathbf{y}} \langle \mathbf{x}_\alpha | a \rangle \langle a | \mathbf{y} \rangle \langle \mathbf{y} | b \rangle \langle b | \mathbf{x}_\zeta \rangle \\ &= (-i)^2 \sum_a e^{-i\epsilon_a(t-t')} \bar{f}(\epsilon_a) \langle \mathbf{x}_\alpha | a \rangle \langle a | \mathbf{x}_\zeta \rangle = -i g^{>}_{\mathbf{x}_\alpha \mathbf{x}_\zeta}(t,t'), \end{aligned} \tag{8}$$

where the completeness of the one-particle basis was used. The sum over $\mathbf{y}$ can be done as integration over space coordinates and spin indices or in any other complete basis. For instance we can perform the sum in the basis of single-particle states, i. e. $\sum_{\mathbf{y}} \to \sum_i + \sum_{\mathbf{k}}$, where i labels bound states and $\mathbf{k}$ is the continuum state labeled by its momentum. Cutting a greater non-interacting GF amounts to specification of an intermediate state, e. g. $g^{>}_{\mathbf{x}_\alpha\mathbf{x}_\zeta}(t,t') \to g^{>}_{\mathbf{x}_\alpha\mathbf{k}}(t,0)g^{>}_{\mathbf{k}\mathbf{x}_\zeta}(0,t')$. To the left hand side we see a single GF that describes the propagation of an electron. Its intermediate state is not determined, whereas the product of two GFs at the right hand side describes observation of the intermediate state in the detector. This procedure can be used to generate double photoemission diagrams starting from the diagrams for the single photoemission accompanied by energy losses. Considering diagrams on Fig. 2 we see that they in particular transform into the diagrams 4A, 4A′, and 4C, respectively, as shown on Fig. 3. This figure contains a number of other topologically distinct species which can be obtained by considering in addition to extrinsic losses, the diagrams describing intrinsic losses, and interference of these two mechanisms. The outlined procedure can be generalized to obtain SPE diagrams from that of the irreducible density-density response function, DPE from SPE, and so on:

$$Z^{(1)}_{\mathbf{k}}(t,t') = \int d(\alpha\zeta)\frac{\delta P^{<}(t,t')}{\delta g^{>}_{\mathbf{x}_\alpha\mathbf{x}_\zeta}(\tau_\alpha,\tau_\zeta)} g^{>}_{\mathbf{x}_\alpha\mathbf{k}}(\tau_\alpha,0)g^{>}_{\mathbf{k}\mathbf{x}_\zeta}(0,\tau_\zeta), \tag{9a}$$

$$Z^{(2)}_{\mathbf{k}\mathbf{k}'}(t,t') = \int d(\gamma\upsilon)\frac{\delta Z^{(1)}_{\mathbf{k}}(t,t')}{\delta g^{>}_{\mathbf{x}_\gamma\mathbf{x}_\upsilon}(\tau_\gamma,\tau_\upsilon)} g^{>}_{\mathbf{x}_\alpha\mathbf{k}'}(\tau_\gamma,0)g^{>}_{\mathbf{k}'\mathbf{x}_\zeta}(0,\tau_\upsilon). \tag{9b}$$

In this way the 4C-type diagram is a descendant of the SPE diagram in Fig. 4(b) which, in turn, traces back to the density-density response function in Fig. 4(a).

It is more than just graphs operations that is illustrated here. In ref. [39] and in a sequence of works [30, 31] it has been shown that cutting procedure, which splits diagrams representing the lesser or the greater response function into *half-diagrams*, can be used to construct approximations with the positive definiteness property. These are the approximations for response functions possessing positive spectral functions in the whole frequency range. Explicit construction outlined there shows that in this case the spectral function can be written in the FGR form, that is, as a sum of terms containing squared modulus of the matrix element of a certain scattering process multiplied with corresponding δ-function assuring overall energy conservation. The matrix element then is expressed in terms of GFs on only one branch of the Keldysh contour, i. e., all its vertices are labeled with either "+" or with "−".

For scattering problems as considered here the Fermi Golden rule approach is particularly useful. It allows to describe arbitrary scattering processes in leading order without referring to complicated mathematical apparatus of Feynman diagrams. It is also the only choice for more complicated processes if perturbative expressions cannot readily be generated. Therefore, it is interesting to identify the ingredients of diagrams on Fig. 3, i. e., the half-diagrams yielding corresponding expressions in the FGR form. Independently whether second-order density-density response diagrams (many of them are listed by Uimonen *et al.* [31]) are used as a starting point, or single photoemission diagrams listed for instance by Caroli *et al.* [18], or the one shown here on Fig. 3, the result will be topologically identical. The difference is in the treatment of dangling lines: in the former case their quantum numbers are summed

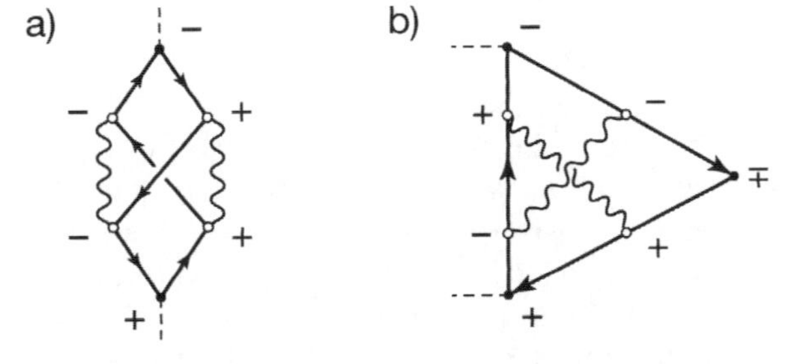

Figure 4. Application of cutting procedure to the lesser density-density response function (left) yields a SPE diagram (right) which, in turn, leads to the 4C-type diagram shown on Fig. 3.

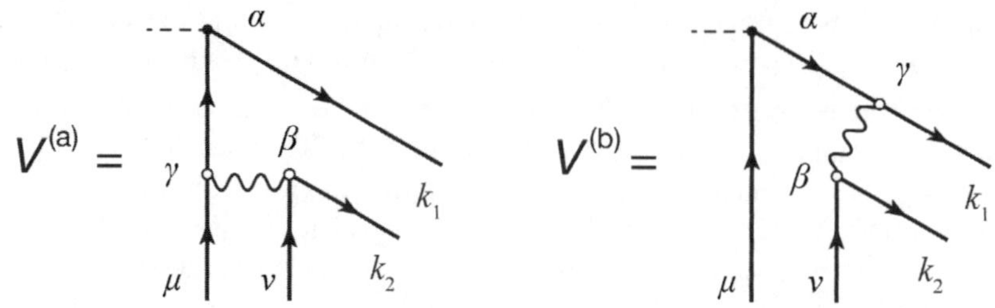

Figure 5. Half-diagrams corresponding to Fig. 3. Same labeling is preserved. All vertices have time arguments on "−"-branch of the Keldysh contour. $V^{(a)}$ has interaction line connected to the lesser propagator for primary electron, the scattering reflects intrinsic losses in the ground state. In AMO physics this process is also known as *shake off* mechanism. In contrast, $V^{(b)}$ has interaction line connected to the greater propagator for primary electron, the scattering reflects extrinsic losses in excited state. In AMO physics this process is also known as *knock out* mechanism. There two more topologically equivalent structures with exchanged $\mathbf{k}_1$ and $\mathbf{k}_2$ indices. They will be denoted as $\bar{V}^{(a)}$ and $\bar{V}^{(b)}$.

over, whereas in SPE one dangling line is observed in the detector and therefore has a fixed index. For emission of n-particles by one photon there should be n dangling lines corresponding to momenta registered (in coincidence) in the detectors. Rather surprisingly, the result of such analysis (Fig. 5) is quite simple and can be recovered from early works of Amusia and Kazachkov [38]; they can also be identified as diagrams in Fig. 7(b) in Uimonen *et al.* [31]. In order to make explicit parallels with works of AMO physics community, we consider Goldstone diagrams (see Chang and McDowell [40] for a short summary of this technique in application to atomic photoionization and L'Huillier *et al.* [41] for more complicated multiphoton examples) that are commonly used in scattering theory. To make this work self-contained the corresponding lowest order DPE Goldstone diagrams are shown in Fig. 6 (cf. Fig. 8.1 in ref. [34]).

In fact, Goldstone diagrams carry more information than standard NEGF ones: the direction of arrow on each fermionic line specifies whether particle or hole propagator should be used. For the latter, "+" and "−" assigned to the vertices only specify the branch of Keldysh contour that they belong to; however, all possible mutual ordering within the branch have to be considered. That is why there are twice as many diagrams in Fig. 6 as compared to Fig. 5: the two possibilities correspond to the two terms in the time-ordered GF of internal line.

Let us verify that all triangular DPE diagrams at Fig. 3 can be obtained by gluing half-diagrams in

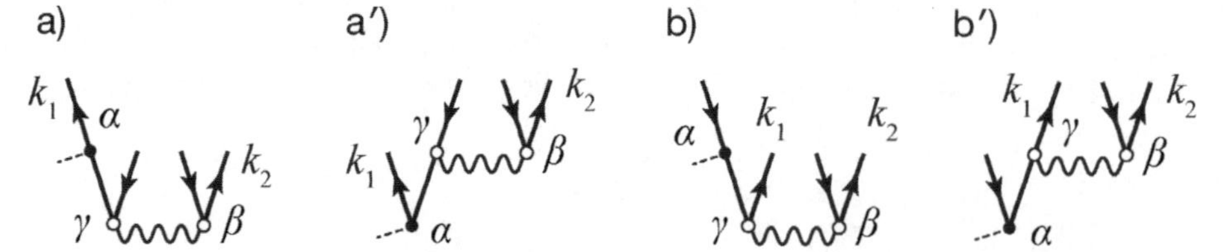

Figure 6. Goldstone diagrams corresponding to the matrix elements of double photoemission in Fig. 5. For NEGF diagrams + and − assigned to the vertices partly determine the time ordering; here, the direction of arrows specifies whether particle (arrow up) or hole (arrow down) propagators should be used. Therefore, each scattering matrix element shown in Fig. 5 is represented by two diagrams: $V^{(a)}$ by (a) and (a'), $V^{(b)}$ by (b) and (b'). This reflect the fact that in addition to dangling lines which are described by either particle or hole propagators, $V^{(a)}$ and $V^{(b)}$ contain one internal g^{--} line. It is given by *time-ordered* propagator comprising both the particle and the hole terms.

Fig. 5. The simplest identities would be

$$\begin{aligned} Z^A &= \sum_{\mu\nu} |V^{(a)} + V^{(b)}|^2_{\mu\nu}, \\ Z^{A'} &= \sum_{\mu\nu} |\bar{V}^{(a)} + \bar{V}^{(b)}|^2_{\mu\nu}, \end{aligned}$$

where Z^A denotes the sum of contributions from column "A" of Fig. 3, the indices μ and ν label dangling lines, and momentum dependence, i. e. $\mathbf{k}_{1,2}$, is not shown explicitly. This is not the full set because, as explained in refs. [30, 31] permutations of dangling lines must also be considered:

$$Z^B = \sum_{\mu\nu} (V^{(a)} + V^{(b)})_{\mu\nu} (V^{(a)} + V^{(b)})^*_{\nu\mu}.$$

Collecting all contributions the full set at Fig. 3, i. e., $Z^{(2)} = Z^A + Z^{A'} + Z^B + Z^{B'} + Z^C + Z^{C'} + Z^D + Z^{D'}$ can be obtained as follows:

$$Z^{(2)} = \sum_{\mu\nu} (V + \bar{V})_{\mu\nu} (V + V^{\mathrm{T}} + \bar{V} + \bar{V}^{\mathrm{T}})^*_{\mu\nu}, \tag{10}$$

where the sum of diagrams in Fig. 5 is denoted by $V = V^{(a)} + V^{(b)}$. There are 32 terms in this expression. The construction guarantees the positivity of corresponding two-particle current.

4. Generalizations

This construction can be generalized to the process of emission of arbitrary number of particles. It is clear that higher order diagrams need to be considered. Already an example of classical system of particles with two-body hard sphere interactions shows that the emission of n particles requires at least $n - 1$ collisions. Quite generally, in order to achieve n-particle emission Goldstone diagrams must contain at least $n - 1$ interaction lines; the scattering cross-section is being proportional to $v^{2(n-1)}$ as shown in ref. [35]. However, what can be said about topological structure of these diagrams?

One possibility to characterize diagrammatic complexity is to enumerate possible topological structures. For equilibrium many-body perturbation theory the problem of counting diagrams was initiated in the work of Molinari [42] who suggested to use the Hedin's system of functional equations [43, 44] that relates the electron GF, the proper self-energy Σ, vertex function Γ, polarization P, and the effective potential W as the generating functions for corresponding diagram counting. The transformation from integral-functional equations to much simpler differential equations for generating functions is done by collapsing all space, spin and time coordinates into one point, i. e., considering zero-dimensional case. These equations have been subsequently solved analytically [45]. Besides pure mathematical interest, the solution can be used as a basis for method development [46, 47] and to get insight into the convergence of approximate theories [48, 49].

We have seen above that arbitrary photoemission process can be described using irreducible lesser density-density response function as a starting point. Corresponding time-ordered response (P^{--}) is the polarization in Hedin's system of equations [44]. Thus, it seems natural to use the nonequilibrium extension of these equations [50, 51] in order to enumerate *decorated* Feynman diagrams (with "+" and "−" assigned to the vertices). Below, these equations are explicitly written discarding the time and space

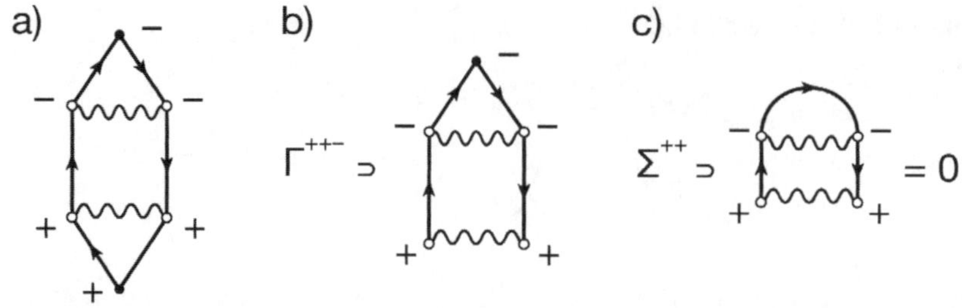

Figure 7. (a) Example of polarization that cannot be obtained from Eq. (11b). Corresponding vertex function (b) can be constructed by applying Eq. (11c) to self-energy (c).

dependence of all quantities (in zero dimensions), but preserving the labeling of vertices:

$$\Sigma^{\eta\bar{\eta}} = i \sum_{\eta_3\eta_4} \bar{\eta}_3\bar{\eta}_4 W^{\eta\eta_3} G^{\eta\eta_4} \Gamma^{\eta_4\bar{\eta},\eta_3}, \tag{11a}$$

$$P^{\eta\bar{\eta}} = -i \sum_{\eta_3\eta_4} \bar{\eta}_3\bar{\eta}_4 G^{\eta\eta_3} G^{\eta_4\eta} \Gamma^{\eta_3\eta_4,\bar{\eta}}, \tag{11b}$$

$$\Gamma^{\eta_1\eta_2,\eta_3} = \delta_{\eta_1\eta_2}\delta_{\eta_2\eta_3} + g^{\eta_4\eta_3} g^{\eta_3\eta_5} \frac{\delta\Sigma^{\eta_1\eta_2}}{\delta g^{\eta_4\eta_5}}, \tag{11c}$$

$$G^{\eta\bar{\eta}} = g^{\eta\bar{\eta}} + g^{\eta\bar{\eta}}\Sigma^{\bar{\eta}\bar{\eta}}G^{\bar{\eta}\bar{\eta}} - g^{\eta\eta}\Sigma^{\eta\bar{\eta}}G^{\bar{\eta}\bar{\eta}} + g^{\eta\eta}\Sigma^{\eta\eta}G^{\eta\bar{\eta}}, \tag{11d}$$

$$W^{\eta\bar{\eta}} = vP^{\eta\eta}W^{\eta\bar{\eta}} - vP^{\eta\bar{\eta}}W^{\bar{\eta}\bar{\eta}}. \tag{11e}$$

Here $\eta = \pm 1$ denotes branches of Keldysh contour, $\bar{\eta} = -\eta$. In equations for the screened interaction some terms are not present because the bare interaction is instantaneous, i. e. $v^{\eta\bar{\eta}} = 0$. One more assumption was incorporated: in Dyson's equations for GFs it was explicitly used that in equilibrium (or in steady state) and at zero temperature the following convolutions vanish $\int d\bar{t}\, X^{\eta\bar{\eta}}(t,\bar{t})Y^{\bar{\eta}\eta}(\bar{t},t') = 0$ (see discussion by Ness *et al.* [51]). Under this assumption equations for time-ordered and anti-time-ordered functions are reduced to the usual form. From the structure of equations (11) the vertices are necessarily decorated such that no isolated islands of pluses or minuses appear [30, 31].

Let us now skip $\pm i$ prefactors and signs (not essential for the diagram enumeration), introduce the prefactor ℓ into eq. (11b) in order to keep count of fermionic loops and write the power series solution for $P^{<}$:

$$\begin{aligned} P^{-+} &= 3\ell v g^{-+}g^{+-}\{[g^{--}]^2 + [g^{++}]^2\} \\ &+ v^2 g^{-+}g^{+-}\Big\{(5\ell^2 + 15\ell)\big([g^{--}]^4 + [g^{++}]^4\big) \\ &+ 6\ell[g^{--}]^2[g^{++}]^2 + (8\ell^2 + 8\ell)g^{--}g^{-+}g^{+-}g^{++}\Big\} + \mathcal{O}(v^3). \end{aligned}$$

The $(8\ell^2 + 8\ell)$ term numbers all diagrams in Fig. 3; however, three other terms of the second order (they contain a single greater propagator and therefore contribute only to SPE, but not to DPE) are missing $[3\ell v^2 g^{-+}g^{+-}[g^{--}]^2[g^{++}]^2$; one such example is shown in Fig. 7(a)]. This can be traced back [Fig. 7(c)] to the self-energy $\Sigma^{\eta\eta}$ in the vertex equation (11c). Per construction (recall that we impose $X^{\eta\bar{\eta}}Y^{\bar{\eta}\eta} = 0$) it only contains $g^{\eta\eta}$ lines and is therefore insufficient to obtain the vertex function shown in Fig. 7(b). Thus, a mere extension of Hedin's equations to nonequilibrium case does not allow to enumerate photoemission diagrams. Spurious diagrams appear if restriction $X^{\eta\bar{\eta}}Y^{\bar{\eta}\eta} = 0$ is lifted, otherwise valid diagrams are missing. There is, actually, a logical explanation to this fact: due to asymmetric position of the vertex function, the Hedin's expressions for the self-energy and polarization do not respect the FGR form pertinent to exact $\Sigma^{\eta\bar{\eta}}$ and $P^{\eta\bar{\eta}}$.

5. Conclusions

In this work the NEGF diagrams for single- and multi-electron emission were treated consistently by using the lesser irreducible density-density response ($P^<$) as a common generating function. In this way n-particle emission is described by diagrams containing n lesser electron propagators with well defined momenta. Alternatively, the photoemission cross-section can be recasted in the Fermi Golden rule form. It is common to write corresponding matrix elements in terms of Goldstone diagrams. Gluing such diagrams that relate to the time-propagation in forward and backward directions yields the NEGF construction. This correspondence was illustrated by considering all possible double photoemission processes to the leading order in Coulomb interaction. Equivalence between the shake off and knock out mechanisms known in AMO physics and intrinsic and extrinsic losses in condensed matter physics is demonstrated.

Since $P^<$ is a central quantity of this study it is desirable to have a set of functional relations that would allow to recursively generate its perturbative expansion in a manner similar to Hedin's equations for time-ordered quantities. An extension of these equations to nonequilibrium case generates a larger set of diagrams, some of them being zero at zero temperature and in a steady state. Because Hedin's equations for self-energy and polarization are not compatible with FGR, these spurious diagrams cannot be easily eliminated.

Acknowledgments

It is a pleasure to thank the organizers Claudio Verdozzi, Andreas Wacker and Carl-Olof Almbladh for an excellent workshop on *Progress in Nonequilibrium Green's Functions* (PNGF6). This work is supported by the DFG grants No. SFB762 and No. PA 1698/1-1.

References

[1] Hüfner S 2003 *Photoelectron spectroscopy: principles and applications* 3rd ed Advanced texts in physics (Berlin; New York: Springer)
[2] Berglund C N and Spicer W E 1964 *Phys. Rev.* **136** A1030–A1044
[3] Chang J J and Langreth D C 1972 *Phys. Rev. B* **5** 3512–3522
[4] Chang J J and Langreth D C 1973 *Phys. Rev. B* **8** 4638–4654
[5] Kane E O 1966 *Phys. Rev.* **147** 335–339
[6] Chulkov E V, Borisov A G, Gauyacq J P, Sánchez-Portal D, Silkin V M, Zhukov V P and Echenique P M 2006 *Chem. Rev.* **106** 4160–4206
[7] Verdozzi C, Durham P J, Cole R J and Weightman P 1997 *Phys. Rev. B* **55** 16143–16158
[8] Verdozzi C, Cini M and Marini A 2001 *Journal of Electron Spectroscopy and Related Phenomena* **117-118** 41–55
[9] Aryasetiawan F, Hedin L and Karlsson K 1996 *Phys. Rev. Lett.* **77** 2268–2271
[10] Pavlyukh Y, Rubio A and Berakdar J 2013 *Phys. Rev. B* **87** 205124
[11] Story S M, Kas J J, Vila F D, Verstraete M J and Rehr J J 2014 *Phys. Rev. B* **90** 195135
[12] Lundqvist B I 1969 *Phys. Kondens. Mater.* **9** 236–248
[13] Langreth D C 1970 *Phys. Rev. B* **1** 471–477
[14] Penn D R 1977 *Phys. Rev. Lett.* **38** 1429–1432
[15] Penn D R 1978 *Phys. Rev. Lett.* **40** 568–571
[16] Schaich W L and Ashcroft N W 1971 *Phys. Rev. B* **3** 2452–2465
[17] Mahan G D 1970 *Phys. Rev. B* **2** 4334–4350
[18] Caroli C, Lederer-Rozenblatt D, Roulet B and Saint-James D 1973 *Phys. Rev. B* **8** 4552
[19] Stefanucci G and van Leeuwen R 2013 *Nonequilibrium Many-Body Theory of Quantum Systems: A Modern Introduction* (Cambridge: Cambridge University Press)
[20] Fujikawa T and Arai H 2002 *Journal of Electron Spectroscopy and Related Phenomena* **123** 19–46
[21] Almbladh C O 1985 *Phys. Scr.* **32** 341
[22] Almbladh C O 1986 *Phys. Rev. B* **34** 3798–3806
[23] Krasovskii E E, Silkin V M, Nazarov V U, Echenique P M and Chulkov E V 2010 *Phys. Rev. B* **82** 125102
[24] Uimonen A M, Stefanucci G and Leeuwen R v 2014 *J. Chem. Phys.* **140** 18A526
[25] Campbell L, Hedin L, Rehr J J and Bardyszewski W 2002 *Phys. Rev. B* **65** 064107
[26] Guzzo M, Lani G, Sottile F, Romaniello P, Gatti M, Kas J J, Rehr J J, Silly M G, Sirotti F and Reining L 2011 *Phys. Rev. Lett.* **107** 166401

[27] Guzzo M, Kas J J, Sponza L, Giorgetti C, Sottile F, Pierucci D, Silly M G, Sirotti F, Rehr J J and Reining L 2014 *Phys. Rev. B* **89** 085425
[28] Hedin L and Lee J D 2002 *Journal of Electron Spectroscopy and Related Phenomena* **124** 289–315
[29] Feibelman P J and Eastman D E 1974 *Phys. Rev. B* **10** 4932–4947
[30] Stefanucci G, Pavlyukh Y, Uimonen A M and van Leeuwen R 2014 *Phys. Rev. B* **90** 115134
[31] Uimonen A M, Stefanucci G, Pavlyukh Y and van Leeuwen R 2015 *Phys. Rev. B* **91** 115104
[32] Kelly H P 1964 *Phys. Rev.* **136** B896–B912
[33] Lindgren I and Morrison J 1982 *Atomic many-body theory* (*Springer series in chemical physics* no 13) (Berlin: Springer)
[34] Amusia M Y 1990 *Atomic photoeffect* Physics of atoms and molecules (New York: Plenum Press)
[35] Pavlyukh Y, Schüler M and Berakdar J 2015 *Phys. Rev. B* **91** 155116
[36] Bardyszewski W and Hedin L 1985 *Phys. Scr.* **32** 439–450
[37] Hedin L 1990 *Journal of Electron Spectroscopy and Related Phenomena* **51** 91–102
[38] Amusia M Y and Kazachkov M P 1968 *Phys. Lett. A* **28** 27–28
[39] Almbladh C O 2006 *J. Phys. Conf. Ser.* **35** 127–144
[40] Chang E S and McDowell M R C 1968 *Phys. Rev.* **176** 126–136
[41] L'Huillier A, Jönsson L and Wendin G 1986 *Phys. Rev. A* **33** 3938–3955
[42] Molinari L G 2005 *Phys. Rev. B* **71** 113102
[43] Hedin L 1965 *Phys. Rev.* **139** A796–A823
[44] Strinati G 1988 *Riv. Nuovo Cimento* **11** 1–86
[45] Pavlyukh Y and Hübner W 2007 *J. Math. Phys.* **48** 052109
[46] Lani G, Romaniello P and Reining L 2012 *New J. Phys.* **14** 013056
[47] Pavlyukh Y, Berakdar J and Rubio A 2016 arXiv:1601.04285[cond-mat]
[48] Berger J A, Romaniello P, Tandetzky F, Mendoza B S, Brouder C and Reining L 2014 *New J. Phys.* **16** 113025
[49] Stan A, Romaniello P, Rigamonti S, Reining L and Berger J A 2015 *New J. Phys.* **17** 093045
[50] Harbola U and Mukamel S 2006 *J. Chem. Phys.* **124** 044106
[51] Ness H, Dash L K, Stankovski M and Godby R W 2011 *Phys. Rev. B* **84** 195114

Journal of Physics: Conference Series **696** (2016) 012006 doi:10.1088/1742-6596/696/1/012006

Second-order bosonic Kadanoff-Baym equations for plasmon-accompanied optical absorption

Michael Schüler, Yaroslav Pavlyukh

Martin Luther University Halle-Wittenberg, 06099 Halle (Saale), Germany

E-mail: michael.schueler@physik.uni-halle.de

Abstract. The availability of ultra-short and strong light sources opens the door for a variety of new experiments such as transient absorption, where optical properties of systems can be studied in extreme nonequilibrium situations. The nonequilibrium Green's function formalism is an efficient approach to investigate these processes theoretically. Here we apply the method to the light-matter interaction of the magnesium 2p core level accompanied by electron-plasmon interaction due to collective excitations in the conduction band. The plasmons are described as massive bosonic quasi-particle excitations, leading to a second-order equations of motion, requiring a new approach for their propagation.

1. Introduction

Electronic correlations are manifest in a number of spectroscopic experiments. Photoemission [1, 2] and, in particular, double photoemission [3, 4, 5], photoabsorption or electron-energy loss spectroscopy (EELS) represent some of the most direct approaches. Here we exemplary focus on photoabsorption. The central quantity hereby is the density-density response function χ, which is a genuine two-body quantity and thus captures electron (or hole) pair correlations. Computing χ is typically an involved task. Depending on the system at hand, few approaches are suitable. For small systems, full (or truncated) CI methods deliver accurate many-body properties including the complete particle-hole (*e-h*) spectrum. This route becomes inaccessible with growing system size, such that approximative schemes are invoked. One of the most employed approach is many-body perturbation theory (MBPT), where χ is obtained from the four-point *e-h* propagator L [6, 7]. In turn, L is the solution to the Bethe-Salpeter equation (BSE). In combination with the GW approximation [8], the BSE approach has been widely applied to molecules, clusters and solids [9, 10, 11, 12, 13, 14].

An alternative route to access χ is to directly use linear response, as the variation of any one-body observable (e. g. the dipole moment) upon applying a weak perturbation potential $\delta\varphi$ is proportional to the convolution of χ with $\delta\varphi$. Hence, carrying out real-time propagation of the system is required in this framework. One important method solving this task is time-dependent density-functional theory (TDDFT) [15]. Its advantage lies in relatively low computation cost, which allows for efficiently computing linear response properties of molecules up to larger clusters [16, 17, 18, 19].

On the other hand, TDDFT is currently only available within the adiabatic approximation and thus lacks memory effects. The nonequilibrium Green's function (NEGF) formalism is a natural tool for accounting both correlation and memory effects. The equations of motion (EOM) for the Green's functions (GFs), the Kadanoff-Baym equations (KBEs), allow hereby taking the MBPT to the time domain [20, 21, 22], providing important assessment on advanced approximation schemes [23, 24, 25]. A time-dependent approach beyond the adiabatic approximation becomes increasingly important in view

Journal of Physics: Conference Series **696** (2016) 012006 doi:10.1088/1742-6596/696/1/012006

of the impressive advances in generating attosecond laser pulses [26, 27, 28], providing insights in transient electron dynamics in systems ranging from atoms to solids [29, 30, 31]. Pump-probe setups using these light sources hence are able to deliver *transient* optical properties such as transient absorption [29, 32, 33, 34].

In this contribution we study the KBE nonequilibrium dynamics of electrons coupled to bosonic quasi-particle excitations such as phonons, vibrons, quantized photons or plasmons. One prototypical model in this regard is the s-model originally proposed by Lundqvist [35] and solved by Langreth [36]. It was introduced to describe photoemission from a deep core states accompanied by the creation of a plasmon [37]. This inelastic process is reflected in a series of plasmon satellites (PSs) on the left-and side of the quasi-particle peak (QP) in the spectral function. Such electron-boson model follows, e. g., from plasmon-pole approximation for the screened interaction W [8], electron-phonon (or electron-vibron) interaction [38, 39, 40, 41, 42]. Although we focus on electron-plasmon interaction here, the methodology presented here is applicable to other electron-boson systems, as well. Hence, related methods such as molecular pump-probe spectroscopy [43, 44, 45, 46, 47] lie within the scope of our approach.

We present the KBEs for the relevant correlators of coupled electron-boson system in the framework of the MBPT (sec. 2), going beyond the standard frozen-boson approach [48, 49]. We apply our theory to the transient absorption of the Mg 2p core level accompanied by plasmon excitations in the conduction band (sec. 3).

2. Hamiltonian and equations of motion

We consider an electronic system with single-particle (SP) energies ϵ_i with corresponding fermionic annihilation operators $\hat{c}_i$. The electrons are assumed to interact with a single boson mode with energy Ω and bosonic annihilation operator $\hat{b}$. The equivalent coordinate-momentum representation of the bosons, that is

$$\hat{Q} = \frac{1}{\sqrt{2}}\left(\hat{b} + \hat{b}^\dagger\right), \quad \hat{P} = -\frac{\mathrm{i}}{\sqrt{2}}\left(\hat{b} - \hat{b}^\dagger\right) \tag{1}$$

will be used in what follows. A generalization to the multi-boson case is straightforward and will be presented elsewhere. Besides the effective interaction mediated by the electron-boson coupling, no further electron-electron interaction is accounted for. The total Hamiltonian describing the system in equilibrium is taken as

$$\hat{H}_0 = \hat{H}_{\mathrm{el}} + \hat{H}_{\mathrm{el-bos}} + \hat{H}_{\mathrm{bos}}\,, \tag{2}$$

where

$$\hat{H}_{\mathrm{el}} = \sum_i \epsilon_i \hat{c}_i^\dagger \hat{c}_i\,, \quad \hat{H}_{\mathrm{bos}} = \frac{\Omega}{2}\left(\hat{P}^2 + \hat{Q}^2\right) \tag{3}$$

are the Hamiltonians describing the individual subsystems. The coupling Hamiltonian is adopted to the deep core-level scenario:

$$\hat{H}_{\mathrm{el-bos}} = \sum_i \Gamma_i \hat{c}_i \hat{c}_i^\dagger \hat{Q}\,. \tag{4}$$

Note that generalizing to the more general matrix case $\Gamma_i \to \Gamma_{ij}$ does not cause any technical difficulties. Light-matter interaction targeting the electrons is taken into account by

$$\hat{H}_{\mathrm{el-L}}(t) = \sum_{ij} F_{ij}(t)\hat{c}_i^\dagger \hat{c}_j + \mathrm{h.c.}\,, \tag{5}$$

Besides the usual one-electron GF $G_{ij}(t,t') = -\mathrm{i}\langle \mathcal{T}\hat{c}_i(t)\hat{c}_j^\dagger(t')\rangle$, the suitable counterpart describing bosonic properties is the coordinate-coordinate fluctuation correlator

$$D(t,t') = -\mathrm{i}\left[\langle \mathcal{T}\hat{Q}(t)\hat{Q}(t')\rangle - \langle \hat{Q}(t)\rangle\langle \hat{Q}(t')\rangle\right]\,. \tag{6}$$

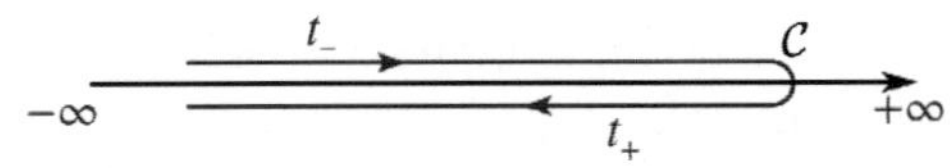

Figure 1. The real contour C, running from $-\infty$ to some finite time and back to $-\infty$. The subscripts + (−) of a contour argument t that it belongs to lower (upper) branch of the contour.

Time arguments lie on the Keldysh contour $t = t_\pm \in C$ shown in fig. 1 (contour ordering operator $\mathcal{T}$).

The derivation of the coupled EOM for $G_{ij}(t,t')$ and $D(t,t')$ follows from the standard Heisenberg EOM. It should be noted though that applying the derivative $\mathrm{i}\partial_t$ to $D(t,t')$ is not yielding a closed equation, as $\partial_t\hat{Q}(t) = \Omega\hat{P}(t)$. Applying the derivative again on the other hand one obtains

$$\partial_t^2\hat{Q}(t) = -\Omega^2\hat{Q}(t) - \sum_i \Gamma_i\hat{c}_i(t)\hat{c}_i^\dagger(t) \, ,$$

which has the form of a driven harmonic oscillator equation. Replacing the occurrence of higher-order correlators by respective convolutions with corresponding self-energies, the EOM for both the electron and the boson GFs attains the form [1]

$$\mathrm{i}\partial_t G_{ij}(t,t') = \delta_{ij}\delta(t-t') + \sum_k h^{\mathrm{MF}}_{ik}(t)G_{kj}(t,t') + \sum_k \int_C \mathrm{d}t''\, \Sigma_{ik}(t,t'')G_{kj}(t'',t') \, , \tag{7}$$

$$-\frac{1}{\Omega}\left(\partial_t^2 + \Omega^2\right)D(t,t') = \delta(t-t') + \int_C \mathrm{d}t''\, \Pi(t,t'')D(t'',t') \, . \tag{8}$$

In the context of electron-phonon interaction, the lowest-order approximations to electron self-energy $\Sigma_{ij}(t,t')$ and the boson self-energy $\Pi(t,t')$ is known as the self-consistent Born approximation [39, 40, 50, 51]. Besides the mean-field term

$$h^{\mathrm{MF}}_{ij}(t) = \epsilon_i\delta_{ij} + F_{ij}(t) - \Gamma_i\delta_{ij}\langle\hat{Q}(t)\rangle \, , \tag{9}$$

one obtains

$$\Sigma^{(2)}_{ij}(t,t') = \mathrm{i}\Gamma_i\Gamma_j G_{ij}(t,t')D(t,t') \, , \tag{10}$$

$$\Pi^{(2)}(t,t') = -\mathrm{i}\sum_{ij}\Gamma_i\Gamma_j G_{ij}(t,t')G_{ji}(t',t) \tag{11}$$

as second-order (in electron-boson interaction) self-energies. Eq. (10)–(11) can be paralleled with the GW approximation to Hedin's equations [7] — $\Gamma_i\Gamma_j D(t,t')$ hereby plays the role of the dynamical part of the screened interaction W. In case W is dominated by one plasmonic excitation, the plasmon-pole approximation applies and allows for reducing a system of interacting electrons to the electron-boson model investigated here. Therefore, we will term eq. (10)–(11) as GW approximation in what follows. Analogously, the GW_0 approximation refers to ignoring $\Pi^{(2)}(t,t')$, such that the boson correlator $D(t,t')$ is entirely non-interacting.

Importantly, the bosonic dynamics is not only captured by the coordinate-coordinate correlator $D(t,t')$ (which contains information on the fluctuations of $\hat{Q}$), but by the boson amplitude $\langle\hat{Q}(t)\rangle$, as well. This is quite distinct from electronic propagators. Because of the explicit appearance of $\langle\hat{Q}(t)\rangle$, a respective EOM needs to be formulated:

$$-\frac{1}{\Omega}\left(\frac{\mathrm{d}^2}{\mathrm{d}t^2} + \Omega^2\right)\langle\hat{Q}(t)\rangle = \mathrm{i}\sum_i \Gamma_i G_{ii}(t^+,t) \, , \tag{12}$$

[1] A rigorous derivation can be done employing a source-field approach similar to Hedin's equations [6] and is the scope of future work.

where t^+ is (with respect to the contour ordering) infinitesimally larger that t.

Inserting the second-order approximation to the self-energy, eq. (10)–(11), eq. (7)–(8) in combination with eq. (9)–(12) leads to a self-consistent set of equations. The solution strategy relies, as usual, on the projection of the contour arguments onto observables times. The specific arrangements $G_{ij}(t_+, t'_-) \equiv G^>_{ij}(t, t')$ ($G_{ij}(t_-, t'_+) \equiv G^<_{ij}(t, t')$) are known as the greater (lesser) GF ($D^\gtrless$ is defined analogously). Note that the time arguments of $G^\gtrless_{ij}$ refer to real, observables times now. By employing the Langreth rules [22, 7], eq. (7)–(8) is then transformed into the KBEs, which are solved with a method similar to ref. [52]. Appropriate modifications for treating the second-order KBE for the boson propagators are incorporated by replacing the Heun method (first order) by Numerov's method (second order).

Note that the contour fig. 1 lacks the imaginary track which required to account for a correlated initial state. Hence, we apply the method of adiabatic switching [22], allowing for working with lesser/greater GFs only.

3. Transient absorption of Mg 2p core level

As a proof-of-principle we apply the theory outlined in sec. 2 to core-level photoabsorption. For concreteness, we consider the transition from the magnesium 2p core level to the conduction band. We construct the electronic part of eq. (2) comprising two electronic levels $|2\mathrm{p}\rangle$ and an excited state $|f\rangle$ located just above the Fermi energy ϵ_F. In principle, a continuum of Bloch states needs to be accounted for when describing photoabsorption, describing (i) transitions within the conduction band including plasmon excitation, and (ii) transitions from the core level to states above the vacuum level (see fig. 2). Both types of processes are reflected in a, possibly large, background signal. We are however interested in the onset attributed to the plasmon satellites originating from the excitation of the core state. In the pump-probe setup detailed below, the pump pulse can be tuned to resonantly create a core hole. The processes (i) and (ii) will thus hardly be affected. Therefore, the mostly static background can be removed. The electron-boson model eq. (2) that lacks this background is then well suited for describing the remaining plasmonic features. Furthermore, we take the continuum of accessible states into account in a approximate fashion by coupling the $|f\rangle$ to an environment. This is accomplished by adding an embedding self-energy $\Sigma^{\mathrm{em},\gtrless}_f(t, t') = \Sigma^{\mathrm{em},\gtrless}_f(t - t')$ to eq. (10). For simplicity, we keep it within the wide-band limit approximation (WBLA) [7], defined by

$$\Sigma^{\mathrm{em},<}_f(t) = \mathrm{i}\,\Gamma_\mathrm{em} \int_{-\infty}^{\epsilon_\mathrm{F}} \frac{d\omega}{2\pi} e^{-\mathrm{i}\omega t} e^{\eta(\omega-\epsilon_\mathrm{F})} , \quad \Sigma^{\mathrm{em},>}_f(t) = -\mathrm{i}\,\Gamma_\mathrm{em} \int_{\epsilon_\mathrm{F}}^{\infty} \frac{d\omega}{2\pi} e^{-\mathrm{i}\omega t} e^{-\eta(\omega-\epsilon_\mathrm{F})} \tag{13}$$

where $\eta > 0$ is a regularization constant (we set $\eta = 0.01$). The embedding coupling constant is set to $\Gamma_\mathrm{em} = 3$ eV, leading to a broadening of $|f\rangle$ comparable to the band width of accessible states below the vacuum energy. The parameters entering the Hamiltonian are estimated in line with experiments [53, 54]. We take $\epsilon_{2\mathrm{p}} = -50$ eV, $\epsilon_f = -3.68$ eV (see fig. 2(a)). Only after the electron was excited from the core state it interacts with plasmons. Hence Γ_f can be assumed as zero, while we take $\Gamma_{2\mathrm{p}} = 5$ eV. The plasmons, in turn, originate from transitions around the Fermi level. They are treated within the small-momentum limit, i. e. the plasmon dispersion is ignored, and only a single plasmon mode $\Omega \simeq 10$ eV is incorporated (see fig. 2(a)). Solving the KBEs in equilibrium yields a spectral function of the core level with a strength of the first PS comparable to what has been observed [53].

In order to investigate the excited-state properties within optical absorption, we apply specific fields. We choose $F_{ij}(t) = f(t) m_{ij}$, where $m_{2\mathrm{p},2\mathrm{p}} = m_{f,f} = 0$ and $m_{2\mathrm{p},f} = m_{f,2\mathrm{p}}$ defines the dipole operator of the system. According to linear response theory [55], the change of the density matrix ρ_{ij} upon applying a weak field $F_{ij}(t)$ is determined by the retarded response function $\chi^\mathrm{R}_{ijkl}(t, t')$:

$$\delta\rho_{ij}(t) = \sum_{kl} \int_0^\infty dt'\, \chi^\mathrm{R}_{ijkl}(t, t') F_{kl}(t') . \tag{14}$$

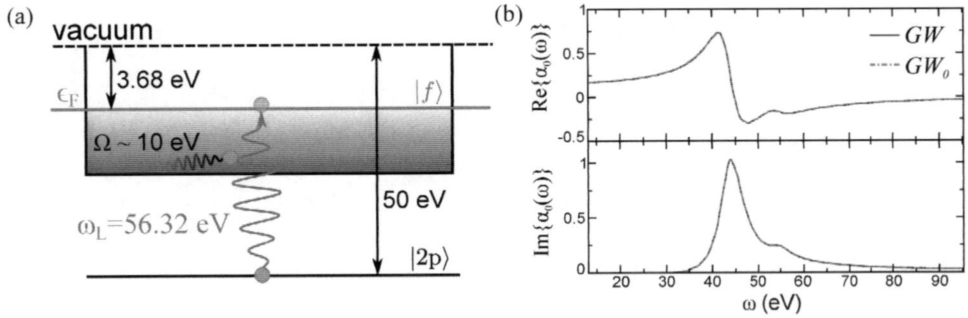

Figure 2. (a) Sketch of model system for describing absorption from the magnesium 2p core level ($|2p\rangle$) to an unoccupied state right above the Fermi level ϵ_F ($|f\rangle$) accompanied by plasmon excitations in the conduction band. (b) Equilibrium polarizability $\alpha_0(\omega)$ in both the GW and the GW_0 approximation, computed by the time-propagation method.

Hence, the dipole response of the system is governed by the polarizability

$$\alpha(t,t') = \sum_{ijkl} m_{ji}\chi^{R}_{ijkl}(t,t')m_{kl}\,.$$

In equilibrium, $\alpha(t,t') = \alpha_0(t-t')$ depends on the time difference only, such that Fourier transformation with respect to $t-t'$ is possible. The convolution eq. (14) is thus turned into multiplication in frequency space. The induced dipole is then just given by $\delta M(\omega) = \alpha_0(\omega)f(\omega)$. The special case $f(t) = f_0\delta(t)$ directly allows for computing $\alpha_0(\omega) = \delta M(\omega)/f_0$. As usual, refractive properties are encoded in the real part, whereas the imaginary part contains information on the absorbed photons.

The equilibrium polarization $\alpha_0(\omega)$ is shown in fig. 2(b), calculated in both the GW as well as in the GW_0 approximations. The results for the two cases are identical — a behavior that is to be expected as the boson self-energy eq. (11) exactly vanishes if the 2p level is fully occupied in the initial state. Therefore, the GW_0 approximation is well suited for studying (general) electron-boson systems close to equilibrium, a fact that is often utilized for simplifying the treatment of electron-phonon interaction [38]. The main feature of the absorption spectrum Im$\{\alpha_0(\omega)\}$ is, besides the strong QP peak, a PS (further satellites are existing but rapidly loosing spectral weight) separated by $+\Omega$ from the QP peak (denoted by PS$^+$). Hence, the absorption spectra resembles the (inverted and shifted) spectral function that is broadened by the environmental coupling of the $|f\rangle$ state.

This picture changes substantially in an excited-state scenario, where the plasmonic dynamics due to the electronic excitation can no longer be ignored. The optical properties encoded in the polarizability of this excited system can, depending on the pulse strength, be very different from fig. 2(b). Experimentally, this dynamical scenario can be studied using transient absorption: a short (but spectrally confined) pulse resonantly induces excitations, whereas a second (spectrally broad) pulse probes the system [29, 32, 33, 34]. In order to reflect this setup we modify the driving field $f(t) = f_{\text{pump}}(t) + f_{\text{probe}}(t)$; $f_{\text{probe}}(t) = f_0\delta(t-\tau)$ is the probe pulse and $f_{\text{pump}}(t) = f_{p,0}\sin(\omega_L t)\sin^2(\pi t/T_p)$ for $0 < t < T_p$ and $f_{\text{pump}}(t) = 0$ otherwise denotes the pump pulse. Here, T_p denotes the pulse duration and ω_L the central frequency. The latter is tuned to $\omega_L = \epsilon_f - \epsilon_{2p}$. The pulse duration is fixed at $T_p = 0.5$ fs, it comprises 5-cycles, but still has a well defined energy. The time τ denotes the time delay between pump and probe; $\tau < 0$ means the probe precedes the excitation, while for $0 < \tau < T_p$ pump and probe overlap and $\tau > T_p$ corresponds to probing the system after the pump pulse has been switched off.

The time-dependent dipole moment $\delta M(t)$ has now two contributions: $\delta M(t) = \delta M_{\text{pump}}(t) + \delta M_{\text{probe}}(t)$. As the pump pules might be strong, linear response theory does not apply here. However, choosing the probe amplitude f_0 sufficiently small allows for characterizing the probe contribution to the total dipole

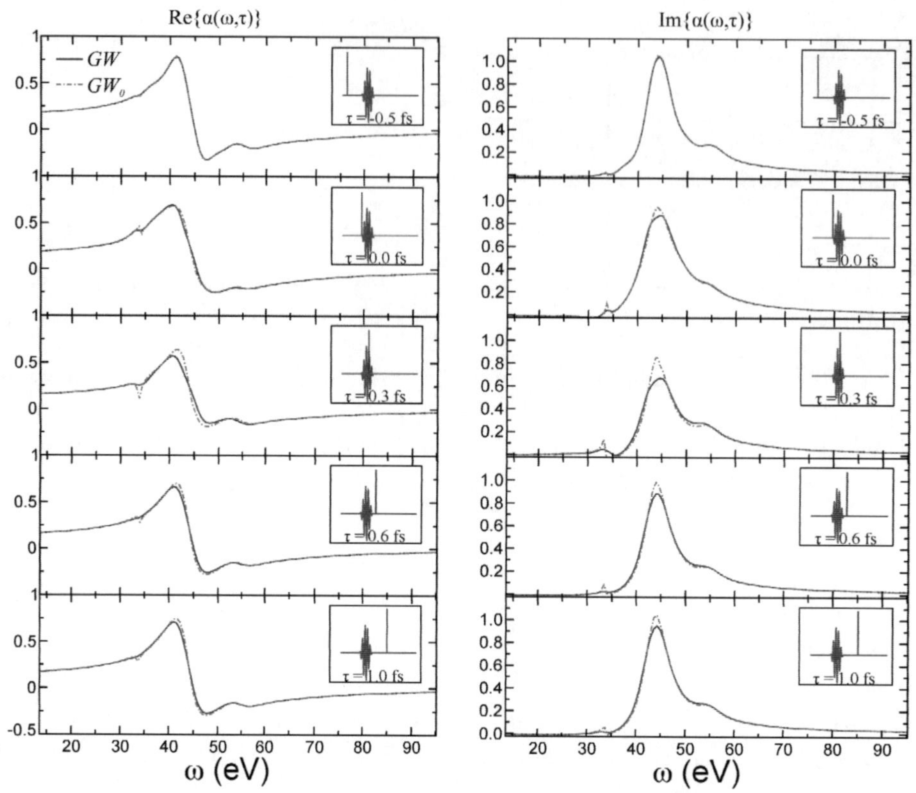

Figure 3. Transient refraction (left panel) and absorption (right) spectra for the GW (solid blue) and GW_0 (red dot-dashed line) approximation. The inset sketches the temporal arrangement of pump (amplitude $f_{p,0} = 0.1$ a. u.) and probe pulses.

moment by $\delta M_{\text{probe}}(t) = f_0\alpha(t,\tau)$. Fourier transforming,

$$\alpha(\omega,\tau) = \frac{1}{f_0}\delta M_{\text{probe}}(\omega) = \int_{-\infty}^{\infty} \mathrm{d}t\, e^{\mathrm{i}\omega t}\delta M_{\text{probe}}(t) = \int_{-\infty}^{\infty} \mathrm{d}t\, e^{\mathrm{i}\omega t}\left[\delta M(t) - \delta M_{\text{pump}}(t)\right], \tag{15}$$

yields the frequency- and delay-dependent polarizability $\alpha(\omega,\tau)$. As performing the time propagation delivers the total dipole $\delta M(t)$ only, two calculations have to be performed for each τ: one with $f_0 = 0$ and one with finite f_0 in order to individually determine the difference in the square brackets in eq. (15).

In fig. 3 we present transient absorption spectra $\alpha(\omega,\tau)$ for a weak pump pulse $f_{p,0} = 3$ eV, corresponding to a depopulation of $|2\mathrm{p}\rangle$ by 0.15. For $\tau < 0$ the behavior of both refraction and absorption are close to the equilibrium case. Increasing τ we observe the transient formation of a second PS located at the left of the QP peak (let us denote it by PS^-) – this is due to the non-zero plasmon occupation resulting from the pump pulse. This picture is consistent with the equilibrium case (no pulses applied), where the spectral function computed with nonzero (but fixed) plasmon occupation displays a PS^- peak, as well (fig. 4(a)). For overlapping pump and probe pulse ($\tau = 0$) the QP peak is suppressed, as finite population of $|f\rangle$ decreases the excitation probability. The QP gains spectral weight again with increasing τ, as the population of $|f\rangle$ vanishes due to the environmental coupling. Generally, the GW and GW_0 approximations agree well also in the transient case, except when pump and probe pulse overlap. This is expected as the system is in a nonequilibrium state due to the laser driving. Furthermore, the GW_0 approximation predicts a spiky PS^- at $\omega \simeq 33$ eV, whereas PS^- is smooth within the full GW approximation.

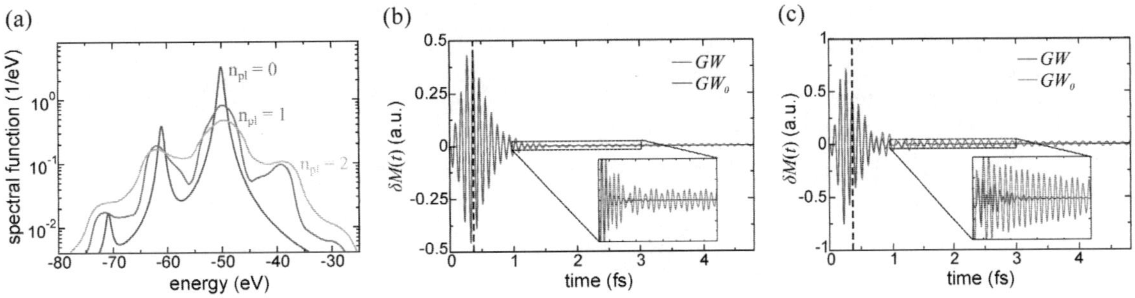

Figure 4. (a) Spectral function of the system in equilibrium with fixed plasmon occupation number $n_{\rm pl} = 0, 1, 2$. (b) Induced dipole oscillations $\delta M(t)$ for $f_{p,0} = 0.1$ a. u., and (c) $f_{p,0} = 0.3$ a. u.; the horizontal dashed black line indicates the time $t = \tau = 0.3$ fs. when the δ-kick is applied.

This is an indication of different damping behavior of $\delta M(t)$ in the two cases: because the full time dependence enters the Fourier transformation in eq. (15), relaxation and dephasing play an important role. For instance, applying the probe pulse long after the system has been excited ($\tau \to \infty$) is only sensitive to the new steady state, while probing immediately after provides insight in the equilibration. Similarly, $\tau \to -\infty$ yields the equilibrium spectrum, whereas $\alpha(\omega, \tau)$ for finite $\tau < 0$ is still influenced by the probe pulse. How rapidly the limit $\tau \to \pm\infty$ is reached depends on the damping strength. For our system, $|f\rangle$ is directly damped by the embedding self-energy, while $|2p\rangle$ experiences damping by correlation due to the electron-plasmon interaction. Additional tests show that $\delta M(t)$ indeed is quickly damped within the GW approximation, while a small oscillation amplitude associated to PS^- remains when employing the GW_0 approximation (see fig. 4(b)). The formation of PS^- hereby is only due to the (first-order) mean-field contribution to eq. (9), meaning that the boson coordinate $\langle \hat{Q}(t) \rangle$ oscillates, but its fluctuation described by $D(t, t')$ is unaffected. Due to the self-energy, energy is transfered from the electronic subsystem to the plasmon subsystem, leading to damping of the electronic excitation. In the GW_0 approximation however, an infinite amount of energy can, in principle, flow into the bosonic degrees of freedom, at variance to the GW approximation.

We proceed by increasing the pump pulse strength to $f_{p,0} = 0.3$ a. u., lowering the occupation of the core level down to 0.17 after the pulse. The results for the dynamical polarizability are depicted in fig. 5. This almost perfect population inversion simulates a strong deviation from the equilibrium. Indeed, increasing τ until pump and probe pulse overlap, $\alpha(\omega, \tau)$ becomes mostly negative – the system does not absorbs photons, but predominantly emits light, mostly at the energy of the QP and PS^- peaks. Hence, the down-conversion to smaller light frequencies occurs in the system. This is a signature of nonlinearities and cannot be observed if the two subsystems are not coupled. For larger τ, the relaxation of the $|f\rangle$ state allows for transferring population to this state again. Since $|2p\rangle$ is almost empty, the optical response is however strongly suppressed. Again we realize that the GW_0 approximation delivers an overall good description, apart from strongly nonequilibrium situations. This is particularly apparent for $\tau = 0.6$ fs, where significant deviations occur. Since the PS^- peak is pertinent to the scenario where the bosonic mode is in nonequilibrium it is poorly described by the GW_0 approximation, as well (similar to fig. 3). Fig. 4(c) shows the dipole $\delta M(t)$ for both approximations. It shows that GW_0 propagation is weaker damped which can be traced back to the form of the imaginary part of the self-energy in the frequency domain. In this case there are no features that would lead to an efficient damping of PS^- state.

4. Conclusions

We studied the equilibrium and transient refraction and absorption properties of a model system describing electron-plasmon interaction of core levels in metals. The plasmon mode was characterized by the collective coordinate $\langle Q(t) \rangle$ and the coordinate-coordinate fluctuation correlator $D(t, t')$. We

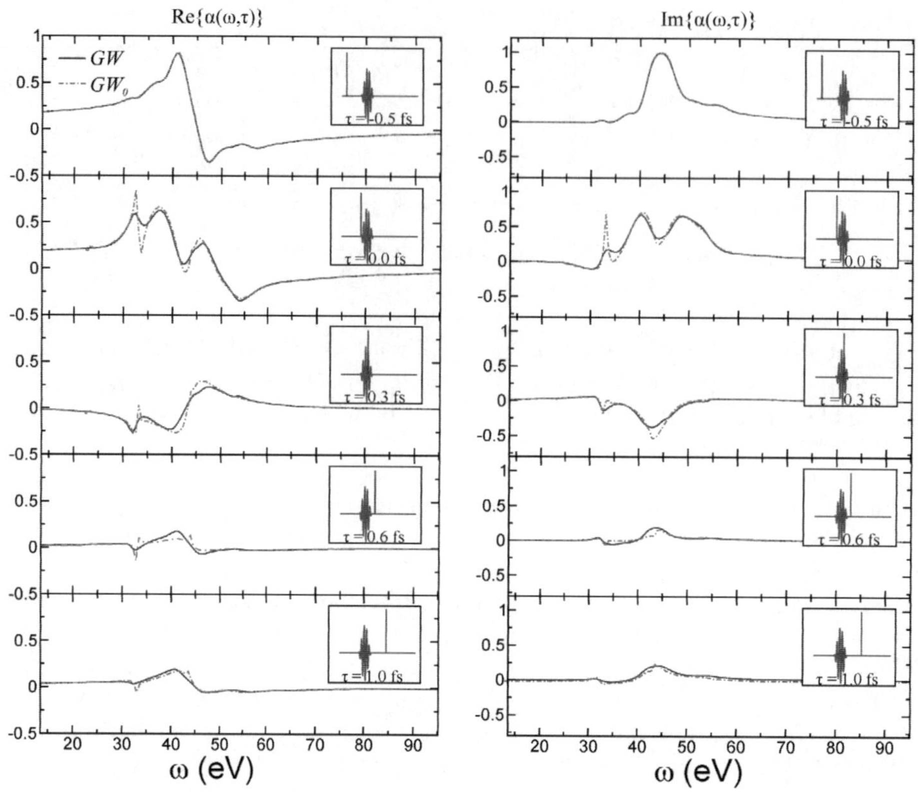

Figure 5. Transient refraction (left panel) and absorption (right) spectra for the GW (solid blue) and GW_0 (red dot-dashed line) approximation. The inset sketches the temporal arrangement of pump (amplitude $f_{p,0} = 0.3$ a. u.) and probe pulses.

presented the equations of motion for the two-times nonequilibrium Green's functions describing both the electronic and bosonic degrees of freedom. These equations were solved to determine the transient optical properties by simulating pump-probe setup. Our full propagation scheme yields, in contrast to the usual frozen-boson approach, accurate absorption spectra even in strong nonequilibrium situations.

Acknowledgments
This work is supported by the German Research Foundation (DFG) Collaborative Research Centre SFB 762 Functionality of Oxide Interfaces (Y. P.) and Grant Number PA 1698/1-1 (M. S. and Y. P.). We would like to thank Claudio Verdozzi, Andreas Wacker, Carl-Olof Almbladh for organizing this inspiring conference.

References

[1] Hüfner S 2003 *Photoelectron Spectroscopy: Principles and Applications* (Springer Science & Business Media)
[2] Uimonen A M, Stefanucci G and Leeuwen R v 2014 *J. Chem. Phys.* **140** 18A526
[3] Berakdar J and Kirschner J 2006 *Correlation Spectroscopy of Surfaces, Thin Films, and Nanostructures* (John Wiley & Sons)
[4] Schattke W and Hove M A V 2008 *Solid-State Photoemission and Related Methods: Theory and Experiment* (John Wiley & Sons)
[5] Pavlyukh Y, Schüler M and Berakdar J 2015 *Phys. Rev. B* **91** 155116
[6] Strinati G 1988 *Riv. Nuovo Cimento* **11** 1–86
[7] Stefanucci G and Leeuwen R v 2013 *Nonequilibrium Many-Body Theory of Quantum Systems: A Modern Introduction* (Cambridge University Press)

Journal of Physics: Conference Series **696** (2016) 012006 doi:10.1088/1742-6596/696/1/012006

[8] Aryasetiawan F and Gunnarsson O 1998 *Rep. Prog. Phys.* **61** 237
[9] Strinati G, Mattausch H J and Hanke W 1982 *Phys. Rev. B* **25** 2867–2888
[10] Albrecht S, Reining L, Del Sole R and Onida G 1998 *Phys. Rev. Lett.* **80** 4510–4513
[11] Benedict L X, Shirley E L and Bohn R B 1998 *Phys. Rev. Lett.* **80** 4514–4517
[12] Rohlfing M and Louie S G 1998 *Phys. Rev. Lett.* **81** 2312–2315
[13] Onida G, Reining L and Rubio A 2002 *Rev. Mod. Phys.* **74** 601–659
[14] Pal G, Pavlyukh Y, Hübner W and Schneider H C 2011 *Eur. Phys. J. B* **79** 327–334
[15] Marques M A L, Maitra N T, Nogueira F M S, Gross E K U and Rubio A 2012 *Fundamentals of Time-Dependent Density Functional Theory* (Springer)
[16] Prodan E and Nordlander P 2002 *Chem. Phys. Lett.* **352** 140–146
[17] Prodan E and Nordlander P 2003 *Nano Lett.* **3** 543–547
[18] Sakko A, Rubio A, Hakala M and Hämaläinen K 2010 *J. Chem. Phys.* **133** 174111
[19] Schüler M, Berakdar J and Pavlyukh Y 2015 *Phys. Rev. A* **92** 021403
[20] Dahlen N E and van Leeuwen R 2007 *Phys. Rev. Lett.* **98** 153004
[21] Myöhänen P, Stan A, Stefanucci G and van Leeuwen R 2009 *Phys. Rev. B* **80** 115107
[22] Balzer K and Bonitz M 2012 *Nonequilibrium Green's Functions Approach to Inhomogeneous Systems* (Springer)
[23] von Friesen M P, Verdozzi C and Almbladh C O 2009 *Phys. Rev. Lett.* **103**
[24] Puig von Friesen M, Verdozzi C and Almbladh C O 2010 *Phys. Rev. B* **82** 155108
[25] Verdozzi C, Karlsson D, Puig von Friesen M, Almbladh C O and von Barth U 2011 *Chem. Phys.* **391** 37–49
[26] Kling M and Vrakking M 2008 *Annu. Rev. Phys. Chem.* **59** 463–492
[27] Krausz F and Ivanov M 2009 *Rev. Mod. Phys.* **81** 163–234
[28] Nagele S, Pazourek R, Feist J, Doblhoff-Dier K, Lemell C, Tőkési K and Burgdörfer J 2011 *J. Phys. B* **44** 081001
[29] Goulielmakis E, Loh Z H, Wirth A, Santra R, Rohringer N, Yakovlev V S, Zherebtsov S, Pfeifer T, Azzeer A M, Kling M F, Leone S R and Krausz F 2010 *Nature* **466** 739–743
[30] Lépine F, Sansone G and Vrakking M J J 2013 *Chem. Phys. Lett.* **578** 1–14
[31] Cavalieri A L, Müller N, Uphues T, Yakovlev V S, Baltuška A, Horvath B, Schmidt B, Blümel L, Holzwarth R, Hendel S, Drescher M, Kleineberg U, Echenique P M, Kienberger R, Krausz F and Heinzmann U 2007 *Nature* **449** 1029–1032
[32] Moskalenko A S, Pavlyukh Y and Berakdar J 2012 *Phys. Rev. A* **86** 013202
[33] Perfetto E, Uimonen A M, van Leeuwen R and Stefanucci G 2015 *Phys. Rev. A* **92** 033419
[34] Perfetto E, Sangalli D, Marini A and Stefanucci G 2015 *Phys. Rev. B* **92** 205304
[35] Lundqvist B I 1969 *Phys. Kondens. Mater.* **9** 236–248
[36] Langreth D C 1970 *Phys. Rev. B* **1** 471–477
[37] Minnhagen P 1975 *J. Phys. C* **8** 1535
[38] Mahan G D 2000 *Many-Particle Physics* (Springer Science & Business Media)
[39] Galperin M, Ratner M A and Nitzan A 2004 *Nano Lett.* **4** 1605–1611
[40] Dash L K, Ness H and Godby R W 2010 *J. Chem. Phys.* **132** 104113
[41] Ness H and Dash L K 2011 *Phys. Rev. B* **84** 235428
[42] White A J and Galperin M 2012 *Phys. Chem. Chem. Phys.* **14** 13809
[43] Gruebele M and Zewail A H 1993 *J. Chem. Phys.* **98** 883–902
[44] Zewail A H 2000 *J. Phys. Chem. A* **104** 5660–5694
[45] Schüler M, Pavlyukh Y and Berakdar J 2013 *J. Phys. Chem. Lett.* **4** 1131–1135
[46] Schüler M, Pavlyukh Y and Berakdar J 2014 *Phys. Rev. A* **89** 063421
[47] Boström E, Mikkelsen A and Verdozzi C 2015 *arXiv:1507.06975 [cond-mat, physics:physics]*
[48] Sentef M, Kemper A F, Moritz B, Freericks J K, Shen Z X and Devereaux T P 2013 *Physical Review X* **3** 041033
[49] Eckstein M and Werner P 2013 *Phys. Rev. Lett.* **110** 126401
[50] Dash L K, Ness H and Godby R W 2011 *Phys. Rev. B* **84**
[51] Säkkinen N, Peng Y, Appel H and Leeuwen R v 2015 *J. Chem. Phys.* **143** 234101
[52] Stan A, Dahlen N E and Leeuwen R v 2009 *J. Chem. Phys.* **130** 224101
[53] Lemell C, Neppl S, Wachter G, Tőkési K, Ernstorfer R, Feulner P, Kienberger R and Burgdörfer J 2015 *Phys. Rev. B* **91** 241101
[54] Canney S A, Sashin V A, Ford M J and Kheifets A S 1999 *Journal of Physics: Condensed Matter* **11** 7507
[55] Giuliani G and Vignale G 2005 *Quantum Theory of the Electron Liquid* (Cambridge University Press)

Journal of Physics: Conference Series **696** (2016) 012007 doi:10.1088/1742-6596/696/1/012007

Nonequilibrium Green's functions and atom-surface dynamics: Simple views from a simple model system.

E. Boström[1], M. Hopjan[1], A. Kartsev [2], C. Verdozzi[1], and C.-O. Almbladh[1]

[1] Mathematical Physics and ETSF, Lund University, Box 118, S-22100 Lund, Sweden
[2] National Research Tomsk State University, Lenina pr. 36, 634050 Tomsk, Russia

E-mail: `emil.bostrom@teorfys.lu.se, miroslav.hopjan@teorfys.lu.se`

Abstract. We employ Non-equilibrium Green's functions (NEGF) to describe the real-time dynamics of an adsorbate-surface model system exposed to ultrafast laser pulses. For a finite number of electronic orbitals, the system is solved exactly and within different levels of approximation. Specifically i) the full exact quantum mechanical solution for electron and nuclear degrees of freedom is used to benchmark ii) the Ehrenfest approximation (EA) for the nuclei, with the electron dynamics still treated exactly. Then, using the EA, electronic correlations are treated with NEGF within iii) 2nd Born and with iv) a recently introduced hybrid scheme, which mixes 2nd Born self-energies with non-perturbative, local exchange-correlation potentials of Density Functional Theory (DFT). Finally, the effect of a semi-infinite substrate is considered: we observe that a macroscopic number of de-excitation channels can hinder desorption. While very preliminary in character and based on a simple and rather specific model system, our results clearly illustrate the large potential of NEGF to investigate atomic desorption, and more generally, the non equilibrium dynamics of material surfaces subject to ultrafast laser fields.

1. Introduction

Since the advent of femtosecond (fs) spectroscopy, the experimental characterisation of ultrafast chemical reactions has grown spectacularly [1]. At the same time, with the development of ever

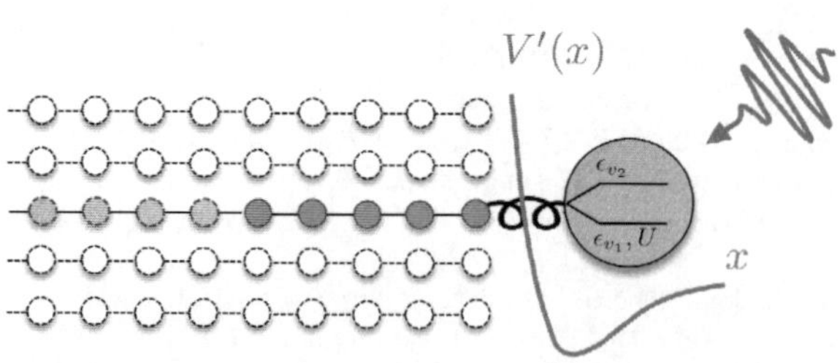

Figure 1. The model system for adsorbate-dynamics investigated in this work, for a finite (5-site) and infinite 1D substrate (a bulk substrate is also shown). The adsorbate is coupled to the substrate via an hopping term $V'(x)$ which depends on the adsorbate-substrate distance x (a schematic effective adsorbate potential is also shown). The interaction U among electrons is present only in the lower valence level of the adsorbate (see Eq. (1) and afterwards for details).

Journal of Physics: Conference Series **696** (2016) 012007 doi:10.1088/1742-6596/696/1/012007

more sophisticated techniques to probe and control atomic and molecular processes far away from equilibrium, there has also been substantial progress in the theoretical description of the ultrafast dynamics of atoms and molecules in the gas phase [2, 1, 3, 4, 5]. At present increasing attention is directed towards the study of surface-adsorbate complexes, since these systems are key to understanding important chemical processes, [6, 7, 8, 9], and also constitute the next natural paradigm for fundamental light-matter processes.

However, for surface-adsorbate systems the level of theoretical understanding is at the moment not on the level comparable with free atoms and molecules [10, 11, 12, 13]. Even though there have been recent steps in this direction [14, 15, 16, 17, 18, 19] using sophisticated methods such as Time-Dependent Density Functional Theory (TDDFT) [20, 21, 22] and Non-Equilibrium Green's Functions (NEGF) [23, 24, 25, 13], the approximate treatment of particle interactions used in practice are not sufficient to accurately describe even the initial stages of desorption. Thus, at present, there is motivation to consider model approaches to ultrafast surface-adsorbate dynamics, that capture the highly non-perturbative situation of correlated electron-nuclear motion and that can be used to gain insight into the outcome of laser induced excitations [26, 27].

Here, we perform an exploratory application of NEGF to adsorbate dynamics, an important physical mechanism at surfaces out of equilibrium. To this end, we apply NEGF to a simplified version of a recently introduced model for adsorbate dynamics induced by ultrashort laser pulses [28]. If solved exactly, the model can deal in a simplified way but on equal footing with electron-electron and electron-nuclear interactions, plasmon screening, real-space nuclear dynamics and core-hole relaxation. Thus it permits to study in a coherent way many competing surface response mechanisms and time scales. Here, for simplicity, core levels in the adsorbate and the plasmon response from the substrate are excluded from the outset.

To solve the model exactly, we must limit ourselves to a finite number of substrate sites, or equivalently a limited number of dissipation channels. As an alternative and a way to overcome this obvious limitation, we also consider a semi-infinite substrate within the framework NEGF, where however approximate treatment of interactions must be introduced. For this approach the finite-size version of our model provides an exact benchmark. To disentangle the effects of our various approximations, we perform them in sequence, starting from the numerically exact solution. The first step is to treat the nuclear dynamics in the Ehrenfest approximation while retaining the exact treatment of the electrons. In the second step the electron dynamics is treated within perturbation theory at the level of the Second-Born approximation (2BA). When both approximations are being employed we are free to extend the substrate to become semi-infinite by standard embedding procedures [29].

It is fair to say that a perturbative treatment of the NEGF (in our case the 2BA) may fail to describe ultrafast real-time dynamics of electrons, especially when dealing with strongly electron correlations. For such systems an accurate and computationally viable first-principles description like NEGF or TDDFT is currently lacking. They both rely on key ingredients [the exchange-correlation (XC) potential for TDDFT and the self-energy Σ for the NEGF] that in general are only approximately known and may be not sufficient in the strongly-correlated fast regimes. Recently, we suggested a step towards the solution to this problem using the strengths of both methods, by mixing a non-perturbative adiabatic-TDDFT XC potential with perturbative-based self-energy of NEGF [30]. Here, we test this hybrid NEGF/TDDFT method at the 2BA level and compare it to the pure NEGF 2BA in our simple model for adsorbate dynamics.

The rest of this paper is structured as follows: the model for adsorbate dynamics is introduced in Sect. 2; the different methods of solutions are described in Sect. 3; the practical details are mentioned in Sect. 4; the results and their discussion are presented in Sect. 5. Some conclusive remarks and outlooks are given in Sect. 6.

Journal of Physics: Conference Series **696** (2016) 012007 doi:10.1088/1742-6596/696/1/012007

2. Model Hamiltonian for adsorbate dynamics

The model we use is a simplified version of the one introduced in Ref. [28]. It consists of a rigid atom chain (the substrate) connected to a mobile adsorbate (Fig. 1). The Hamiltonian is

$$\hat{H}(t) = \hat{H}_a + \hat{H}_s + \hat{H}_{as} + \hat{\Lambda}(t), \tag{1}$$

where $\hat{H}_a$ describes the adsorbate, $\hat{H}_s$ the substrate and $\hat{H}_{as}$ the adsorbate-substrate interaction. To bring the system out of equilibrium a laser field $\hat{\Lambda}(t)$ is applied. The adsorbate part H_a is

$$H_a = \frac{\hat{p}^2}{2M} + \sum_{v=v_1,v_2} \sum_{\sigma} \epsilon_v \hat{n}_{v,\sigma} + U\hat{n}_{v_1,\uparrow}\hat{n}_{v_1,\downarrow} \tag{2}$$

where $\hat{p}$ is the momentum of a mobile atom with mass M. The operator $a^\dagger_{v,\sigma}$ creates an electron with spin σ at the valence orbital v of the adsorbate with energy ϵ_v, and $\hat{n}_{v,\sigma} = a^\dagger_{v,\sigma}a_{v,\sigma}$. In the adsorbate level v_1, valence electrons mutually interact with interaction strength U. The substrate is modelled as a non-interacting tight-binding chain with one orbital per site:

$$H_s = -V \sum_{\langle RR' \rangle,\sigma} c^\dagger_{R,\sigma} c_{R',\sigma} \ , \tag{3}$$

where $c^\dagger_{R,\sigma}$ creates an electron at site R with spin σ. In the following, the hopping amplitude V between neighbouring sites R and R' is taken as the energy unit, i.e. $V = 1$. The adsorbate-substrate interaction Hamiltonian H_{as} is given by

$$H_{as} = \frac{\kappa}{\hat{x}^4} - g e^{-\lambda(\hat{x}-1)} \sum_{v,\sigma} \left(a^\dagger_{v,\sigma} c_{S,\sigma} + h.c. \right) \tag{4}$$

where we denote the surface site of the substrate by S and x is the adsorbate-surface distance. The first term gives a repulsive interaction between the atomic cores, while the second term is attractive and due to the exchange of electrons between the adsorbate and the surface. We consider κ and λ as phenomenological parameters that can be adjusted to give reasonable values for the binding energy E_b, vibrational frequency ω_{ph} and effective hopping amplitude $V' = g\langle e^{-\lambda(\hat{x}-1)} \rangle$. The shape of the field $\hat{\Lambda}(t)$ in Eq.(1) depends on the experiment considered, and is switched on at $t = 0$ so that $\hat{\Lambda} = 0$ for $t \le 0$. In the dipole approximation, and for a field coupling only the adsorbate valence levels, $\hat{\Lambda}(t) = \Lambda(t) \sum_\sigma a^\dagger_{v_1\sigma} a_{v_2\sigma} + h.c.$, where $\Lambda(t)$ is an arbitrary function of time with a vanishing integral.

3. Exact and approximate treatments

3.1. Finite substrate: Exact wavefunction solution

An exact solution of the Schrödinger equation for the model presented above is viable only for a short substrate, since the size $\mathcal{S}$ of the Hilbert space $\mathcal{H}$ grows exponentially with the number of orbitals. In more detail, we note that $\mathcal{H} = \mathcal{H}_{el} \otimes \mathcal{H}_n$, with $\mathcal{H}_{el}$ the electronic and $\mathcal{H}_n$ the atomic (nuclear) Hilbert spaces, respectively. Due the explicit form of H in Eq. (1), the size of $\mathcal{S}_n$ grows linearly, while, for a spin-compensated system with L electronic orbitals, we find $\mathcal{S}_{el} = \binom{L}{L/2}^2 \xrightarrow[L\to\infty]{} 2^{2L+1}/\pi L$, i.e. an exponential scaling of $\mathcal{S}$. With and a space-grid subspace for the nucleus of size $\mathcal{S}_n \simeq 500$, substrate lengths $L \lesssim 6$ can be comfortably afforded.

Specifically, the ground state $|g\rangle$ of the model is then determined with exact diagonalization (ED) , by expanding the state vector in the basis $\{|n_{i\sigma}, x_k\rangle\}$, where $n_{i\sigma}$ is the occupation on site i of electrons with spin σ and x_k is the k:th mesh-point on a uniform grid in the interval $[0, x_{max}]$. Assuming that the system is in the state $|g\rangle$ at $t = 0$, we then propagate the Schrödinger equation for $t > 0$ using the short iterated Lanczos algorithm.

Journal of Physics: Conference Series **696** (2016) 012007 doi:10.1088/1742-6596/696/1/012007

3.2. Finite and semi-infinite substrate: NEGF

If we are not interested in the full wavefunction of the system but we choose to directly look at single particle quantities, NEGF offer a natural general and in-principle-exact way to go beyond size limitations, and to treat electrons and nuclei exactly and on equal footing (in practice approximations must be introduced). For our present system, this would imply a two-component formulation of coupled Kadanoff-Baym (KBE) equations, one component for each type of degree of freedom [19, 25]. However, to just explore the scope of NEGF for adsorbate dynamics, we here adopt a much cruder strategy, namely we use NEGF for electrons and a classical treatment for the nuclei. The most straightforward approximation is then to use the Born-Oppenheimer separation of electronic and atomic degrees of freedom, and for each atomic configuration extract effective potentials generated by the electrons and felt by the atoms, known as Born-Oppenheimer surfaces. Given these potential surfaces, there exists a number of schemes to propagate an initial atomic wave packet; here we consider the Ehrenfest approximation (EA) [31], which amounts to replace the nuclear wavepacket with a delta function, and to let the atomic position evolve under the average force from the electrons, or in other words demoting the operators $\hat{x}$ and $\hat{p}$ to classical variables [1].

For the model in Eq. (1) the mixed quantum-classical equations in the EA are

$$M\ddot{x}(t) = -\frac{\partial}{\partial x}\langle\psi_e|\hat{H}(t)|\psi_e\rangle, \tag{5}$$

$$i\frac{d}{dt}|\psi_e\rangle = \left(\hat{H}(t) - \frac{p^2}{2M} - \frac{\kappa}{x^4}\right)|\psi_e\rangle. \tag{6}$$

Once reformulated in the language of NEGF, these become

$$M\ddot{x}(t) = \frac{4\kappa}{x^5} - 2g\lambda e^{-\lambda(x-1)}\sum_v \Im\left[G^<_{vS}(t,t) + G^<_{Sv}(t,t)\right], \tag{7}$$

$$\left[i\frac{d}{dz_1} - h_{ij}(z_1;x)\right]G_{jk}(z_1,z_2) = \delta_{ik}\delta(z_1,z_2) + \int_\gamma dz\Sigma_{ij}(z_1,z)G_{jk}(z,z_2), \tag{8}$$

where the single particle Green's function is defined via the time contour ordering $\mathcal{T}$ as $G_{ij}(z_1,z_2) = -i\langle\mathcal{T}c_i(z_1)c_j^\dagger(z_2)\rangle$ for a convenient choice of basis $\{\varphi_i\}$, here taken as the site basis of our model. In the coupled equations above, Eqs. (7) governs the nuclear dynamics, whilst Eq. (8) is the standard Kadanoff-Baym equation for G, where dependence of the one-body Hamiltonian matrix h_{ij} on x is shown, and an implicit dependence of Σ and G on x is understood.

If the system is composed of an interacting region connected to a non-interacting reservoir (as for example, the adsorbate and substrate parts in our system), the latter can be removed from the explicit description and still taken into account exactly via an embedding procedure [29]. In this case, the self-energy for the interacting region becomes $\Sigma_{ij}(z_1,z_2) = \Sigma^{em}_{ij}(z_1,z_2) + \Sigma^{MB}_{ij}(z_1,z_2)$. The first term is the embedding self-energy that gives an exact coupling to the reservoir, and the second term is the many-body self-energy that contain all interactions between electrons in the active region. After finding the initial $G_{ij}(t_0,t_0)$ on the imaginary part of the Keldysh contour, the KBE are propagated self-consistently within the same approximation for Σ_{ij} [23]. With NEGF and the embedding self-energies, a much larger number of orbitals can be explicitly taken into account (plus the levels of the macroscopic reservoir) than e.g. with exact diagonalization, and in any dimensionality. The downside is that interactions now have in general to be treated in an approximate way, e.g. perturbatively: $\Sigma^{MB}_{ij}(z_1,z_2) \approx \Sigma^{PT}_{ij}(z_1,z_2)$.

[1] Within the Ehrenfest approximation, the size of the Hilbert-space is determined only by $\mathcal{H}_{el}$, but the substrate-lengths L which can be considered within our exact treatment remain rather short. Hence the usefulness of NEGF, also for a classical treatment of the nuclei.

3.3. A NEGF/TDDFT hybrid scheme

In our numerical calculations for adsorbate dynamics, correlations are treated in the 2nd Born approximation (2BA), $\Sigma^{PT} \equiv \Sigma^{2BA}$. However, starting from this approximation, we also consider a second approach, recently introduced in Ref. [30], which adds to the Kadanoff-Baym equation an exchange-correlation potential $v^{xc,NP}$ of DFT which is nonperturbative in character. At the same time, we subtract a perturbative exchange-correlation potential $v^{xc,PT}$, to avoid double counting of the interactions. The approach is conserving in the Kadanoff-Baym sense, and has the same numerical complexity of standard KBE. In this hybrid NEGF/TDDFT scheme, the modified Kadanoff-Baym equation reads (repeated indices are implicitly summed over)

$$\left[i\frac{d}{dz_1} - h_{ij}(z_1;x) - \left(v_i^{xc,NP}(z_1) - v_i^{xc,PT}(z_1)\right)\delta(z_1,z_2)\delta_{ij}\right] G_{jk}(z_1,z_2)$$
$$= \delta_{ik}\delta(z_1,z_2) + \int_\gamma dz\left(\Sigma_{ij}^{em}(z_1,z) + \Sigma_{ij}^{PT}(z_1,z)\right)G_{jk}(z,z_2). \quad (9)$$

Both exchange-correlation potentials are taken in adiabatic local density approximation (ALDA) [32], i.e. the time-dependent XC-potential is approximated as a potential coming from a reference equilibrium system with the same density $v_i^{xc}(z_1) \approx v_{ref.}^{xc}(n_i(z_1))$. The reference system is chosen according to the model to be solved, as discussed in the next section.

3.4. Additional aspects of the NEGF/TDDFT scheme

In the NEGF/TDDFT method, we need suitable XC potentials $v^{xc,NP}$ and $v^{xc,PT}$ for the lower valence level. The geometrical structure of the model in Fig. 1 does not allow for a direct application of standard 1D ALDA (which is obtained from a 1D homogeneous Hubbard model where each site has a bond on the left and one on the right). However, both in equilibrium and for time-varying perturbations acting only on the adsorbate, symmetry arguments permit to re-interpret the adsorbate as a two-orbital central site connected to the parity-even states (but not to odd ones) of two identical 1D semi-infinite substrates, with a reduced site-substrate(s) coupling $V'_{\rm sym} = V'/\sqrt{2}$. For the lower valence level v_1 of the central site in the symmetric model, we employ the ALDA potential $v^{xc} \approx v_{ref.}^{xc}\left(n_{\nu_1}, U/V'_{\rm sym}\right)$ using the homogeneous 1D Hubbard model as the reference equilibrium system [33, 34, 35]. The corresponding potentials $v^{xc,NP}$ and $v^{xc,PT}$ are shown in Fig. 2: they mutually differ in many ways in their behavior, the key difference being that $v^{xc,NP}$ ($v^{xc,PT}$) has (has not) a discontinuity at half-filling.

4. Practical details

4.1. On the different treatments

In the actual calculations, we consider both finite ($L = 5$) and semi-infinite 1D substrates (see Eq. 1 and Fig. 1). For $L = 5$, the model is solved exactly with a full-quantum mechanical solution for the electrons and the nucleus (Sect. 3.1) and, in the EA, with different degrees of approximations for the electron dynamics (Sects. 3.2, 3.3). In the EA for the nuclei plus the exact dynamics of the electrons (EA+Ex$_e$), the electronic ground state $|g_e\rangle$ and nuclear equilibrium position x_{eq} are found at the stationary points of Eq. (5-6), after which we propagate Eq. (5-6) in time using the Lanczos technique for the electrons and a velocity Verlet-type algorithm for the nuclei. When the EA is used in conjunction with NEGF, the electronic correlations are taken into account at the 2BA level (EA+2BA), or via the hybrid NEGF/TDDFT scheme (EA+Hyb). The ground state Green's function $G_{ij}(t_0,t_0)$ and equilibrium position x_{eq} are now found solving self-consistently Eq. (8) on the vertical track of the contour for G and Eq. (7) for x. For the dynamics, Eqs. (7-8) are propagated in time according to standard procedures on the time-square for G [36, 37, 38] together with a velocity Verlet algorithm for the nucleus.

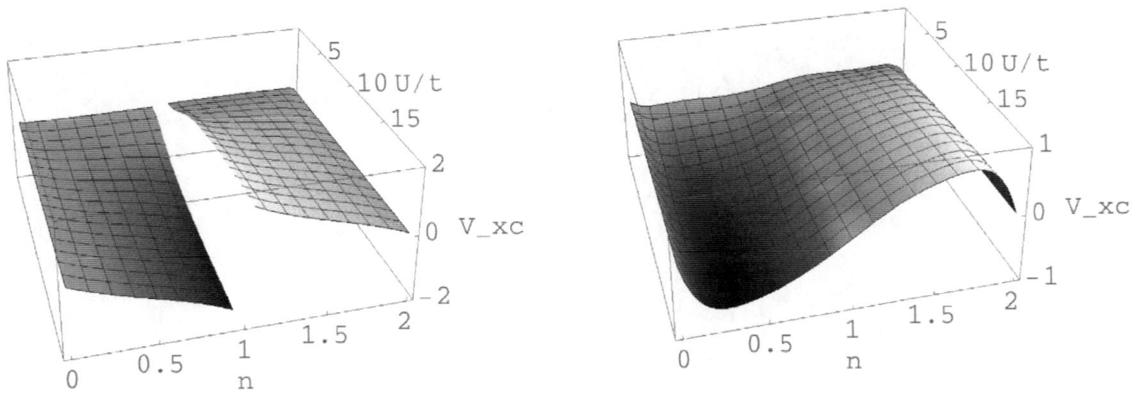

Figure 2. XC potentials for the reference homogeneous one-dimensional Hubbard model $H = t\sum_{\langle ij\rangle,\sigma} a^{\dagger}_{i,\sigma}a_{j,\sigma} + U\sum_i n_{i,\downarrow}n_{i,\uparrow}$. The exact nonperturbative exchange-correlation potential $v^{xc,NP}_{ref.}$ is shown to the left, and to the right is the corresponding perturbative potential $v^{xc,PT}_{ref.}$ in the 2BA. We plot the potentials as a function of uniform density n and the ratio of interaction and hopping $\frac{U}{t}$ for fixed interaction $U = 4$.

Within the NEGF treatments, we also consider a semi-infinite 1D substrate via an embedding self energy [29], with the solution method otherwise unchanged.

4.2. On the external fields

We used two time-dependent fields to perturb the system: $\Lambda(t) \equiv \Lambda_a(t) = Ae^{-(t-t_0)^2/2\sigma^2}$ and $\Lambda(t) \equiv \Lambda_b(t) = \Lambda_a(t)\sin(\omega t + \pi/2)$, with amplitude either $A = 2$ or 4. The electronic parameters of the model are chosen as $V = 1$, $\epsilon_1 = 1$, $\epsilon_2 = 3$, thus giving a cationic adsorbate in the ground state, and we consider the interaction strengths $U = 4, 6$ and 8. For the nucleus, $\kappa = 2.4$, $g = 1.8$ and $\lambda = 1.2$, giving a binding energy $E_b \simeq 1.5$, an harmonic frequency $\omega_{ph} \simeq 0.2$ and an effective hopping amplitude $V' \simeq 1.8$. These values are typical for chemisorption.

The Gaussian pulse Λ_a is an artificial perturbation and serves as a tool to investigate slower dynamics, while the sinusoidally modulated pulse Λ_b represents a realistic pulse with a carrier wavelengh of 400nm (for a hopping amplitude $V \simeq 2$ eV the unit of time is $1/V \simeq 1.3$ fs, which is the period of 400 nm radiation). Both perturbations extend from $t = 0$ to $t = 8$, with time measured in units of the inverse hopping $1/V$. In this time interval the evolution is dominated by the perturbation and all the methods give basically the same result. For later times the dynamics is dominated by the memory of the system, and the solutions differ depending on the approximation used.

5. Results

5.1. Exact vs Ehrenfest dynamics for the nuclei

Let us first look at the evolution of the nuclear subsystem, as shown in Fig. 3. With the full quantum mechanical solution we observe bond stretching for the pulse Λ_a, but for large times the nuclear position returns to a region close to its equilibrium value. For the pulse Λ_b the nuclear coordinate continues to increase also for large times, signalling a finite probability of desorption. This is typical of the full quantum solution, where in general one part of the initial wave packet remains bound and performs oscillations in the potential well, while another part overcomes the potential barrier and thereafter propagates quasi-freely (strictly the potential is zero only for $x \to \infty$). We note that with increasing interaction U and increasing amplitude

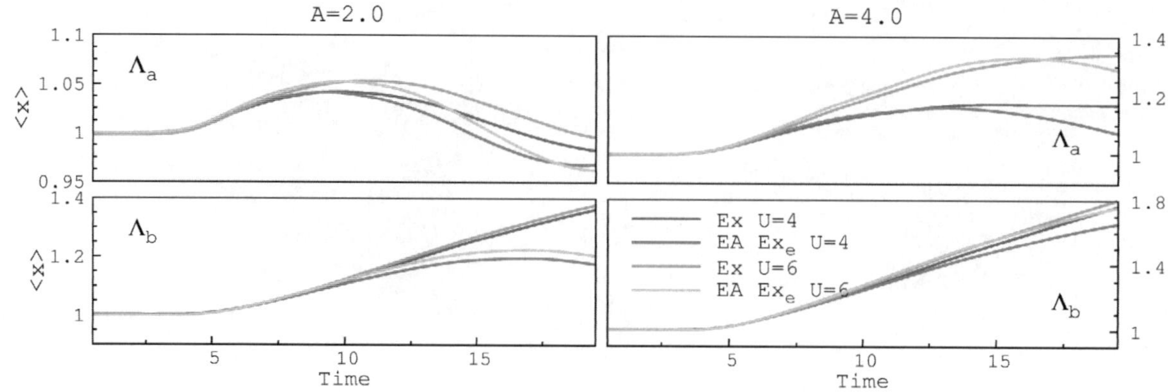

Figure 3. Time-dependent interatomic distances in the exact quantum treatment (red) and the EA+Ex$_e$ (green), for the values $U = 4$ and $U = 6$ of the interaction strength. The first row shows results for the external field $\Lambda_a(t)$ of amplitude $A = 2.0$ and $A = 4.0$, and the second row gives the corresponding results for the field $\Lambda_b(t)$.

A, the desorption probability also increases. The former is a result of the interaction induced quenching of electronic motion between the adsorbate and surface, resulting in a lower average bond kinetic energy and therefore a weaker bond (cf. the form of electron-nuclear coupling in Eq. 4 and the definition $K_{as} = c^\dagger_{v1}c_S + c^\dagger_{v2}c_S + h.c.$ for the bond kinetic energy).

Since the EA is a classical treatment of nuclei the only possible outcomes of an experiment is to have either full or no desorption, and in the intermediate regime the approximation will inevitably fail to reproduce the exact dynamics. This is verified in Fig. 3 where the exact results are best reproduced in the cases where the quantum solution predicts either a very small or very large desorption probability. The agreement is in general better for the pulse Λ_a, which varies more slowly in time and is closer to the adiabatic situations where the EA is known to perform well.

5.2. Exact vs NEGF electronic correlations within the Ehrenfest dynamics.

We now turn our attention to the electron dynamics, as depicted in Fig. 4. When comparing the exact solution with the EA+Ex$_e$ dynamics we see clear differences, that are most pronounced in the case of partial desorption. In particular the EA shows oscillations in the electron density of greater amplitude, that persist even for long time. This can be explained in part by the inability of the EA to transfer energy from the electronic to the nuclear subsystem, due to neglect of electron-nuclear correlations, that in the exact simulation act as a dissipation channel for the electrons. In part it is also due to the fact that the effective adsorbate-surface hopping amplitude is larger in the EA, making is more probable for the electrons to tunnel between the substrate and the adsorbate. With quantum mechanical nuclei, the hopping amplitude (proportional to $e^{-\lambda x}$, see Eq. 4) is decreased by the presence of a split-off part of the nuclear wave packet having propagated to large distances.

Comparing the exact and the EA+2B electron evolution we observe that the oscillations in density seen in the EA are reduced compared to the EA+Ex$_e$ result, and that this difference increases with stronger interactions. The damping of electron oscillations is a generic feature of approximate solutions of the KBE in finite systems, and was first discussed in [38]. Briefly, the reason for damping is that self-consistent approximations (like the 2B) include infinitely many electron-hole excitations (through the infinite set of diagrams contained in the self-energy), which cannot occur in finite systems. Thus, even though the agreement between the exact and the

EA+2B solution seems better than EA+Ex$_e$, it should be kept in mind that this result is due to unphysical effects.

With the EA+Hyb method we observe a slightly better agreement in phase with the EA+Ex$_e$ treatment, but for the model considered here the improvement of local correlations brought by the exchange-correlation potential have a rather small overall effect. This points to the importance of non-local correlations for the desorption process, and to capture more closely the exact dynamics approximations beyond 2B (or even inclusion of non-perturbative effects) are necessary. As a first step in this direction we are currently working on including correlations in the T-matrix approximation.

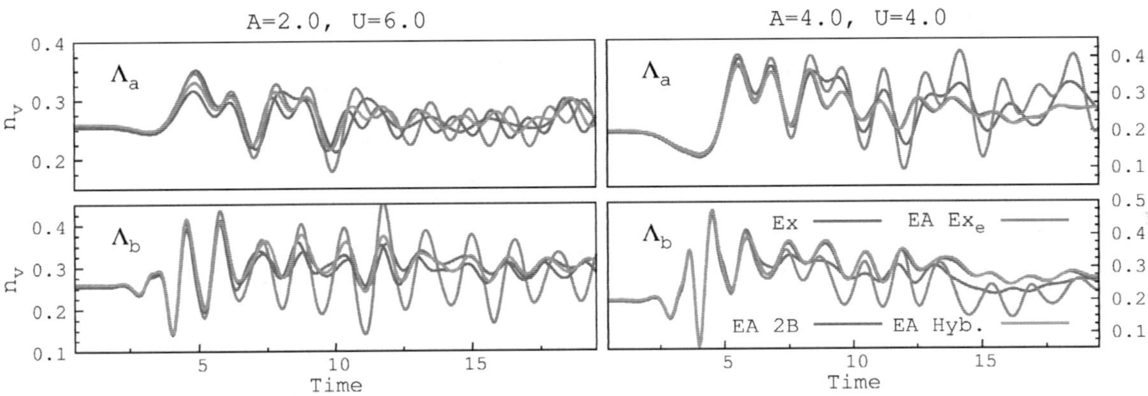

Figure 4. Time-dependent densities of the lower valence level in the exact quantum treatment (red) and the approximate schemes EA+Ex$_e$ (green), EA+2B (blue) and EA+Hyb (orange).The first row shows results for the external field $\Lambda_a(t)$ of amplitude $A = 2.0$ with interaction strength $U = 6$ and $A = 4.0$ with interaction strength $U = 4$, and the second row gives the corresponding results for the field $\Lambda_b(t)$.

5.3. The effect of a semi-infinite substrate

Finally, we contact the system to a semi-infinite lead and compare the resulting dynamics within EA+2B and EA+Hyb to the corresponding finite system simulations (see Fig. 5). Interestingly, the addition of the particle reservoir seems to have a rather small impact on the short-time dynamics, both for the electronic and nuclear subsystem. At larger times the desorption probability decreases in the presence of a lead, since there is now an additional energy dissipation channel for the electrons that reduce the amount of energy transferred to the nuclei. However, for the nuclear part there is not much difference between the EA+2B and EA+Hyb methods.

The electron dynamics of the finite and extended systems show a qualitative agreement, the major difference being that the amplitude of the density oscillations decay faster in the latter case. For a contacted system the damping of electron oscillations is an inherent physical effect, in contrast to the artificial behavior discussed in the previous section. For the faster pulse Λ_b we also observe a slight dephasing of the EA+2B and EA+Hyb solutions, favoring the EA+Hyb method in both the finite and contacted case.

The difference between the EA+2B (or EA+Hyb) and the exact solution is typically greater than the difference between the finite and contacted solution within a given approximation scheme. This might imply that to model adsorbate-surface systems, and in particular non-perturbative processes such as desorption, it is more important to properly account for the electron-electron and electron-nuclear correlations close to the surface, than for the infinite extent of the system. This conclusion is strongly dependent on the parameters chosen, but for a

Journal of Physics: Conference Series **696** (2016) 012007 doi:10.1088/1742-6596/696/1/012007

strong electron-nuclear coupling (so that the system is in the surface molecule limit) we expect it to hold.

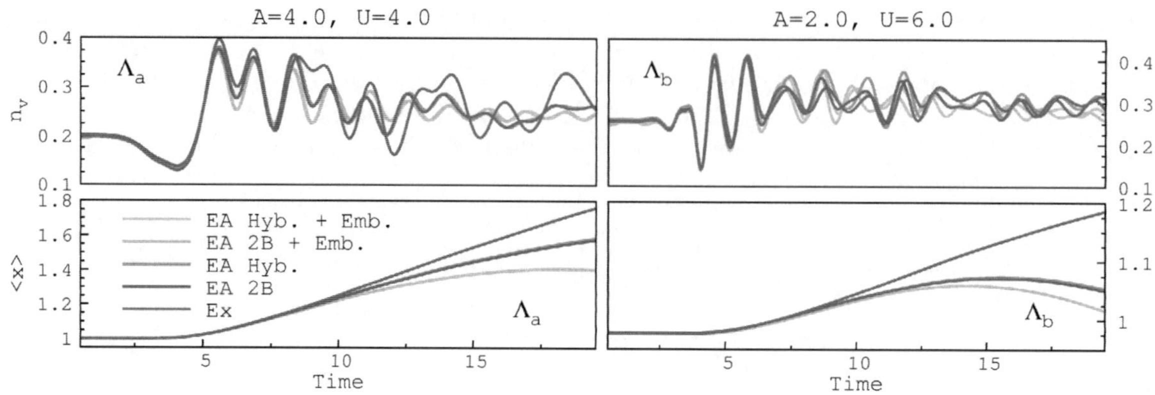

Figure 5. Time-dependent densities of the lower valence level (top) and interatomic distance (bottom), for a finite chain in the exact quantum treatment (red) and the approximate schemes EA+2B (blue) and EA+Hyb (orange), as well as for a contacted system within EA+2B (light blue) and EA+Hyb (light orange). The first column shows results for the external field $\Lambda_a(t)$ of amplitude $A = 4.0$ with the interaction strength $U = 4$, and the second column gives the results for the external field $\Lambda_b(t)$ of amplitude $A = 2.0$ with the interaction strength $U = 6$.

6. Conclusions

We present a model to treat real-time surface-adsorbate system dynamics induced by ultrafast laser pulses, and solve it exactly numerically in its finite size realisation. The exact results are compared to a treatment within the framework of NEGF, on the level of the second Born approximation and also to a recently introduced hybrid NEGF/TDDFT approach. By the standard embedding procedure this allows us to also solve the semi-infinite version of our model. In the finite size case we treat the nuclear coordinate exactly using first quantisation, and compare it to the semi-classical case of having exact electronic dynamics and nuclear motion within the Ehrenfest approximation.

For the process of desorption, interesting due to its non-perturbative character, we find good qualitative agreement between the exact and Ehrenfest nuclear dynamics when the exact result predicts either a very small or a very large desorption probability. In the intermediate case, characterized by a splitting of the nuclear wave packet into a bound and a quasi-free part of comparable magnitude, the Ehrenfest treatment fails to reproduce the exact results. For the case of chemisorption considered here, we further find that the inclusion of a semi-infinite substrate have a limited effect but predicts a slightly lower desorption probability. Our results also indicate the importance of non-local correlation effects beyond second Born.

References

[1] Zewail A H 2000 *J. Phys. Chem.* A **104** 5660
[2] F. Remacle *et al* 1999 *J. Phys. Chem.* A **103** 10149
[3] Miron C 2012 *et al Nature Physics* **8** 135
[4] Månsson E P *et al* 2014 *Nature Physics* **10** 207
[5] García-Vela A and Henriksen N E 2015 *J. Phys. Chem. Lett.* **6** 824
[6] Somorjai G A and Li Y 2010 *Introduction to Surface Chemistry and Catalysis* (Wiley)
[7] Henderson M A 2011 *Surf. Sci. Rep.* **66** 185
[8] Hornett S M *et at* 2014 *Phys. Rev.* B **90** 081401(R)

[9] Linic S *et al* 2015 *Nature Materials* **14** 567
[10] Petek H and Ogawa S 2002 *Annu. Rev. Phys. Chem.* **53** 507
[11] Frischkorn C and Wolf M 2006 *Chem. Rev.* **106** 4207
[12] Dahlström J M and Lindroth E 2014 *J. Phys.* B **47** 124012
[13] Perfetto E, Uimonen A-M, van Leeuwen R, Stefanucci G 2015 *Phys. Rev.* A **92** 033419
[14] Craig C F, Duncan W R and Prezhdo O V 2005 *Phys. Rev. Lett.* **95** 163001
[15] Butriy O *et al* 2007 *Phys. Rev.* A **76** 052514
[16] Tapavicza E, Tavernelli I and Rothlisberger U 2007 *Phys. Rev. Lett.* **98** 023001
[17] Abedi A, Maitra N T and Gross E K U 2010 *Phys. Rev. Lett.* **105** 123002
[18] Säkkinen N, Peng Y, Appel H and R. van Leeuwen 2015 *Preprint* arXiv:1403.2968
[19] Säkkinen N, Peng Y, Appel H and R. van Leeuwen 2015 *Preprint* arXiv:1507.04726
[20] Runge E and Gross E K U 1984 *Phys. Rev. Lett.* **52** 997
[21] Crawford-Uranga A *et al* 2014 *Phys. Rev.* A **90** 033412
[22] Hellgren M, Räsänen E and Gross E K U 2013 *Phys. Rev.* A **88** 013414
[23] Stefanucci G and van Leeuwen R 2013 *Nonequilibrium Many-Body Theory of Quantum Systems* (Cambridge University Press)
[24] Dahlen N E and van Leeuwen R 2007 *Phys. Rev. Lett.* **98** 153004
[25] Pavlyukh Y, Schüler M and Berakdar J 2015 *Phys. Rev.* B **91** 155116
[26] Kim K H *et al.* 2011 *Phys. Rev. Lett.* **107** 047401
[27] Díez Muiño R *et al.* 2011 *Proc. Nat. Acad. Sci. U.S.A.* **108** 971
[28] Boström E, Mikkelsen A and Verdozzi C 2015 *Preprint* arXiv:1505.07364
[29] Myöhänen P, Stan A, Stefanucci G and van Leeuwen R 2008 *Europhys. Lett.* **84** 67001
[30] Hopjan M, Karlsson D, Ydman S, Verdozzi C, Almbladh C-O *Preprint* arXiv:1508.03089
[31] Ehrenfest P 1927 Z. Phys. **46** 455
[32] Zangwill A and Soven P 1980 *Phys. Rev.* A **21** 1561
[33] Schönhammer K, Gunnarsson O and Noack R M 1995 *Phys. Rev.* B **52** 2504
[34] Lima N A, Silva M F, Oliveira L N and Capelle K 2003 *Phys. Rev. Lett.* **90** 146402
[35] Verdozzi C 2008 *Phys. Rev. Lett.* **101** 166401
[36] Köhler H S, Kwong N H and Yousif H A 1999 *Comp. Phys. Comm.* **123** 123
[37] Stan A, Dahlen N E and van Leeuwen R 2009 *J. of Chem. Phys.* **130** 224101
[38] Puig von Friesen M, Verdozzi C and Almbladh C-O 2009 *Phys. Rev. Lett.* **103** 176404

Progress in Non-equilibrium Green's Functions (PNGF VI) IOP Publishing
Journal of Physics: Conference Series **696** (2016) 012008 doi:10.1088/1742-6596/696/1/012008

The time-dependent generalized active space configuration interaction approach to correlated ionization dynamics of diatomic molecules

S Bauch[1], H R Larsson[2], C Hinz[1] and M Bonitz[1]

[1] Institute for Theoretical Physics and Astrophysics, Christian-Albrechts-Universität Kiel, 24098 Kiel, Germany

[2] Institute for Physical Chemistry, Christian-Albrechts-Universität Kiel, 24098 Kiel, Germany

E-mail: [1] bauch@theo-physik.uni-kiel.de

Abstract. In this contribution, we review the time-dependent generalized-active-space configuration interaction (TD-GAS-CI) approach to the photoionization dynamics of atoms and molecules including electron correlation effects. It is based on the configuration interaction (CI) expansion of the many-body wave function and the restriction of the determinantal space to a reduced subspace. For its numerically efficient application to photoionization, a partially-rotated basis set is used which adopts features of a localized basis with a good reference description and a grid representation for escaping wave packets. After reviewing earlier applications of the theory, we address the strong-field ionization of a one-dimensional model of the four-electron LiH molecule using TD-GAS-CI and demonstrate the importance of electron-electron correlations in the ionization yield for different orientations of the molecule w.r.t the peak of the linearly polarized laser field. A pronounced orientation-dependent variation of the yield with the pulse duration and the level of considered electron-electron correlations is observed.

1. Introduction

With the emergence of ultrashort and strong laser pulses in the femtosecond and subfemtosecond regimes [1], a time-dependent theory for electron dynamics with the ability of treating strong fields and electron correlations is needed. Among the processes are the time-resolved observation of post-collision interaction [2, 3], time-delays in photoionization [4], the time-resolved observation of strong-field tunneling [5], and many more. See also the review articles [6, 7].

Often, only numerically costly methods are able to describe the ultrafast motion of the interacting electrons in the external fields. Pioneering work was based on the numerical solution of the time-dependent Schrödinger equation (TDSE) in the context of strong-field processes, e.g. [8, 9], for one-electron systems. Despite the tremendous evolution of computing architectures over the past decades, the numerical solution of the TDSE still remains a challenging task for extended systems with more than one active electron and is limited to helium, H_2 and similar systems, e.g., [10, 11, 12]. Therefore, approximative methods are inevitable. Perturbation theory in the electrical field of the excitation is extremely successful by considering the fully-correlated ground-state wave function for calculating spectra, but not applicable to the high-intensity or time-resolved regime by construction. The alternative approach is the approximative treatment of the electron interaction. Within a wave-function based perspective, this leads to the concept of configuration interaction (CI) and self-consistent-field (SCF) methods, see [13] for a review. Time-dependent methods emerging from the latter are the widely applied multi-configurational time-dependent Hartree-Fock (MC-TDHF) theory, e.g. [14, 15, 16] and time-dependent restricted and complete active space (TD-R/CAS) SCF methods [17, 18, 19]. Similar approaches are the TD-renormalized natural orbital [20] and the TD orbital-adaptive coupled-cluster [21] theories. All these methods are based on a time-dependent optimization of the orbital space, thereby leading to a highly non-linear propagation scheme with convergence issues still under investigation, see also the article by *Hinz et al.* [22] in this volume. The CI expansion on the other hand is based on a time-independent orbital picture and conserves the linearity of the underlying TDSE, although much larger determinantal spaces are required for an accurate description. This deficiency is addressed within the presented TD-GAS-CI approach.

2. Time-dependent generalized-active space configuration interaction

The underlying equation of motion is the time-dependent N-electron Schrödinger equation,

$$i\frac{\partial}{\partial t}|\Psi(t)\rangle = H(t)|\Psi(t)\rangle. \tag{1}$$

with a time-dependent Hamilton operator

$$H(t) = \sum_{i=1}^{N} t_i + v_i(t) + \sum_{i<j}^{N} w_{ij} \,. \tag{2}$$

$t_i \equiv t(\boldsymbol{r}_i)$ and $v_i \equiv v_{\text{ion}}(\boldsymbol{r}_i) + E(t)\boldsymbol{r}_i$ describe the kinetic and potential energies of particle i at position $\boldsymbol{r}_i$, and $w_{ij} \equiv w(\boldsymbol{r}_i, \boldsymbol{r}_j)$ denotes the interaction between electrons i and j. The excitation with an external field, e.g. a laser, is coupled via the dipole operator in length gauge, $\boldsymbol{r}$, with the time-dependent electrical field amplitude $E(t)$. In this contribution, we consider only linear polarization.

2.1. The generalized-active space method

The generalized-active space (GAS) concept to calculate approximate solutions to Eq. (1), as many recently developed methods, is inspired from time-independent quantum chemistry considerations [23, 24]. It is founded on the configuration interaction (CI) expansion of the N-electron wave function,

$$|\Psi_N(t)\rangle = \sum_{I\in \text{GAS}} c_I(t)|I\rangle \,, \tag{3}$$

where $|I\rangle$ denotes an N-electron Slater determinant constructed from $2N_b$ time-independent single-particle spin orbitals $\chi_i(\boldsymbol{x}) \equiv \phi_i(\boldsymbol{r})s(\sigma)$ with $\boldsymbol{x} = (\boldsymbol{r}, \sigma)$ referring to the spatial orbital $\phi_i(\boldsymbol{r})$ and the spin part $s(\sigma)$ with $s \in \{|\uparrow\rangle, |\downarrow\rangle\}$. $c_I(t)$ are the time-dependent complex expansion coefficients. Typically, $|I\rangle$ is constructed from HF orbitals or HF-like pseudo orbitals, which describe the excitations of the system more accurately than common HF virtual orbitals [25].

The basic idea of GAS is the reduction of the exponentially large N-electron Hilbert space to an active subspace, which is capable of describing the physics, but to much less numerical cost than the full CI (FCI) method. This is achieved by partitioning the single-particle space into subspaces with fixed number of electrons. Thereby, it is possible to freeze certain parts of the many-body wavefunction which are unlikely to take part in the dynamics (e.g. by considering energy scales). Details of the idea, technical aspects of the GAS construction and illustrative examples are given in Refs. [25, 26]. A schematic of examples for GAS partitions is drawn in

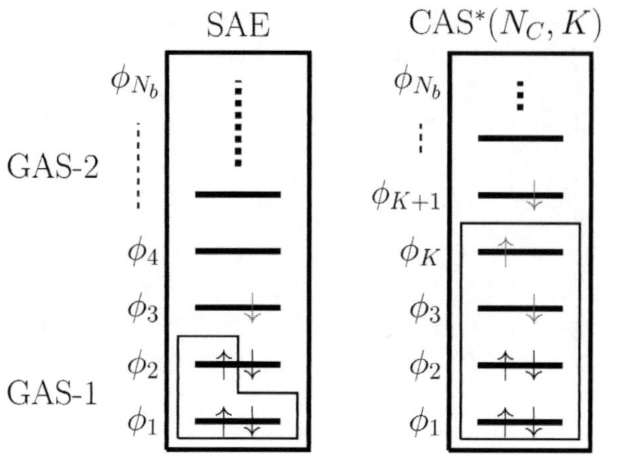

Figure 1. Schematic of GAS definitions for SAE (left) and correlated CAS (right) calculations with four electrons. ϕ_i are the spatial orbitals, which can be occupied by ↑ and ↓ electrons. For SAE, three electrons in GAS-1 are frozen and the outer ↓ electron is allowed to be excited, $i > 2$ (gray arrows). For the CAS, $N_C = 4$ electrons occupy K orbitals in any possible configuration. For $i > K$, only one electron is allowed to occupy higher orbitals. In the picture, the example of ↓ is sketched (gray arrows indicate excited electrons). In the calculations, every realization is possible.

Fig. 1. The time-dependent realization of GAS-CI was first introduced in the context of time-dependent restricted-active-space (RAS) CI in Ref. [26] and applied to the ionization of helium, beryllium and neon [27]. Throughout, we use the historically motivated acronyms "RAS" and "GAS" as synonyms of the same method. A comprehensive study of convergence and a detailed layout of the method are discussed in [25], further studies using TD-GAS-CI comprise correlation effects in the enhanced-ionization phenomena of molecules [28] and a study of the four-electron LiH molecule in prolate spheroidal coordinates [29]. For more information see also the review article [13].

Within the GAS formalism, several well-established approximation schemes are contained. In this contribution, we will use the single-active electron (SAE), the CI singles (CIS) and special cases of the complete active space (CAS) approximations. For the SAE case, expansion (3) consists only of excitations of one electron with all other $N-1$ electrons kept frozen in their HF reference orbitals, see Fig. 1 (left) for a schematic of a four-electron system. In contrast to common SAE calculations, no additional pseudo potential needs to be constructed in the GAS framework and all interactions are contained on the same level in the Hamiltonian, Eq. (2). For CIS, the restriction to only one active electron is loosened, and all electrons may be excited, however, only one at a time.

The important CAS case is more involved, as it can be tuned toward the fully converged FCI case containing all electron correlation contributions. We will use the specially adapted notation CAS*(N_C, K), which corresponds to $N_C \leq N$ electrons occupying $K \ll N_b$ spatial orbitals ($2K$ spin orbitals) and on top of that single-excitations out of the CAS, which is referred to by the * symbol. In this notation, the familiar CIS method is denoted by CAS*$(N_C, N_C/2)$ [N_C electrons occupy always $N_C/2$ spatial orbitals and one electron may be excited]. Therefore, in the quantum chemistry community, also the term "multi-reference CIS" is widely used for this kind of GAS with $K > N_C/2$. These CAS* approximations are especially useful for photoionization, as they describe the correlated motion of all (or some) electrons within a subspace and one electron is allowed to occupy every spin orbital, such as continuum states. See Fig. 1 (right) and the corresponding caption for an example. Further information on notations and illustrative examples are provided in Refs. [25, 29].

2.2. Mixed-basis set approach

An essential ingredient to a successful application of truncated CI expansions, i.e., the efficient application of the GAS-CI concept, is a good reference state. Typically in usual CI calculations, the Hartree-Fock (HF) or a similar optimized state is chosen, from which excited Slater

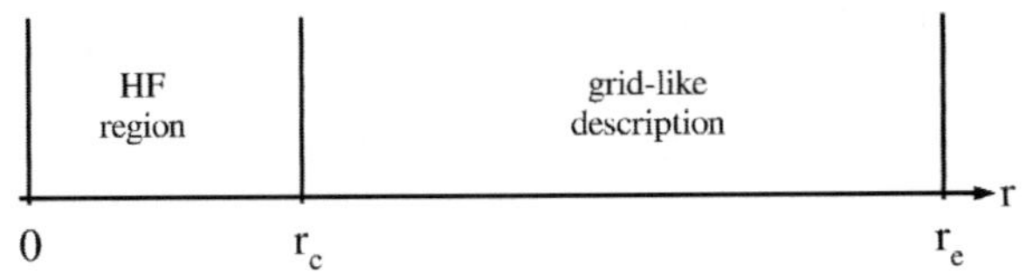

Figure 2. Layout of the mixed basis. Close to the atom or molecule, $r < r_c$, a localized basis (HF-like) is constructed to allow for good CI-reference properties. For $r \geq r_c$, a grid-like representation is employed. For 1d systems, a similar boundary appears for $r < 0$.

determinants are created. However, especially the two-electron integrals in the HF basis comprise a four-indexed object, w_{ijlk}, which prohibits the application for large basis sets because $\mathcal{O}(N_b^4)$ two-electron integrals need to be calculated and stored. Note that N_b is large ($\gg 100$) for photoionization calculations due to the discretized continuum description.

Therefore, we employ a mixed basis set, allowing for an accurate description of the bound part and the scattering states, which combines the best of both worlds: the good reference properties in vicinity to the nucleus and a grid-like discretized continuum with sparse interaction matrices. Technically, a finite-element discrete-variable-representation (FE-DVR) basis is used and partially transformed to a HF-like basis within a central region of the computational grid close to the nuclei, $r < r_c$, see Fig. 2. The great benefit originates from the fact that the two-electron integrals in FEDVR representation, e.g. [30], are extremely sparse, $w_{ijkl} \propto \delta_{ij}\delta_{kl}$, and under the partial rotation, this property is conserved for the largest part of the basis ($r \geq r_c$) [25].

In the context of time-dependent CI calculations, this technique was introduced by *Hochstuhl and Bonitz* in Ref. [26]. The details of the transformation from the full grid to the mixed basis and technical aspects for its efficient implementation are presented and discussed in Ref. [25], the application to atoms in spherical coordinates is addressed in Refs. [26, 27] and in prolate spheroidal coordinates for diatomic molecules in [29]. A similar partitioning technique is also applied in the successful time-dependent R-Matrix theory [31, 32], in a recent formulation even with two-electron continua [33].

3. Application to four-electron molecules

To demonstrate TD-GAS-CI, we consider in the following the four-electron LiH molecule in one spatial dimension and the fixed-nuclei (Born-Oppenheimer) approximation. Hence, only the electronic degrees of freedom are treated in a time-dependent fashion, which is applicable due to the large separation of time scales caused by the large mass differences between the heavy nuclei and the electrons.

3.1. 1d model of LiH

We consider a regularized binding potential for the electrons (R is the internuclear distance, $Z_i \in (1, 3)$ are the nuclear charges) of

$$v_{\text{ion}}(x) = -\frac{Z_1}{\sqrt{(x - R/2)^2 + 1}} - \frac{Z_2}{\sqrt{(x + R/2)^2 + 1}}\,, \tag{4}$$

and an electron-electron interaction term of

$$w(x_1, x_2) = \frac{1}{\sqrt{(x_1 - x_2)^2 + 1}}\,, \tag{5}$$

to remove the Coulomb singularities. Such forms of the potentials are well-established models in the literature, see, e.g. Ref. [25] and references therein. A similar model has been used for

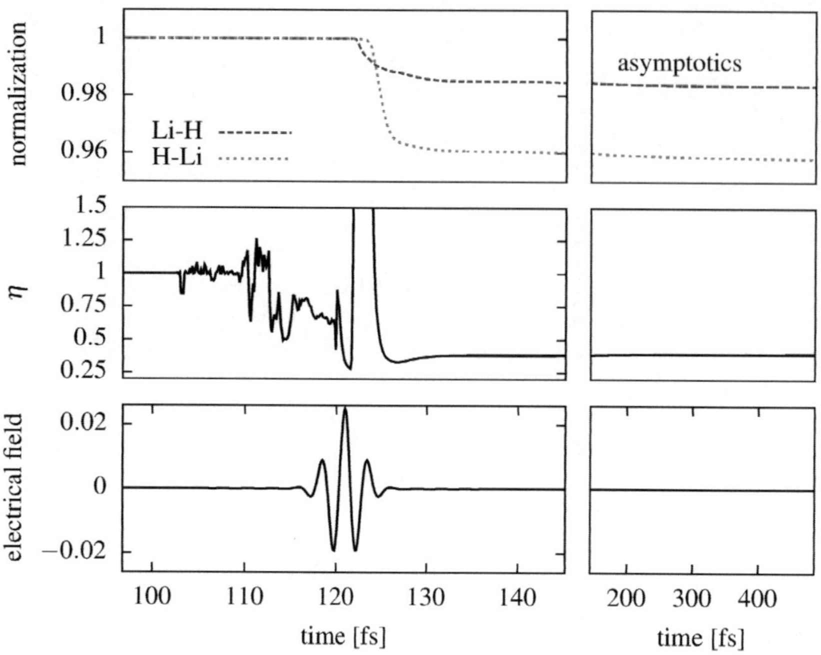

Figure 3. (color online). Normalization of the wave function (top) for both orientations of the LiH molecule with respect to the peak of the electrical field (bottom) and the ionization parameter η, Eq. (8) (center). The asymptotic behavior for long times is shown in the right panels, the dynamics during the pulse in the left panels. Data is for $\tau = 75$ and $E = 0.025$ and $t_0 = 125$ in SAE approximation.

the investigation of enhanced ionization phenomena in molecules, also using the TD-GAS-CI formalism [28].

In such a model, the laser field is polarized along the internuclear axis. Throughout, we consider fields in Eq. (2) of the form

$$E(t) = f(t)E_0 \cos\left[\omega(t - t_0) + \varphi_{\text{CEP}}\right] , \tag{6}$$

with a Gaussian-shaped envelope

$$f(t) = \exp\left[(t - t_0)^2/(2\tau^2)\right] . \tag{7}$$

The photon energy is denoted by ω, the maximum field strength by E_0, which relates to the peak intensity of the field via $I_0[\text{W/cm}^2] = 3.56 \cdot 10^{16}\ \text{W/cm}^2 \times E_0^2$. φ_{CEP} denotes the phase between the envelope $f(t)$ and the carrier wave, the pulse duration is given by τ. In the following, we will choose $\varphi_{\text{CEP}} = 0$. Due to the asymmetric potential, Eq. (4), a strong dependence of the orientation w.r.t. to the dominant field cycle is expected [25]. We will denote the configuration $Z_1 = 1, Z_2 = 3$ with Li-H and $Z_1 = 3, Z_2 = 1$ with H-Li. For the former, the direction of the force acting on the electrons due to the external electric field at its maximum [$t = t_0$ in Eq. (7)] points from H to the Li, thus ionization happens from the Li end (and vice versa for the opposite configuration).

3.2. Calculation of ionization yields

To calculate total ionization probabilities, we add a complex absorbing potential (CAP) to the hamiltonian, which is located at the boundary region of the computational box. During the time evolution, the normalization $n(t) = |\langle\Psi(t)|\Psi(t)\rangle|^2$ is therefore not conserved and the absorbed part, $P(t) = 1 - n(t)$, is a measure for the ionization yield. This procedure was introduced by *Kulander* [8]. Details of the employed CAP can be found in Refs. [25, 28].

Journal of Physics: Conference Series **696** (2016) 012008 doi:10.1088/1742-6596/696/1/012008

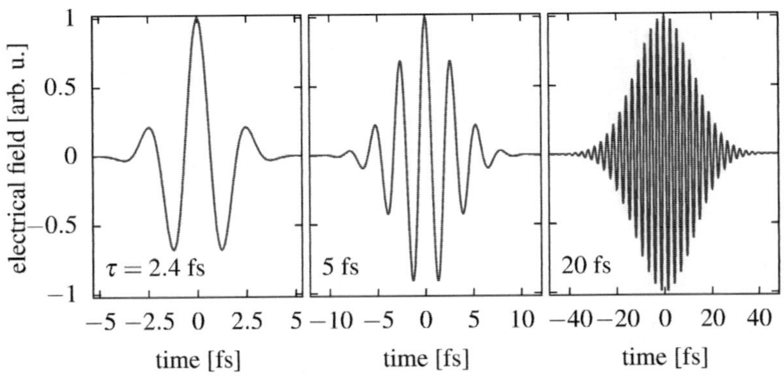

Figure 4. Shape of the Gaussian pulses, Eq. (6), used in this work ($\varphi_{\mathrm{CEP}} = 0$). In the lower corner, the FWHM duration of the intensity ($I(t) \propto E^2(t)$) is printed. The situation $t_0 = 0$ in Eq. (6) is chosen to emphasize the asymmetry of the pulse w.r.t. the sign of the electrical field amplitude.

To investigate the orientation dependence of the ionization yield in our 1d model, we follow Refs. [25, 29] and introduce an orientation parameter

$$\eta = \frac{\mathcal{P}^{\text{H-Li}}(t \to \infty)}{\mathcal{P}^{\text{Li-H}}(t \to \infty)}, \tag{8}$$

which describes the ratio of ionization w.r.t. the orientation of the molecule. Values of $\eta < 1$ correspond to a more favourable ionization of Li-H, whereas $\eta > 1$ to H-Li.

To illustrate the methodology, the temporal behavior of the ionization yields are given in Fig. 3 for both orientations of the LiH molecule along with the parameter η for an intermediate electrical field strength $E = 0.025$ (throughout, atomic units are used if not stated otherwise), a pulse duration of $\tau = 75$ (3 fs FWHM of intensity) and using the SAE approximation. After a fast dynamics in the observables during the pulse (left panels), a convergence for large times is found (right panels in the figure), as expected. The ionization probabilities entering in Eq. (8) are extracted from the asymptotic value of the ionization yield, typically at $t > 500$ fs. We checked for convergence of the results with the size of the simulation box, the grid spacing, and the position of the CAP and found no significant variation of the observables.

3.3. Pulse duration

We first turn our attention to the behavior of the ionization parameter η [Eq. (8)] as a function of the pulse duration τ at a fixed electrical field strength of $E_0 = 0.025$, which corresponds to an intensity of 2.2×10^{14} W/cm^2 and a photon energy of $\omega = 0.057$ (800 nm). By tuning the pulse duration from 2.4 fs to about 40 fs, qualitatively different regimes from single-cycle pulses to long continuous-wave like pulses are covered, see Fig. 4 for an illustration. Due to the asymmetric target molecule and the chosen CEP of the pulse, a strong influence of the orientation of the molecule w.r.t. the peak electrical field strength ($t = 0$ in Fig. 4) is expected for short pulses (left panel), which should vanish for rather long pulses (right panel).

The corresponding ratio of the ionization yield η at different levels of GAS approximations as a function of the pulse duration is given in Fig. 5. Let us first discuss the uncorrelated SAE results (red dotted line). The strongest asymmetry is observed for the shortest pulses ($\tau < 4$ fs), which results from the strong directional character of the single-cycle pulse with one dominant cycle pointing at either the Li or the H nucleus. $\eta < 1$ corresponds to a preferred configuration of ionization of Li-H, which shows in the limit of the shortest possible pulse at $\omega = 0.057$ a factor of five ($\eta \approx 0.2$) higher yields than the configuration H-Li. As expected, an increase of the pulse duration converges η to unity due to the symmetry of the long pulse. The observed convergence is not monotonic but oscillatory with the pulse duration with a period of about 8 fs.

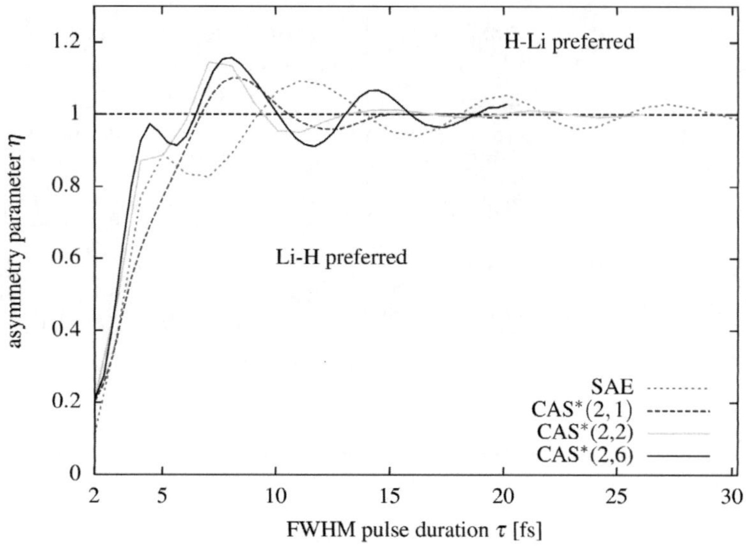

Figure 5. (color online). Ionization parameter η, Eq. (8), as a function of the FWHM pulse duration of the pulse, Eqs. (6), at an electrical field strength of $E_0 = 0.025$ (2.2×10^{13} W/cm^2). $\eta > 1$ ($\eta < 1$) corresponds to a more favourable ionization of H-Li (Li-H).

The reviewed TD-GAS-CI formalism allows us to investigate the influence of electronic correlations. The first improvement is the CAS$^*(2,1)$ [or equivalently TD-CIS, see Sec. 2.1] approximation (blue dashed line), which shows similar behavior as the SAE result, but with a drastic phase shift of the observed oscillations in η. Further inclusion of correlations (green and black solid lines) shifts the oscillations such that the preferred configuration is opposite to that of the uncorrelated SAE calculation (black solid vs. red dashed line). This means that for certain regimes of pulse durations, correlation decides which configuration is easier to ionize and correlated and uncorrelated results are "flipped".

To check whether this oscillatory behavior is only present at the chosen intensity, we calculated η as a function of the pulse duration for different electric field strength in SAE approximation, see Fig. 6. For small electric fields, only the first minimum is found (red dotted and blue dashed lines) and by successively increasing the intensity more and more pronounced oscillations in the ratio appear. This lets us conclude that the observed structures are especially important at higher intensities.

3.4. Validity of the frozen-core approximation

The above calculations considered only CAS space with two active electrons ($N_C = 2$) thereby freezing the inner shell. This should be a valid scheme since the core level located nearby the lithium atom is much more tightly bound than the valence orbital. Nevertheless, the TD-GAS-CI formalism allows us to test the validity of this frozen core approximation within computational limits. We in the following compare CAS calculation with equivalent sizes, i.e., CAS$^*(2,K)$ and CAS$^*(4,K+1)$, where the latter include four active electrons.

For the TD-CIS-like calculations, this results in CAS$^*(2,1)$ with a frozen core and two active electrons and CAS$^*(4,2)$ with four active electrons. The results are presented in Fig. 7, and nearly no difference is observed (green fine dotted vs black solid curves) from which we conclude that the frozen-core approximation is valid for the considered regimes of intensity and photon energy. We note, however, that for intense pulses or high photon energies, where inner-shell ionization becomes possible, it is necessary to include the electrons in the dynamics.

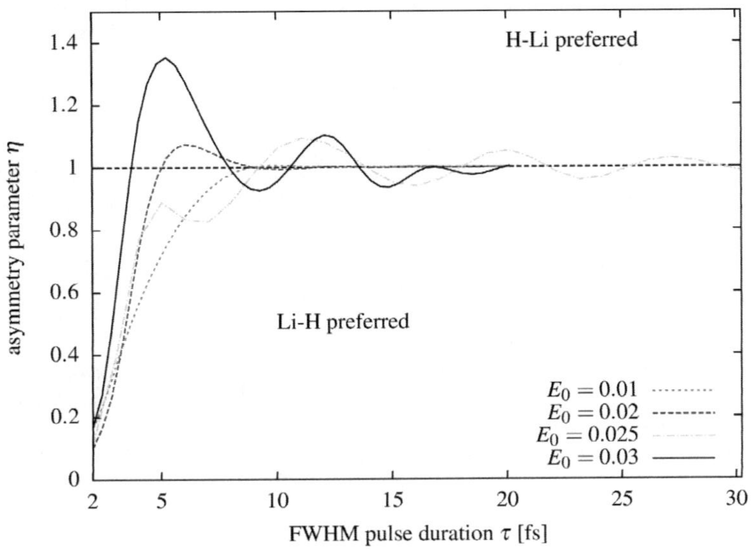

Figure 6. (color online). The same as Fig. 5 but for different pulse intensities at SAE level.

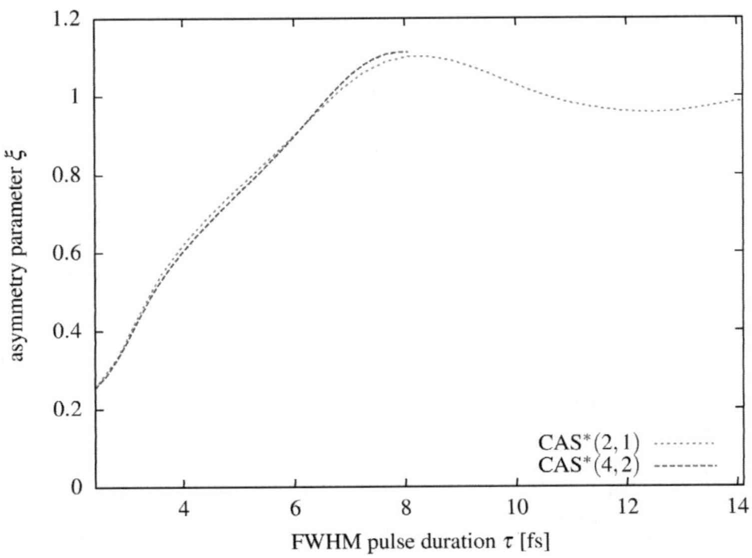

Figure 7. (color online). Comparison of a frozen-core, CAS*(2, 1), and an all-active electrons, CAS*(4, 2), calculation on the CIS level to test for the validity of the frozen-core approximation. Due to the increased computational cost for four active electrons, CAS*(4, 2) is shown only up to 8 fs.

4. Conclusions

In this contribution, we reviewed the time-dependent generalized active space configuration interaction (TD-GAS-CI) approach to the photoionization of atoms and molecules. We introduced the basic idea of selecting important Slater determinants in the CI expansion and addressed the mixed basis set as an essential ingredient to a numerically tractable scheme. We then demonstrated the theory by applying it to the ionization of a heteronuclear 1d model molecule in a strong field with its strongest peak pointing in one direction. For such an excitation scenario the total ionization yields of the two possible orientations in one spatial dimension have been compared and strong influences of electronic correlations have been found. Even the opposite configuration w.r.t. the electric peak for correlated vs. uncorrelated calculation can be found. Ongoing work is in the direction of 3d molecules and larger systems. One possible approach by means of a prolate spheroidal basis is pursued in Ref. [29], with the focus on the

3d LiH molecule. Following these extensions of the TD-GAS-CI theory, it will become possible to validate the presented correlation-driven observations of a one-dimensional model in more realistic systems which can then be tested by experiment.

Acknowledgments
The authors gratefully acknowledge funding by the German Federal Ministry for Education and Research in the frame of the Verbundprojekt "FSP-302" and computing time at HLRN via grants shp00006 and shp00013.

[1] Brabec T and Krausz F 2000 *Reviews of Modern Physics* **72** 545–591
[2] Schütte B, Bauch S, Frühling U, Wieland M, Gensch M, Plönjes E, Gaumnitz T, Azima A, Bonitz M and Drescher M 2012 *Physical Review Letters* **108** 253003
[3] Bauch S and Bonitz M 2012 *Physical Review A* **85** 053416
[4] Schultze M, Fieß M, Karpowicz N, Gagnon J, Korbman M, Hofstetter M, Neppl S, Cavalieri A L, Komninos Y, Mercouris T, Nicolaides C A, Pazourek R, Nagele S, Feist J, Burgdörfer J, Azzeer A M, Ernstorfer R, Kienberger R, Kleineberg U, Goulielmakis E, Krausz F and Yakovlev V S 2010 *Science* **328** 1658–1662
[5] Uiberacker M, Uphues T, Schultze M, Verhoef A J, Yakovlev V, Kling M F, Rauschenberger J, Kabachnik N M, Schröder H, Lezius M, Kompa K L, Muller H G, Vrakking M J J, Hendel S, Kleineberg U, Heinzmann U, Drescher M and Krausz F 2007 *Nature* **446** 627–632
[6] Kling M F and Vrakking M J 2008 *Annu. Rev. Phys. Chem.* **59** 463–492
[7] Krausz F and Ivanov M 2009 *Reviews of Modern Physics* **81** 163–234
[8] Kulander K C 1987 *Physical Review A* **35** 445–447
[9] Krause J L, Schafer K J and Kulander K C 1992 *Physical Review Letters* **68** 3535–3538
[10] Taylor K T, Parker J S, Dundas D, Meharg K J, Doherty B J S, Murphy D S and McCann J F 2005 *Journal of Electron Spectroscopy and Related Phenomena* **144–147** 1191–1196
[11] Parker J S, Doherty B J S, Taylor K T, Schultz K D, Blaga C I and DiMauro L F 2006 *Physical Review Letters* **96** 133001
[12] Bauch S, Balzer K and Bonitz M 2010 *EPL (Europhysics Letters)* **91** 53001
[13] Hochstuhl D, Hinz C M and Bonitz M 2014 *The European Physical Journal Special Topics* **223** 177–336
[14] Meyer H D, Manthe U and Cederbaum L S 1990 *Chemical Physics Letters* **165** 73–78
[15] Hochstuhl D, Bauch S and Bonitz M 2010 *Journal of Physics: Conference Series* **220** 012019
[16] Hochstuhl D and Bonitz M 2011 *The Journal of Chemical Physics* **134** 084106–084106–10
[17] Miyagi H and Madsen L B 2013 *Physical Review A* **87** 062511
[18] Sato T and Ishikawa K L 2013 *Physical Review A* **88** 023402
[19] Sato T and Ishikawa K L 2015 *Physical Review A* **91** 023417
[20] Brics M and Bauer D 2013 *Physical Review A* **88** 052514
[21] Kvaal S 2012 *The Journal of Chemical Physics* **136** 194109
[22] Hinz C, Bauch S and Bonitz M 2015 *J. Phys.: Conf. Ser.: this volume*
[23] Olsen J, Roos B O, Jørgensen P and Jensen H J A 1988 *The Journal of Chemical Physics* **89** 2185–2192
[24] Sorensen L K, Fleig T and Olsen S 2010 *Zeitschrift für Physikalische Chemie International journal of research in physical chemistry and chemical physics* **224** 671–680
[25] Bauch S, Sørensen L K and Madsen L B 2014 *Physical Review A* **90** 062508
[26] Hochstuhl D and Bonitz M 2012 *Physical Review A* **86** 053424
[27] Hochstuhl D and Bonitz M 2013 *Journal of Physics: Conference Series* **427** 012007
[28] Chattopadhyay S, Bauch S and Madsen L B 2015 *arXiv:1508.01629* , accepted for publication in Phys. Rev. A (2015)
[29] Larsson H R, Bauch S and Bonitz M 2015 *arXiv:1507.04107*
[30] Balzer K, Bauch S and Bonitz M 2010 *Physical Review A* **81** 022510 URL `http://link.aps.org/doi/10.1103/PhysRevA.81.022510`
[31] van der Hart H W, Lysaght M A and Burke P G 2007 *Physical Review A* **76** 043405
[32] Lysaght M A, van der Hart H W and Burke P G 2009 *Physical Review A* **79** 053411
[33] Wragg J, Parker J S and van der Hart H W 2015 *Physical Review A* **92** 022504

Journal of Physics: Conference Series **696** (2016) 012009 doi:10.1088/1742-6596/696/1/012009

Instabilities and inaccuracies of multi-configuration time-dependent Hartree-Fock

C M Hinz, S Bauch and M Bonitz

Institut für Theoretische Physik und Astrophysik, Christian-Albrechts-Universität zu Kiel, Leibnizstrasse 15, D-24098 Kiel, Germany

E-mail: hinz@theo-physik.uni-kiel.de

Abstract. We demonstrate that the widely used multi-configuration time-dependent Hartree-Fock method is restricted to a certain class of applications and fails for scenarios where periods of low entanglement occur during the propagation. By using illustrative and physically relevant examples, based on the Hubbard model of solid state physics, we show the existence of serious instabilities in the method itself and demonstrate that the method does not converge with respect to electron correlations. Possible cures of the approach are discussed.

1. Introduction

In recent years, time-dependent processes in many-particle systems have seen increased attention since their more complex structure leads to the emergence of previously unobserved effects. While it is theoretically possible to describe these systems exactly with methods like time-dependent full configuration interaction (FCI) [1], it is not viable in practice for most of them due to the exponential growth of the configuration space with respect to the particle number, the so called exponential wall.

To mitigate the influence of the exponential wall, it is necessary to employ approximations which is usually possible due to the fact that not all parts of the entire Hilbert space are accessed during the time propagation. This is, for example, exploited by the multi-configuration time-dependent Hartree-Fock (MCTDHF) method [2–5], which is a reformulation of the the multi-configuration time-dependent Hartree (MCTDH) method [6, 7] for systems with exchange correlation. This family of methods is particularly suited for systems which require a large number of basis functions.

However, it turns out that MCTDHF suffers from serious instabilities if it is applied to a certain class of systems. This issue is especially severe if the simulation needs to be very accurate or encompasses a long timescale. This makes the validity of applying MCTDHF to such a system uncertain. These difficulties are known in the mathematical literature [8, 9] and recently mentioned in the quantum chemistry literature [10] but are not considered seriously in the physics community.

In this contribution, we will demonstrate the problem on the basis of a physically relevant example. Finally, we will discuss several possible approaches to cure this problem.

Journal of Physics: Conference Series **696** (2016) 012009 doi:10.1088/1742-6596/696/1/012009

2. Theory

Prior to discussing the actual problem, we will give a short overview about the principles of MCTDHF. A more elaborate review can be found in Ref. [1]. Throughout this contribution, we will assume an N-particle Hamiltonian of the form

$$\hat{H}(t) := \sum_{pq} h_{pq}(t)\, \hat{a}_p^\dagger \hat{a}_q + \frac{1}{2} \sum_{pqrs} w_{pqrs} \hat{a}_p^\dagger \hat{a}_r^\dagger \hat{a}_s \hat{a}_q \tag{1}$$

with a (time-dependent) single-particle Hamiltonian $\hat{h}(t)$ and an interaction $\hat{w}$, respectively their matrix elements h_{pq} and w_{pqrs} in a given basis set.

In the MCTDHF framework, the wave function is expressed as a superposition of time-dependent Slater determinants (or configurations),

$$|\,\Psi\,\rangle = \sum_I C_I(t)\,|\,I, t\,\rangle, \tag{2}$$

which are written in occupation number representation with respect to the time-dependent orbitals $\{|\,\phi_i(t)\,\rangle\}$. The equations of motion for the expansion coefficients $C_I(t)$ and the orbitals $|\,\phi_i(t)\,\rangle$ can be obtained by using the Lagrange formulation of the time-dependent variational principle which involves the minimization of the action functional,

$$S\left[\{C_I\}, \{|\,\phi_i\,\rangle\}\right] = \int dt \left\{ \langle\,\Psi\,|\hat{H} - \mathrm{i}\frac{\partial}{\partial t}|\,\Psi\,\rangle - \sum_{kl} \mu_{kl}(t)\left(\langle\,\phi_k\,|\,\phi_l\,\rangle - \delta_{kl}\right) \right\}, \tag{3}$$

with respect to the expansion coefficients, the orbitals and the Lagrange multiplier $\mu_{kl}(t)$. For example, the resulting conditions for the orbitals would read

$$0 \equiv \frac{\delta}{\delta\langle\,\phi_i\,|} S\left[\{C_I\}, \{|\,\phi_i\,\rangle\}\right]. \tag{4}$$

After taking the functional derivatives and simplifying the results, one obtains the wave function equation,

$$\mathrm{i}\dot{C}_I(t) = \sum_J \langle\,I\,|\hat{H}(t)\,|\,J\,\rangle C_J(t), \tag{5}$$

and the orbital equation,

$$\mathrm{i}|\,\dot{\phi}_n\,\rangle = \hat{P}\left\{ \hat{h}(t)\,|\,\phi_n\,\rangle + \sum_{pqrs} \left(\mathbf{D}^{-1}\right)_{np} d_{pqrs} \langle\,\phi_r\,|\hat{w}|\,\phi_s\,\rangle |\,\phi_q\,\rangle \right\}, \tag{6}$$

with the single-particle density matrix $D_{pq} := \langle\,\Psi\,|\hat{a}_p^\dagger \hat{a}_q|\,\Psi\,\rangle$ and the two-particle density matrix $d_{pqrs} := \langle\,\Psi\,|\hat{a}_p^\dagger \hat{a}_r^\dagger \hat{a}_s \hat{a}_q|\,\Psi\,\rangle$. $\hat{P} := \hat{1} - \sum_m |\,\phi_m\,\rangle\langle\,\phi_m\,|$ is the projection operator onto the orthogonal complement of the subspace spanned by the orbitals. An additional equation–ensuring the orthogonality of the orbitals–can be obtained via the Lagrange multipliers. However, this condition is already implicitly fulfilled by the orbital equation and is usually omitted.

As pointed out in the introduction, it is usually not necessary to consider the entire FCI Hilbert space at each point of time. The MCTDHF approximation exploits this by only trying to describe this subspace at each point of time, making it possible to use a relatively small number of orbitals M. At the same time, the *time dependence* of the orbitals allows to continuously adapt the spanned subspace to hopefully imitate the evolution of the exact subspace as closely

Journal of Physics: Conference Series **696** (2016) 012009 doi:10.1088/1742-6596/696/1/012009

as possible. As the number of orbitals restricts the possible entanglement contained in the N-particle state, the MCTDHF approximation is only applicable for systems with small to medium entanglement. Conversely, one can significantly reduce the computational complexity, especially if it is necessary to use a large single-particle basis to describe the system. As the number of orbitals is increased, the result should converge to the exact FCI result.

From a numerical point of view, the restriction to a smaller subspace can be interpreted as performing a low-rank approximation with respect to the density matrix. While the scheme provides the ability to specify the *highest possible* rank for any calculation, nothing prohibits it to actually use a smaller rank. This results in two problems: First, the scheme does not need to be improved by the addition of an orbital as the algorithm might simply exclude it from the calculation. This might result in a difficult convergence which is actually observed in practice. The second issue is even more severe and is the focus of this contribution. It turns out that the MCTDHF equations (5) and (6) do not uniquely describe the evolution of the N-particle state for a large class of systems. Moreover, the resulting numerical instability actually leads to larger errors for an increasing number of orbitals instead of a convergence.

This flaw is not a problem of the numerical implementation but a fundamental one since it stems directly from the definition of the action functional (3). Since unoccupied (or virtual) orbitals do not give a contribution to the N-particle state $|\Psi\rangle$ the action as per definition (3) stays invariant under rotation of these orbitals. A direct consequence of this is that the functional derivatives (4) with respect to these orbitals vanish as those are postulated to be orthonormal. Hence, the evolution of these orbitals is not uniquely described by equation (6). As the cause is an inherent property of its derivation, it is in general not possible to rectify this flaw within the MCTDHF framework.

One might argue that this additional freedom should not change the result of the calculation, since the N-particle state does not depend on the virtual orbitals. However, any orbital might be (re-)occupied during the calculation. At this point of time, the state of the orbital might influence not only the physical state of the system–which is described by the N-particle state–but also all orbitals including the virtual ones. Therefore, an important insight is that the entire mathematical state of the simulation–described by the wave function coefficients and all orbitals–is very important for the behavior of the MCTDHF equations. Even the validity of a certain MCTDHF approximation could be changed by the choice of virtual orbitals. Another crucial point is that the MCTDHF equations are non-linear equations, hence even slight modifications of the virtual orbitals may lead to substantial errors. It should be noted however that all of this is irrelevant if the virtual orbitals are never occupied. This property mitigates the problem at least in the case of time-dependent Hartree-Fock (TDHF).

While it is so far not clear if the virtual orbitals actually influence the simulation in practice or if the MCTDHF equations are able to correctly adjust the virtual orbitals upon occupation, we will demonstrate in this contribution that the former is the case. Even from a purely theoretical point of view, this leads to some serious problems. One consequence is that MCTDHF calculations are not easily reproducible. For example, it is not sufficient to specify that the initial state is the ground state of the system as the set of obtained virtual orbitals may vary between different methods to calculate it. Furthermore, even the existence of several valid solutions makes this method, strictly speaking, mathematically ill-posed which might affect the applicability of certain numerical methods to solve the equations of motion.

In the case of a numerical implementation, all the above problems become even more critical. Due to the finite accuracy of floating point calculations, implementation varieties may lead to small differences in the virtual orbitals and as a result to divergent behavior between these implementations. If an implementation is parallelized, even different runs of the same simulation could show this behavior due to changes in the order of execution. Furthermore, orbitals with very low occupation need to be counted as unoccupied since the difference is not distinguishable within the accuracy of the implementation. Hence, one might see the same problems even for a wider class of problems.

Numerically, the most crucial calculation is the inversion of the single-particle density matrix $\mathbf{D}$ in the orbital equation (6), respectively, the solution of the corresponding system of linear equations. One way to quantify the numerical stability of this operation is the calculation of the singular value decomposition (SVD),

$$\mathbf{D} = \mathbf{U}\boldsymbol{\Sigma}\mathbf{V}^{\dagger}, \tag{7}$$

and subsequent analysis of the singular values,

$$\sigma_i := \Sigma_{ii}. \tag{8}$$

From a physical point of view, the singular values are nothing more than the occupations of the natural orbitals and the matrices $\mathbf{U}$ and $\mathbf{V}$ specify the transformation between the time-dependent orbitals and the natural orbitals. Using the singular values, one can now calculate the condition number of the single-particle density matrix,

$$\kappa(\mathbf{D}) := \|\mathbf{D}\|\|\mathbf{D}^{-1}\|, \tag{9}$$

which can be expressed via the singular values as,

$$\kappa(\mathbf{D}) = \frac{\sigma_{\mathbf{max}}}{\sigma_{\mathbf{min}}}, \tag{10}$$

if one chooses the spectral norm $\|.\|_2$ in equation (9). $\sigma_{\mathbf{max}}$ and $\sigma_{\mathbf{min}}$ are the largest respectively the smallest singular value. If one solves the linear system

$$\mathbf{D}x = b, \tag{11}$$

the relative error in the solution $\frac{\|\Delta x\|}{\|x\|}$ due to a change Δb in the right-hand side is bounded by

$$\frac{\|\Delta x\|}{\|x\|} \leq \kappa(\mathbf{D})\frac{\|\Delta b\|}{\|b\|}. \tag{12}$$

Such a change might be introduced by a modification of the virtual orbitals or a general numerical inaccuracy during the calculation of the right-hand side. A similar bound can be found for the error with respect to a change in the matrix $\mathbf{D}$, though we will not discuss this here as the effect is quite similar.

Due to its occurrence as a factor on the right-hand side of the inequality (12), the condition number $\kappa(\mathbf{D})$ can be interpreted as an 'error amplification factor'. Hence, for a large condition number even small implementation variances may lead to significantly different results. This condition is fulfilled if at least one of the natural orbitals is significantly less occupied than the other ones. The worst case is, by construction, a Hartree-Fock-like state, as some orbitals are exactly unoccupied. On the contrary, if all natural orbitals are equally occupied–which corresponds to a state of maximal entanglement–this stability problem never manifests itself. From this observation, we can deduce another consequence. Since the added orbitals will be used to approximate ever smaller portions of the exact state, their occupation will inevitably decrease. The only imaginable system where this might not occur at some point would be a system with extremely large entanglement. But since the MCTDHF approximation only works for moderately entangled systems anyway without degenerating into a full CI calculation this point is rather moot. This leaves two choices: (1) Either one tries to reach convergence which will eventually be prevented by the invariably increasing instability, or (2) one stays at a low number of orbitals and accepts the approximation error. Ultimately this means, that MCTDHF can not converge in most cases. While there is an optimal parameter for the number of orbitals w.r.t. to the introduced error, it is debatable if this suffices for most applications, especially if one has high accuracy requirements.

In practice, this numerical problem is usually suppressed by regularizing the single-particle density matrix. A possible workaround has recently been formulated by using a different integration scheme [11], which however introduces other shortcomings [10]. In our implementation we use the regularization

$$\mathbf{D} \to \mathbf{D} + \epsilon e^{-\frac{\mathbf{D}}{\epsilon}} \tag{13}$$

proposed in [7] with a regularization parameter ϵ. All possible choices have in common that they will inevitably introduce a systematic error. To avoid this, the parameter is set to a sufficiently small value, a typical choice is $\epsilon = 10^{-10}$. While small enough to be neglectable as a source of error, it turns out that this amount of regularization is insufficient in practice. Hence, it is arguable if regularization is a suitable solution for most applications of MCTDHF.

3. Numerical Results

To illustrate the problem in practice, we focus on a simple, but physically relevant system. Let us consider a small Hubbard cluster with strongly confined initial state. For a demonstration of the breakdown in atomic and molecular systems, we refer to a forthcoming publication [12].

We use a small Hubbard cluster with nearest-neighbor hopping and on-site interaction which is described by the Hamiltonian

$$\hat{H} = -\sum_{<ss'>}\sum_{\sigma=\uparrow,\downarrow} a^{\dagger}_{s\sigma}a_{s'\sigma} + U\sum_{s} a^{\dagger}_{s\uparrow}a_{s\uparrow}a^{\dagger}_{s\downarrow}a_{s\downarrow}. \tag{14}$$

The first summation in equation (14) goes over all combinations of neighboring sites. A more comprehensive discussion of Hubbard clusters can be found in Ref. [13]. For this benchmark, we use two electrons, either one spin up and spin down, in eight sites. The effective interaction strength U is set to 1.

For times $t < 0$ the electrons are confined to the left-most (first) site by applying an infinitely large potential to all other sites. Conveniently, this implies that the initial state of the simulation can be analytically specified. Since only the occupation of first site leads to a finite energy contribution, all configurations except

$$\left|\,\Psi_0\,\right\rangle := \left|\,\phi_{0\uparrow}\phi_{0\downarrow}\,\right\rangle \tag{15}$$

must vanish. Therefore, the initial state $\left|\,\Psi_0\,\right\rangle$ is a Hartree-Fock state.

This fact has far reaching consequences for the time-dependent propagation. Since all time-dependent orbitals except for the first one are initially not occupied and as a result do not take part in the determination of the physical state, the choice of these virtual orbitals is arbitrary. The only restriction is that they must form a basis together with the first orbital.

One way of exploiting these additional degrees of freedom is to apply a unitary transformation to all orbitals which leaves the occupied ones invariant. In general, such a transformation takes the form of a block diagonal matrix,

$$\mathbf{U}_{\text{virt}} := \begin{pmatrix} \mathbb{1} & \mathbf{0} \\ \mathbf{0} & \mathbf{U} \end{pmatrix}, \tag{16}$$

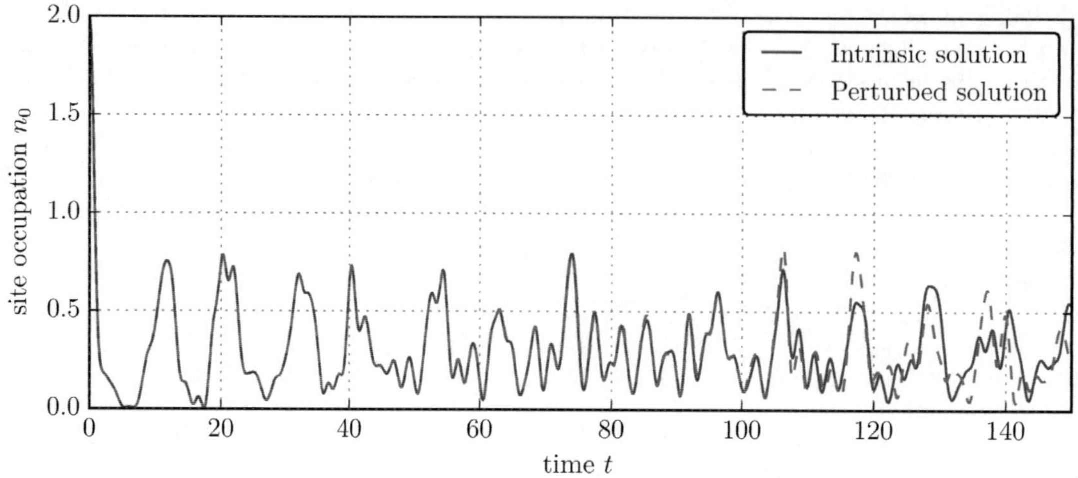

Figure 1. (color online) Occupation of the left-most Hubbard site for the intrinsic solution (blue, solid) and the perturbed solution (green, dashed). The Hubbard cluster consists of two particles in eight sites.

where the first block is the identity matrix and the second block is an arbitrary unitary matrix. In equation (16), we assume that the orbitals are sorted in descending order with respect to their occupation and that $\mathbf{U}$ only acts on the virtual orbitals. For his calculation, we have chosen

$$\mathbf{U} = e^{\mathrm{i}\mathbf{K}} \tag{17}$$

as the unitary transformation which is applied to the virtual orbitals. The matrix $\mathbf{K}$ in equation (17) is an arbitrary Hermitian matrix which we define as

$$\mathbf{K} = \mathbf{J} + \mathbf{J}^{\dagger} \tag{18}$$

with a random complex matrix $\mathbf{J}$. Figure 1 depicts what happens if one applies the matrix $\mathbf{U}_{\mathrm{virt}}$ to the orbitals prior to the time-dependent propagation. The perturbed solution (dashed line) deviates significantly from the intrinsic solution even though the initial physical states are equivalent. Since all other parts of the simulation remained unchanged between the two runs, one has to conclude that the choice of virtual orbitals may have a substantial effect on the solution.

It should be noted that this problem occurs although the implementation was regularizing the single-particle density matrix with a parameter $\epsilon = 10^{-10}$. Hence, regularization is unable to solve the stability problem in general. While it might be possible to improve the stability using a larger regularization parameter, this has its own issues as one introduces another source of errors. Since it is practically impossible to predict or even measure the impact of these additional errors due to the non-linear nature of the MCTDHF equations, we have to assume that it will change the result significantly. As we will see below, even a small error might be amplified to a relevant deviation. However, the simulation would not be possible without some sort of regularization due to an arbitrarily large condition number during the initial phase of the simulation. If the resulting error grows large enough, the density matrices loose their defining properties, for example they cease to be positive semidefinite and self-adjoint, violating the fundamentals of quantum mechanics.

During the second benchmark, we will investigate the influence of implementation variations on the simulation results for a Hubbard cluster. To emulate the differing behavior of distinct implementations, we introduce small artificial errors into our simulation in each time step. Using equation (12), one can define a set of possible deviations for a given relative error in the right-hand side of the linear system $\frac{b}{\|b\|}$ and a given exact solution x. All of these deviations are contained in a sphere of radius

$$\Delta := \frac{b}{\|b\|}\|x\|. \tag{19}$$

For this demonstration, we have chosen to use deviations which are uniformly distributed over the surface of this sphere. Such deviations can be calculated via the formula

$$D_r := \Delta \sqrt[M]{U}\frac{d_r}{\|d_r\|_2}. \tag{20}$$

U is chosen to be one while d_r is a vector of random numbers with a standard normal distribution. Another possible way to obtain a deviation would be to sample the variable U in definition (20) from a uniform real distribution in the range $[0, 1]$ which results in a uniform distribution within the sphere. However we have not pursued this here as we wanted to analyze the worst possible case.

To calculate the exact solution, we have solved the system with 256-bit precision which should give us a sufficiently good approximation. In the case of the right-hand side of Eq. (11), this is unfortunately not possible since the computational costs would be to high. Instead, we assume that the double precision solution is accurate enough and choose an relative error in units of the machine epsilon for double precision $\epsilon_M = 1.1 \cdot 10^{-16}$. Given that the computation of the right-hand side consist of several million floating point operations even a relatively high error is feasible. Even if the double precision solution differs significantly from the analytical one, our results still hold since in this case the error would be larger resulting in an even less stable propagation. After determining the deviation, it is simply added to the exact solution.

To investigate the effect of these artificial variances, we consider a Hubbard cluster with two particles in ten sites. To decrease the entanglement for the demonstration, the particles are initially confined by a harmonic oscillator potential $\frac{1}{2}\Omega^2\hat{x}^2$ with a trap frequency $\Omega = 0.75$. As the relative error of the right-hand side, we choose $\frac{b}{\|b\|} = 2 \cdot 10^9\,\epsilon_M$. This large error was chosen to bypass the problem that for Hubbard-like systems MCTDHF becomes extremely instable with increasing number of orbitals. Even small changes in the parameters lead to a totally different behavior including program termination due to violated assumptions in the algorithm. To avoid this erratic behavior and provide reproducible results, we have chosen to use only four orbitals and increase the error of the right-hand side accordingly to obtain roughly the same error in the solution as for 6 orbitals. This avoids the failure while still allowing us to investigate the issue in a reasonable setup. Besides, a relative error in the order of magnitude of 10^{-8} is quite realistic for most numerical calculations.

As depicted in Fig. 2, the deviations between solutions corresponding to two different seeds can be much larger than the introduced error due to error amplification. Moreover, not only do the solutions differ quantitatively but also qualitatively at certain times of the simulation. It should be also noted that the occupation of the natural orbitals increases quickly which lessens the effect for later times. Thus, it is even more surprising that the effect is fairly pronounced.

While this effect is a problem for lowly entangled systems, Hubbard-like systems are strongly entangled in many important setups, thus avoiding this issue. However, as we will show below, the MCTDHF approximation even fails in those cases. To demonstrate this, we examine a Hubbard cluster with 8 particles in 8 basis functions. Similar to the first benchmarking system, the particles are confined to the four left-most sites for times $t < 0$. The effective interaction strength U is set to 0.1. The parameters were taken from Ref. [14] which discusses the system in the context of several other methods.

Journal of Physics: Conference Series **696** (2016) 012009 doi:10.1088/1742-6596/696/1/012009

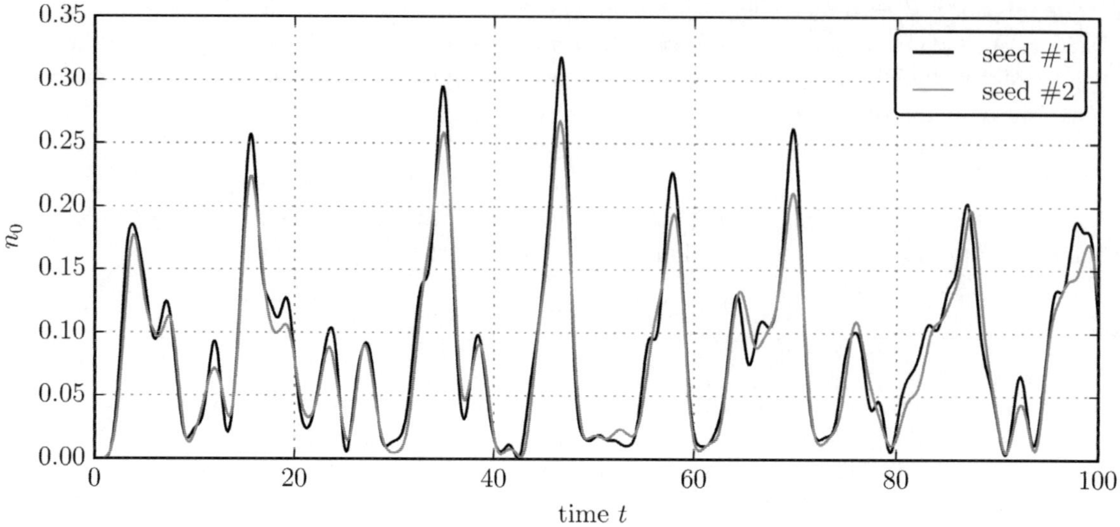

Figure 2. Occupation of the left-most Hubbard site for a cluster consisting of two electrons and ten sites for two different seeds. The electrons are initially confined by a harmonic oscillator potential with an angular frequency $\omega = 0.75$.

While the system starts in a Hartree-Fock-like state, and thus produces a comparably large deviation during the first few time steps, the strong excitation rapidly produces a strongly entangled state and as a result averts the stability problem for later times. Therefore, the system has not enough time to accumulate a significant error. Though, care must be taken to avoid that the error within the first few time steps is small enough. To avoid this pitfall, we have applied a regularization of $\epsilon = 10^{-10}$ to the single-particle density matrix.

It is obvious from Fig. 3 that MCTDHF is unsuitable even in this favorable case. For times prior to $t = 20$, the Hartree-Fock result almost matches the exact solution and the application of MCTDHF does not lead to definite improvements. Depending on one's accuracy requirements the much simpler TDHF might already suffice. At the same time, both TDHF and the best MCTDHF approximation ($M = 7$) fail to describe the behavior of the system for later times. Not only do they fail to describe the properties of the system quantitatively but also qualitatively. Again, the obtained improvement of MCTDHF beyond Hartree-Fock is uncertain as both methods show quite similar deviations at times. In conclusion, it is obvious that MCTDHF is not suitable to describe neither lowly nor strongly entangled systems.

4. Discussion

In this contribution, we have demonstrated that the practicalness of the MCTDHF method is severely restricted by an inherent instability caused by unoccupied orbitals. Due to error amplification, even small errors in the initial state or any intermediate step can cause significant changes in the obtained results. In the end, the convergence of MCTDHF against the exact FCI result is defeated by the increasing error with respect to the number of orbitals.

These problems are only avoided if the system exhibits strong entanglement throughout the entire simulation. However, as shown in Section 3, this would imply necessity of a full CI calculation making the usage of MCTDHF disputable. Apart from that, MCTDHF is only applicable to problems with low accuracy requirements, since one can stay at a lower number of orbitals, or very short calculations. Even in this case, one should carefully monitor the stability of the calculations–for example via the condition number–and reduce the number of time-dependent

Journal of Physics: Conference Series **696** (2016) 012009 doi:10.1088/1742-6596/696/1/012009

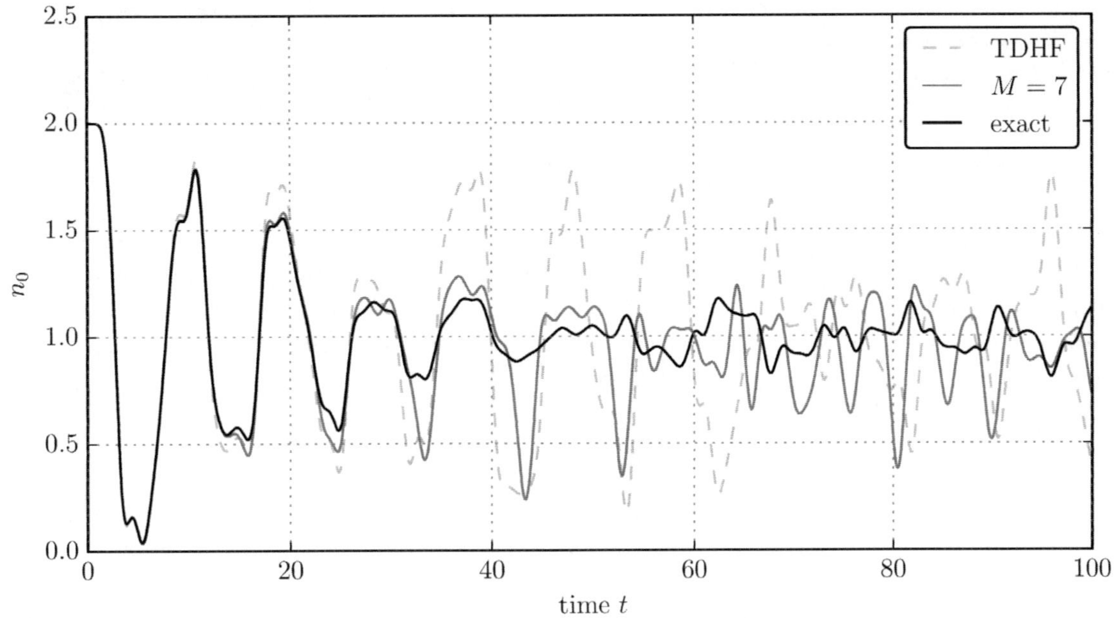

Figure 3. Occupation of the left-most Hubbard site for TDHF (light-gray, dashed), MCTDHF with $M = 7$ (medium-gray, solid) and FCI (black, solid). The system consists of 8 particles in 8 sites.

orbitals accordingly.

One possible solution would be to solve the orbital equation with sufficiently high accuracy to suppress any error amplification. Apart from the issue of consuming vastly more computational resources in itself, this has another severe limitation. From our experience, the condition number, and, therefore, the instability, tends to increase exponentially with respect to the number of orbitals making this practically intractable as one would need to expend an also exponentially increasing amount of resources. In the end, this would just create another exponential wall making the method ineffectual. Another solution which one might think of is an adaptive scheme which adds or removes orbitals as their occupation changes. Unfortunately, this idea is flawed since without a governing principle one would need to guess new orbitals and the inherently arbitrary choice would change the result of the calculation. One way to circumvent this is to specify another set of equations which determine the evolution of the virtual orbitals. A possible choice for such a system of equations was recently proposed by U. Manthe [10]. As of now, it is not obvious if this method can remove the instability or just hides it. Since the equations and their implementation will be reflected in the results, the effects of the intrinsic instability of the orbital equation can only be avoided if it is possible to solve these equations with sufficiently high accuracy.

Another way to avoid this instability would be the choice of an entirely different method. Unfortunately, the variety of candidates is quite restricted as it is most likely that other MCTDH-like methods, for example, time-dependent complete-active-space self-consistent-field [15, 16], time-dependent restricted-active-space self-consistent-field [17] and ultimately MCTDH itself, will exhibit the same instability. Moreover, we suspect that other methods using time-dependent orbitals, like the time-dependent renormalized natural orbital theory [18] and orbital-adaptive time-dependent coupled-cluster [19], might show similar problems as all of these methods involve an inversion of the single-particle density matrix or equivalent operations. One contender for a replacement are CI-based methods with time-independent orbitals, like the time-dependent generalized-active-space configuration interaction (TD-GAS-CI) method [1, 20–22], due to their static Hilbert space. Another class of methods which have proven to be especially suited for Hubbard-like systems are based on the non-equilibrium Green function (NEGF) approach[14, 23–25]. However, the associated increase in the computational complexity necessitates the development of more powerful computational facilities.

Acknowledgments
This work was supported by the BMBF in the frame of the Verbundprojekt FSP 302, computing time at the HLRN via the grants `shp00006` and `shp00013`. We would also like to thank Haruhide Miyagi and Lars Bojer Madsen for a helpful discussion of the depicted issue.

References

[1] D. Hochstuhl, C. M. Hinz, and M. Bonitz, *The European Physical Journal Special Topics* **223**, 177–336 (2014).
[2] J. Caillat, J. Zanghellini, M. Kitzler, O. Koch, W. Kreuzer, and A. Scrinzi, *Physical Review A* **71**, 012712 (2005).
[3] J. Zanghellini, M. Kitzler, T. Brabec, and A. Scrinzi, *Journal of Physics B: Atomic, Molecular and Optical Physics* **37**, 763 (2004).
[4] T. Kato and H. Kono, *Chemical Physics Letters* **392**, 533–540 (2004).
[5] M. Nest, T. Klamroth, and P. Saalfrank, *The Journal of Chemical Physics* **122**, 124102 (2005).
[6] H.-D. Meyer, U. Manthe, and L. S. Cederbaum, *Chemical Physics Letters* **165**, 73–78 (1990).
[7] U. Manthe, H.-D. Meyer, and L. S. Cederbaum, *The Journal of Chemical Physics* **97**, 3199–3213 (1992).
[8] D. Conte and C. Lubich, *ESAIM: Mathematical Modelling and Numerical Analysis* **44**, 759–780 (2010).
[9] C. Bardos, I. Catto, N. Mauser, and S. Trabelsi, *Archive for Rational Mechanics and Analysis* **198**, 273–330 (2010).
[10] U. Manthe, *The Journal of Chemical Physics* **142**, 244109 (2015).
[11] C. Lubich, *Applied Mathematics Research eXpress* **2015**, 311–328 (2015).
[12] C. M. Hinz, S. Bauch, and M. Bonitz, *to be published.*
[13] S. Hermanns, N. Schlünzen, and M. Bonitz, *Physical Review B* **90**, 125111 (2014).
[14] D. Lacroix, *Physical Review B* **90** (2014).
[15] T. Sato and K. L. Ishikawa, *Physical Review A* **88**, 023402 (2013).
[16] T. Sato and K. L. Ishikawa, *Physical Review A* **91**, 023417 (2015).
[17] H. Miyagi and L. B. Madsen, *The Journal of Chemical Physics* **140**, 164309 (2014).
[18] M. Brics and D. Bauer, *Physical Review A* **88**, 052514 (2013).
[19] S. Kvaal, *The Journal of Chemical Physics* **136**, 194109 (2012).
[20] D. Hochstuhl and M. Bonitz, *Physical Review A* **86**, 053424 (2012).
[21] S. Bauch, L. K. Sørensen, and L. B. Madsen, *Physical Review A* **90**, 062508 (2014).
[22] S. Bauch, C. M. Hinz, and M. Bonitz, *submitted to Journal of Physics: Conference Series* (2015).
[23] K. Balzer, S. Bauch, and M. Bonitz, *Physical Review A* **81**, 022510 (2010).

[24] K. Balzer, S. Bauch, and M. Bonitz, *Physical Review A* **82**, 033427 (2010).

[25] N. Schlünzen, S. Hermanns, M. Bonitz, and C. Verdozzi, *arXiv:1508.02947 [cond-mat]* (2015).

Journal of Physics: Conference Series **696** (2016) 012010 doi:10.1088/1742-6596/696/1/012010

Towards Quantum Transport for Central Nuclear Reactions

Pawel Danielewicz

National Superconducting Cyclotron Laboratory, Michigan State University, East Lansing, MI 48824, USA

E-mail: danielewicz@nscl.msu.edu

Arnau Rios

Department of Physics, University of Surrey, Guildford, GU2 7XH, United Kingdom

E-mail: a.rios@surrey.ac.uk

Brent Barker

Biological, Chemical and Physical Sciences Department, Roosevelt University, Chicago, IL 60605, USA

E-mail: bbarker@roosevelt.edu

Abstract. Nonequilibrium Green's functions represent a promising tool for describing central nuclear reactions. Even at the single-particle level, though, the Green's functions contain more information that computers may handle in the foreseeable future. In this study, we explore slab collisions in one dimension, first in the mean field approximation and demonstrate that only function elements close to the diagonal in arguments are relevant, in practice, for the reaction calculations. This bodes well for the application of the Green's functions to the reactions. Moreover we demonstrate that an initial state for a reaction calculation may be generated through adiabatic transformation of interactions. Finally, we report on our progress in incorporating correlations into the dynamic calculations.

1. Introduction

Expansion in the scope of nuclear investigations is largely owed to the construction of new accelerator facilities, in particular those accelerating secondary exotic beams, leading to the growth in discovered nuclides and nuclides that can be used in further experimentations. A second high-intensity generation of the latter facilities is now under construction and these include the Facility for Rare Isotope Beams (FRIB) at the Michigan State University, at the moment the largest investment in the low-energy nuclear physics in North America. The expansion on the experimental side is paralleled by a growth in the theory for nuclear structure that, till now largely phenomenological, has increasingly become fundamentally based, even down to QCD, as an effective theory. Not surprisingly, expectations rise then with respect to the reaction theory. The theory for central reactions has been till now phenomenological to a much greater extent than the structure theory and, in particular, largely semiclassical. Many

Journal of Physics: Conference Series **696** (2016) 012010 doi:10.1088/1742-6596/696/1/012010

particles participate in central reactions, so the theory needs to be statistical in nature. For dynamics, the natural general way forward is the quantum transport.

2. Description of Central Nuclear Reactions

Historically, just a handful of methods have been developed to describe central nuclear reactions. The time-dependent Hartree-Fock (TDHF) method has been exhaustively employed in describing low-energy reactions [1]. Nowadays, the TDHF simulations can be performed in full 3D and involve nuclei as heavy as Uranium [2]. However, the validity of TDHF requires correlations to play a negligible role in the dynamics [3]. At low energies, one might argue that the role of correlations is minimized by wavefunction antisymmetrization. Conversely, one would expect correlations to dominate at higher energies, where role of the Pauli principle weakens. With increase of collision energy the limitation of TDHF is exhibited in the fact that nuclei in TDHF simulations become transparent to each other, see Fig. 1. No such transparency is observed experimentally.

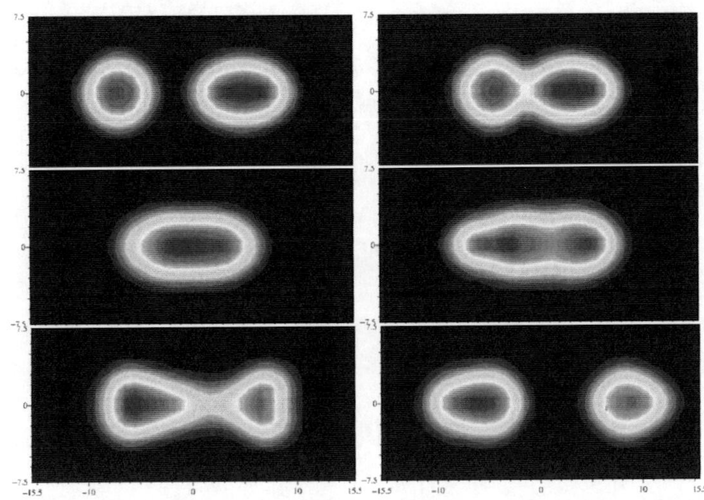

Figure 1. Head-on collision of $^{16}O + {}^{22}Ne$ at $E_{\rm cm} = 95\,{\rm MeV}$ in the TDHF calculation by Umar and Oberacker [4]. In head-on collisions at high enough energies, the nuclei pass through each other within TDHF.

Intermediate and high energy central nuclear reactions have been commonly described in terms of the Boltzmann-equation (BE) models [5]. In these models the evolution of the phase-space distribution functions $f(\boldsymbol{r},\boldsymbol{p},t)$ of nucleons and other particles is followed. BE models have been fairly successful in describing many aspects of higher-energy reactions, see e.g. Fig. 2. However, the use of BE in reactions has been criticized on theoretical grounds. BE relies on the quasiparticle picture and simple estimates [6] indicate that particle scattering rates are comparable to particle energies, which undermines that picture. In this context, it is theoretically difficult [7] to separate collisional effects, described with cross-sections and entering the collisional integrals in BE from mean-field effects entering the quasiparticle energies.

3. Kadanoff-Baym Equations

The TDHF and BE approaches to central nuclear reactions are fundamentally tied with the single-particle density matrix

$$\rho(\boldsymbol{r}_1\,\boldsymbol{r}_1'\,t) = \langle\Phi|\psi^\dagger(\boldsymbol{r}_1'\,t)\,\psi(\boldsymbol{r}_1\,t)|\Phi\rangle\,. \tag{1}$$

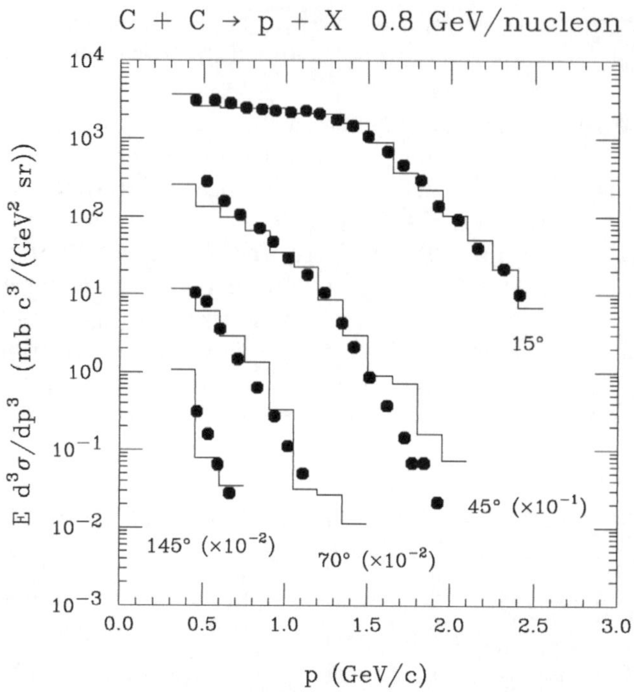

Figure 2. Proton spectra from an 800 MeV/nucleon ^{12}C + ^{12}C reaction. Dots represent data of Ref. [8] and histograms represent BE calculations of Ref. [9].

The density matrix yields all single-particle observables. E.g. the particle density represents the diagonal of this matrix as the expectation value of the density operator:

$$n(\boldsymbol{r}\,t) = \rho(\boldsymbol{r}\,\boldsymbol{r}\,t) = \langle\Phi|\psi^\dagger(\boldsymbol{r}\,t)\,\psi(\boldsymbol{r}\,t)|\Phi\rangle\,. \tag{2}$$

The Wigner function, that may be viewed as the quantal version of the classical phase-space distribution, results from evaluating a Fourier transformation of the equal-time Green's function, i.e. density matrix, in relative coordinates:

$$f(\boldsymbol{p}\,\boldsymbol{r}\,t) = \int \mathrm{d}(\boldsymbol{r}_1 - \boldsymbol{r}_1')\,\mathrm{e}^{-i\boldsymbol{p}(\boldsymbol{r}_1-\boldsymbol{r}_1')}\,\rho(\boldsymbol{r}_1\,\boldsymbol{r}_1'\,t)\,. \tag{3}$$

Here, the spatial argument of the Wigner function is $\boldsymbol{r} = (\boldsymbol{r}_1 + \boldsymbol{r}_1')/2$. It can be seen that the doubled spatial argument in the density matrix accounts for position and momentum in the classical limit. By extension, one can expect that the doubled time argument in the Green's function corresponds to time and energy in the classical limit, i.e. the density in momentum *and* energy at a given position and time should be:

$$-iG^<(\boldsymbol{p}\,\epsilon\,\boldsymbol{r}\,t) = \int \mathrm{d}(\boldsymbol{r}_1 - \boldsymbol{r}_1')\,\mathrm{d}(t_1 - t_1')\,\mathrm{e}^{i[\epsilon(t_1-t_1')-\boldsymbol{p}(\boldsymbol{r}_1-\boldsymbol{r}_1')]}\,(-i)G^<(\boldsymbol{r}_1\,t_1\,\boldsymbol{r}_1'\,t_1')\,. \tag{4}$$

Indeed, for the static case of a Hartree-Fock state, where the Green's function is generally a superposition of products of occupied orbitals ϕ_α,

$$-iG^<(\boldsymbol{r}_1\,t_1\,\boldsymbol{r}_1'\,t_1') = \sum_\alpha \phi_\alpha(\boldsymbol{r}_1\,t_1)\,\phi_\alpha^*(\boldsymbol{r}_1'\,t_1')\,; \tag{5}$$

Journal of Physics: Conference Series **696** (2016) 012010 doi:10.1088/1742-6596/696/1/012010

Eq. (4) yields

$$-iG^<(\boldsymbol{p}\,\epsilon\,\boldsymbol{r}\,t) = \sum_\alpha f_\alpha(\boldsymbol{p}\boldsymbol{r})\,\delta(\epsilon-\epsilon_\alpha)\,. \tag{6}$$

Here, ϵ_α are single-particle energies of the orbitals. In nuclear physics, the spectral function represented by Eq. (4) is explored for ground-state nuclei in inelastic electron scattering.

Outside of the mean-field approximation, an intrinsically consistent dynamics for the Green's function (1), in nonequilibrium or even finite temperature situation, can only be arrived when considering different orderings of single-particle operators in an expectation value, encompassed in the Green's function

$$i\,G(1,1') = \langle\Phi|T\left\{\psi(1)\,\psi^\dagger(1')\right\}|\Phi\rangle\,. \tag{7}$$

Here, the time arguments of the operators are assigned to either side of a time contour that runs first forward and then backward in time, representing the evolution of the ket in the expectation value, on one hand, and of the bra, on the other [6, 10]. The superoperator T orders operators according to their order on the time contour. Depending on the assignment of time arguments to the contour branches, different actual ordering of the operators can result, with the superscript '<' in the Green's function referring to the order in (1) and '>' to the opposite order.

A perturbative expansion of the evolution operators in the Green's function (7), followed by resummations, yields a Dyson equation familiar from other contexts

$$G = G_0 + G_0\,\Sigma\,G\,. \tag{8}$$

Here, the time integrations run over the contour forward and then backward in time, reaching above the maximal time of operators in any considered expectation value. The self energy Σ can be expressed in terms of source operators j,

$$\left(i\,\frac{\partial}{\partial t_1}+\frac{\boldsymbol{\nabla}_1^2}{2m}\right)\psi(1) = j(1)\,, \tag{9}$$

with

$$i\,\Sigma(1,1') = \langle\Phi|T\left\{j(1)\,j^\dagger(1')\right\}|\Phi\rangle_{\rm irr}\,. \tag{10}$$

The integral Dyson equation may be converted into an integro-differential equation through application of the differential inverse operator G_0^{-1} to both sides of the equation. For a specific order of operators in G (7), one arrives at the set of Kadanoff-Baym equations [10]:

$$\left(i\,\frac{\partial}{\partial t_1}+\frac{\boldsymbol{\nabla}_1^2}{2m}\right)G^\lessgtr(1,1') = \int \mathrm{d}1''\,\Sigma^+(1,1'')\,G^\lessgtr(1'',1') + \int \mathrm{d}1''\,\Sigma^\lessgtr(1,1'')\,G^-(1'',1')\,. \tag{11}$$

Implicit, in the expansion and resummation leading to the Dyson equation and, thus, to the Kadanoff-Baym (KB) equations, is the presumption that the impact of any multi-particle correlations dies off with time.

In different limits, for a variety of cases of Σ, the K-B equations yield a variety of useful known results and moreover they interpolate and extrapolate beyond those limits and associated results. Thus, when the mean field contribution dominates the self-energy, $\Sigma_{\rm mf} >> \Sigma^\lessgtr$, in a highly degenerate system, the TDHF description for the system follows. When the scales of variation for the Green's functions and self-energies are significantly larger for the average than relative function arguments, then the quasiparticle approximation follows, with evolution governed by the Boltzmann equation,

$$-i\,G^<(1,1') \approx \int \mathrm{d}\boldsymbol{p}\,f(\boldsymbol{p},1)\,\mathrm{e}^{i\,\boldsymbol{p}(\boldsymbol{x}_1-\boldsymbol{x}_{1'})-i\,\epsilon_{\boldsymbol{p}}(t_1-t_{1'})}\,. \tag{12}$$

Given the two limiting approaches contained in the KB equations, already employed in the practice of central nuclear reactions, it is natural to ponder a direct application of the nonequilibrium Green's functions to the reactions, covering the variety of circumstances that are possibly or clearly out of reach of those limiting approaches. The problem with a direct solution of the KB equations, though, is the 8-dimensional nature of the functions that can easily overwhelm the current and near-future computational power. By contrast, the TDHF equations involve 4 dimensions, or 5 if one counts the large number of orbitals, and these already tax the current power.

4. Towards Reaction Simulations

General issues that must be considered when coping with nonuniform systems, described in terms of nonequilibrium Green's functions, include the space-time matrix form of the dynamics, preparation of an initial state and the abundance of matrix elements that need to be considered. Thus, for a discretization involving sensible 50 values in any space-time direction, the number of matrix elements that would need to be considered is $50^8 = 4\times10^{13}$ for every function, of the order of 100 TB of data!

In order to progress on those issues, we start [11] with mean-field dynamics in one dimension. Our primary goals include assessing whether the initial state for nuclear reaction could be prepared within the same calculational scheme as the reaction simulations and whether all matrix elements of the Green's functions are equally important in calculations of the dynamics. If many could be discarded, it might be possible to reduce the computational task for realistic calculations to an acceptable scale.

When the self-energies $\Sigma^{\lessgtr}$ are suppressed, the KB equation for $G^<$ becomes

$$\left(i\frac{\partial}{\partial t_1}+\frac{\nabla_1^2}{2m}-\Sigma_{\rm mf}\left(-iG^<(1,1)\right)\right)(-i)G^<(1,1')=0\,. \qquad (13)$$

Here, we presume that the self energy $\Sigma_{\rm mf}$ is local and depends only on local density, which is commonplace in nuclear physics. In that situation, the kinetic and potential parts of single-particle Hamiltonian are, respectively, diagonal either in momentum or configuration space. The equation may be solved in that situation by employing the split-operator method

$$\begin{aligned} G^<(t_1+\Delta t,t_{1'}) &= \mathrm{e}^{-i\Delta t(K+\Sigma)}\,G^<(t_1,t_{1'}) \\ &= \left(\mathrm{e}^{-i\Delta t\,\Sigma/2}\,\mathrm{e}^{-i\Delta t\,K}\,\mathrm{e}^{-i\Delta t\,\Sigma/2}+\mathcal{O}\left((\Delta t)^3\right)\right)G^<(t_1,t_{1'})\,, \end{aligned} \qquad (14)$$

and by employing the Fast-Fourier-Transform (FFT) to move back-and-forth between the configuration and momentum representations. In each representation, application of the respective portion of the evolution operator amounts to a multiplication of the Green's function by a phase factor, in itself assuring unitarity. As a final simplification, as the mean-field self-energy depends only instantaneous values of the Green's function, the KB equation can be closed at the level of equal time-arguments of the function, i.e. density matrix, cf. Eqs. (7) and (1).

In exploring the preparation of the initial state, we test the utility of adiabatic switching of the interactions to arrive at interacting ground state [12, 13]. Specifically, when starting from the density matrix for a Harmonic Oscillator (HO), we evolve the Hamiltonian from that of HO to the mean-field one, using a switching profile function $f(t)$:

$$\mathcal{H}(t)=\mathcal{H}_{\rm HO}\,f(t)+\mathcal{H}_{\rm mf}(t)\,\left(1-f(t)\right)\,. \qquad (15)$$

The function is required to reach the limits of $f\to1$ as $t\to-\infty$ and of $f\to0$ as $t\to+\infty$. Most often we use:

$$f(t)=\frac{1}{1+\exp\frac{t-\tau_0}{\tau}}\,. \qquad (16)$$

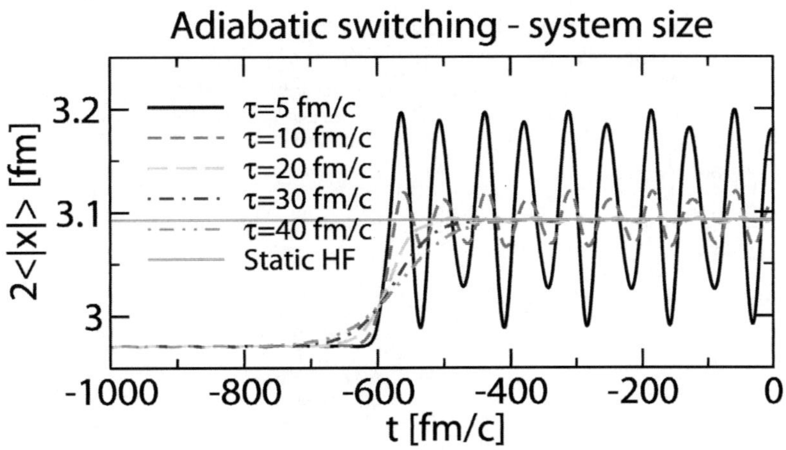

Figure 3. Time evolution of slab width when starting from an HO configuration with $N_s = 2$ filled shells at time $t = -1000\,\text{fm}/c$ and transforming the single-particle potential from the HO to the mean-field form following the Fermi-Dirac switching function with different values of the τ parameter.

Arriving at energy close to minimal is relatively straightforward [11]. More of a challenge is ensuring close to stationary nature of the evolved state. Figure 3 illustrates how a stationary interacting final state is gradually obtained when making the transition in the Hamiltonian more and more adiabatic. Figure 4 shows the changes in the density profile of a 1-dimensional slab when adiabatically switching over from the HO to the mean-field Hamiltonian. We find here that it is indeed possible in practice to arrive at a satisfactory interacting ground state through adiabatic switching within the framework intended for studying reactions. At submission of this paper, we were alerted to an exploration of the sensitivity of physics results in adiabatic switching, to switching function [14].

We next turn to the issue of reaction dynamics and the relative importance of various elements in the density matrix and, by proxy, in the Green's functions. To initiate a collision of slabs in one dimension, we construct the net density matrix by combining density matrices of ground-state slabs boosted to opposing momenta in the c.m. system:

$$\rho(x, x', t = 0) \to \mathrm{e}^{ipx}\, \rho(x, x', t = 0)\, \mathrm{e}^{-ipx'} . \tag{17}$$

Figure 5 shows evolution of slab reaction at incident c.m. energy of 0.1 MeV/nucleon. In the absence of Coulomb interactions, fusion occurs at this energy. At energies $15\,\text{MeV/nucleon} \gtrsim E_{\text{cm}} \gtrsim 0.6\,\text{MeV/nucleon}$, the slabs pass through each other, but get excited and may break up. At still higher energies, $E_{\text{cm}} \gtrsim 15\,\text{MeV/nucleon}$, the identity of original slabs is largely lost, see Fig. 6 and the system breaks into a multitude of fragments. In the nuclear reaction terminology, this is called multifragmentation.

Figure 7 shows next the intensity plot for the real part of the density matrix for the system of slabs colliding at the c.m. energy of 25 MeV/nucleon. The density matrix is always largest along the $x = x'$ diagonal and the real components tend to marginally dominate over the imaginary components. Values of the matrix along the diagonal represent density, i.e. Fig. 6. Early on in the reaction, significant values of the density matrix are limited to square-like regions of which the diagonals are colinear with the diagonal of the matrix and of which the size represents the support of nucleon wavefunctions within the individual slabs. When the system breaks up into pieces, the region of significant values for the density matrix expands. Besides regions adjacent to the diagonal, islands and streaks of significant values emerge, moving away, with time, from

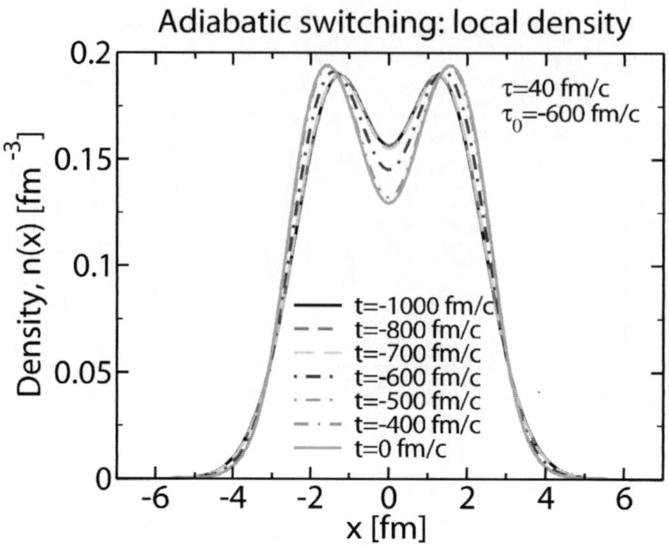

Figure 4. Evolution of the density profile, when starting from an HO configuration with $N_s = 2$ filled shells at time $t = -1000\,\mathrm{fm}/c$ and transforming the single-particle potential to the mean-field form using $\tau = 40\,\mathrm{fm}/c$.

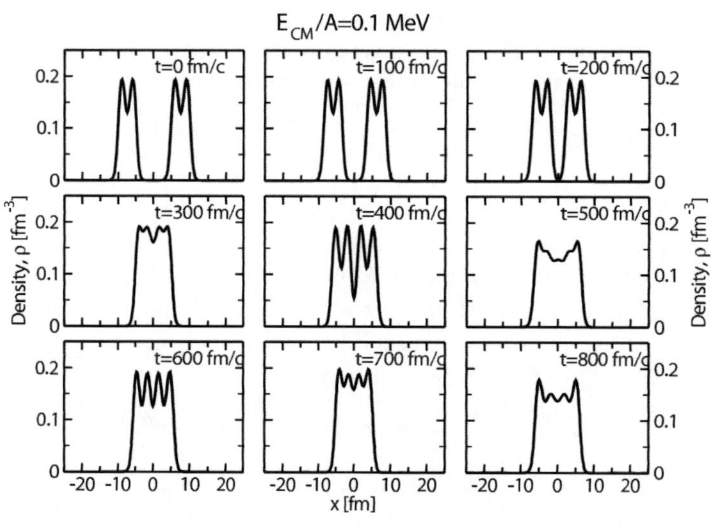

Figure 5. Evolution of density for slabs colliding at the c.m. energy of 0.1 MeV/nucleon.

the diagonal. These are associated with the fragmentation of the original wavefunctions, into pieces taking off with different fragments. The regions of significant values for $\rho(x, x', t)$ represent phase correlations between amplitudes for the nucleons, from the individual original states, to move away with one or another fragment.

5. Tinkering with Evolution

The need to monitor far off-diagonal values of the Green's functions would have been devastating for the capability of carrying reaction simulations in 3 dimensions. However, if the fragments from the break-up are hardly likely to ever to meet again, then the phase information in

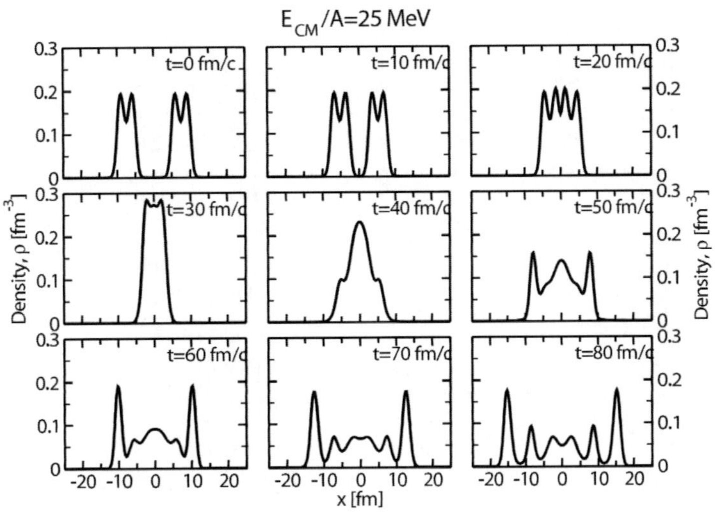

Figure 6. Evolution of density for slabs colliding at the c.m. energy of 25 MeV/nucleon.

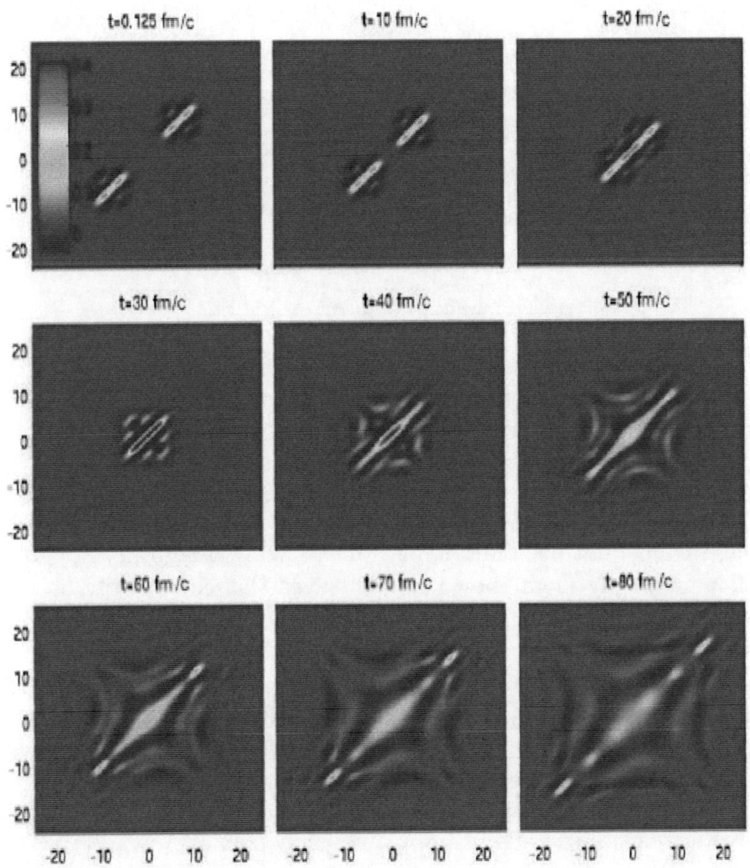

Figure 7. Intensity plots for the real part of the density matrix $\rho(x, x', t)$ in the collision of slabs at $E_{\rm cm} = 25\,{\rm MeV/nucleon}$. The horizontal and vertical axes represent, respectively, the arguments x and x' of the matrix, in fm.

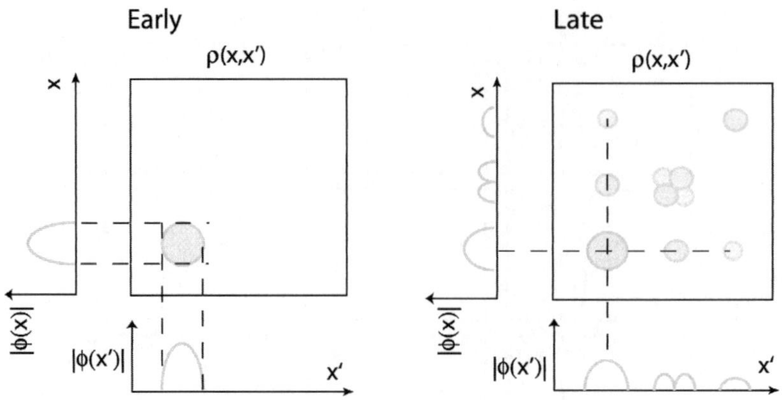

Figure 8. Schematic illustration for the origin of significant off-diagonal elements in the density matrix $\rho(x,x',t) = \sum_\alpha n_\alpha \varphi_\alpha(x,t)\,\varphi_\alpha^*(x',t)$. When nuclear fragments separate, the individual nucleon wavefunctions fragment, maintaining phase coherence between possibilities for individual nucleons moving out with one or another fragment.

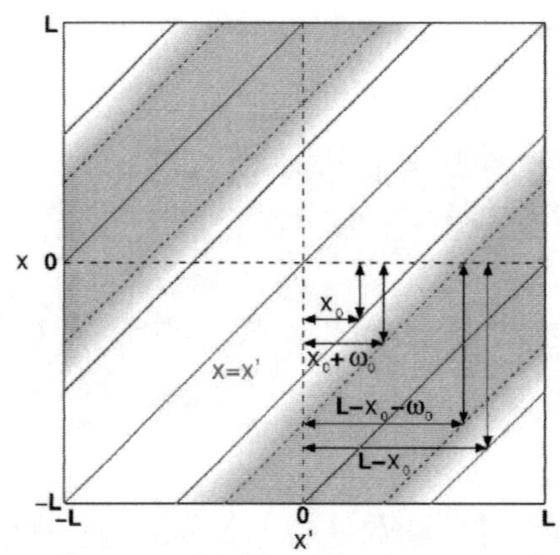

Figure 9. Manipulation of the elements of the density matrix $\rho(x,x',t)$ in testing. Dark regions represent elements that get suppressed.

the far off-diagonal values should not matter. To test the importance of the far off-diagonal elements, we repeat the calculations employing now a strong imaginary superoperator potential that suppresses elements away from the axis and leaves the vicinity of the axis in the density matrix intact, see Fig. 9. Note that the periodic boundary conditions for the system imply a tile-like periodicity of the density matrix ρ in the x-x' plane, with the values on the diagonal repeated next on the lines passing through the the corners at $(-L, L)$ and $(L, -L)$. With the periodicity imposed onto the superoperator, the suppression of the matrix elements occurs in valleys in the x-x' plane, that are separated by ridges where the elements remain intact.

Figure 10 compares results for the density at different times, obtained in the calculations of slab collisions at $E_{\rm cm} = 25\,{\rm MeV/nucleon}$, for different retained regions of the matrix elements along the diagonal. Only when the matrix elements are suppressed from $|x - x'| = 5\,{\rm fm}$ on, some changes in the density, compared to the standard evolution, begin to emerge at late times. Even these changes are subtle as the system may still reach about the same final state, just slightly later. Less severe trimmings of the elements leave no visible signs in the evolution of density for times relevant for reaction dynamics. Figure 11 shows further the evolution of system size and

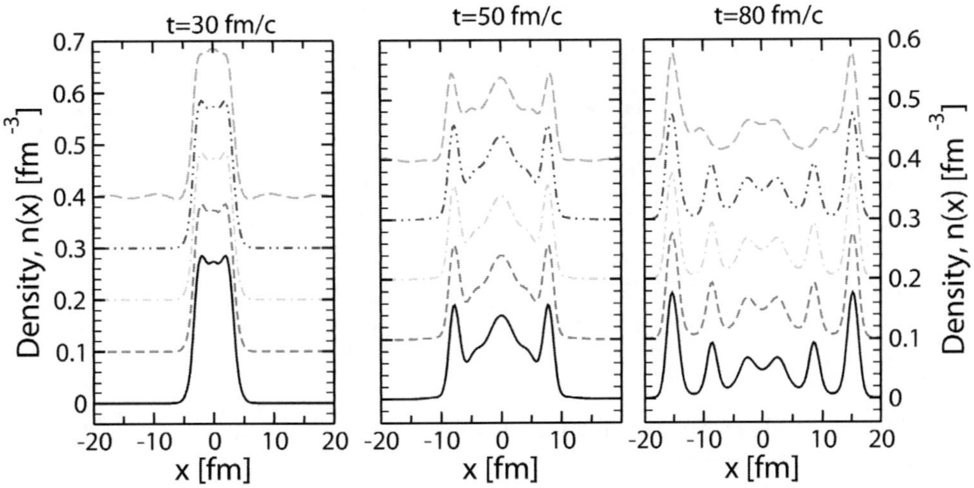

Figure 10. Density at different stages of evolution of the system of slabs at the incident c.m. energy of 25 MeV/nucleon. The bottom curves represent the density from standard evolution. The other curves, from bottom up, represent the density obtained from evolution where the matrix elements at $|x - x'| > 20$, 15, 10, and 5 fm, respectively, are suppressed. For clarity, the results for the density from the different evolutions are staggered by 0.1 fm^{-3}.

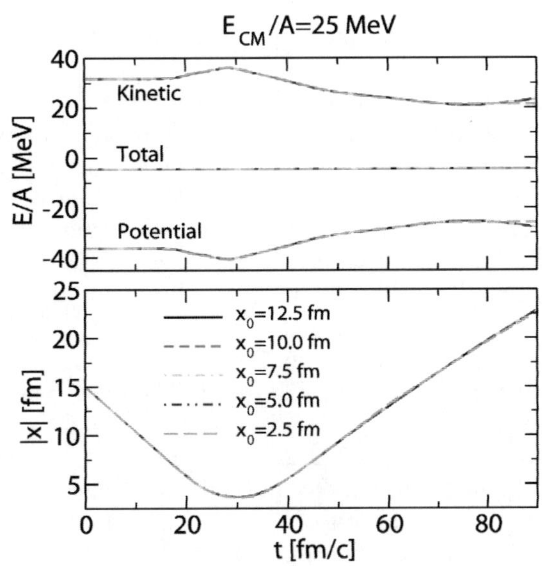

Figure 11. Net energy components (top) and system size (bottom) for elements of density matrix suppressed at $|x - x'| > 2x_0$, at different values of x_0.

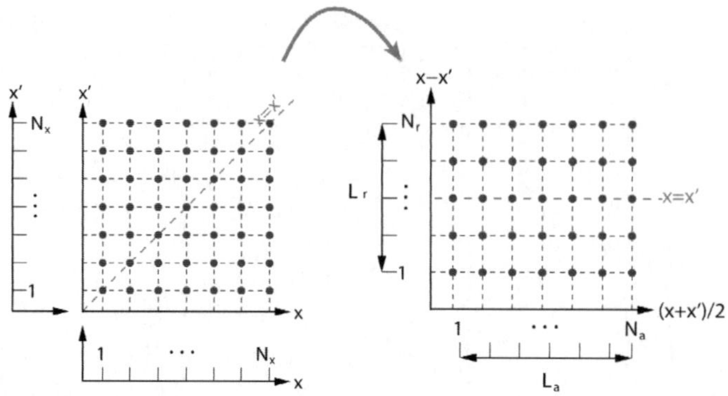

Figure 12. Change in discretization for Green's functions, when turning from the traditional to relative and average arguments.

of components of net energy for different suppressions of matrix elements in the density matrix. Only for elements suppressed at $|x - x'| > 5\,\mathrm{fm}$, a small change in the energy breakdown begins to emerge at late times.

Insensitivity of the dynamics to the removal of far-away matrix elements of the density matrix bodes well for applying nonequilibrium Green's functions to central nuclear reactions. In 3 dimensions + time, such a removal would dramatically reduce the computational cost. Importantly, it appears that one can make a compromise between the demands of low computational cost and of accuracy. On more way of reducing the cost of calculations in 3 dimensions is in exploiting the fact that local momentum distributions are never far away from isotropy in central nuclear reactions. With this, we hope to be able expand the functions in relative positions in terms of the cartesian spherical harmonics. The tensorial properties of the harmonics are preserved under Fourier transformation [15] and the harmonics transform in a simple fashion under shift of the reference point [16]. For weak anisotropies only a few harmonics may need to be retained.

6. Further Advances

Far away elements in the density matrix pertain to a fine momentum resolution for the Wigner function. Elements close to the diagonal pertain to coarse characteristics in momentum. When standard cartesian discretization employed for the Green's functions and density matrix, the momentum and position resolution scale are tied together. However, there is no reason for coupling those. Change of variables to relative and average arguments and discretization of the latter variables, cf. Fig. 12, allows for a decoupling of the scales and also facilitates discarding of far-away matrix elements of the functions. The decoupling should facilitate application of the nonequilibrium Green's functions to intermediate energy central nuclear reactions, as momenta grow with energy, but scales for spatial nonuniformities stay about the same. The rotated numerical meshes have been explored within the Ph.D. Thesis of B. Barker [17] and have been shown to work well.

When incorporating correlations, off-diagonal structure of the Green's functions needs to be considered [18], that augments the storage and calculational complexity in solving the KB equations [19, 20]. With regard to the current round of efforts, aiming at building up the infrastructure for modelling of nuclear reactions, we examined the effects of both abrupt [21] and gradual the build-up of the correlations in a homogenous and inhomogeneous system within an oscillator trap. As an example, Fig. 13 shows the evolution of density in the configuration space and occupation for a slab starting in an uncorrelated state, for an abrupt build-up of

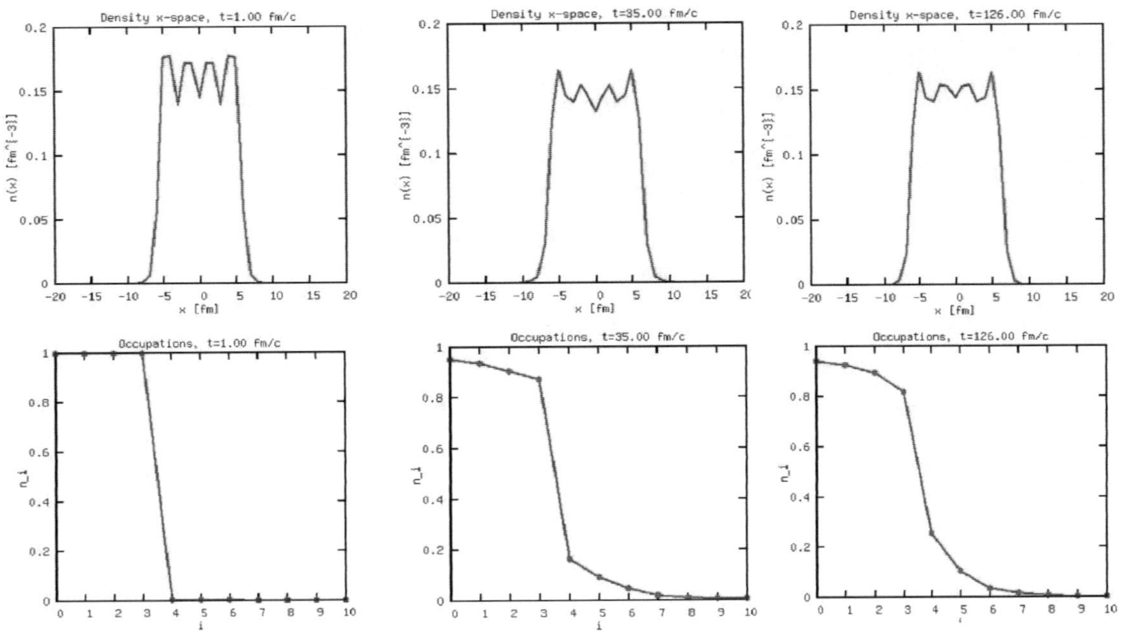

Figure 13. Density as a function of position (top panels) and occupation (bottom), for states diagonalizing the density matrix, in a correlated evolution of an $A = 16$ state starting as uncorrelated, at $t = 0$, in a HO trap.

the correlations. The self-energies $\Sigma^{\lessgtr}$ are taken here in the Born approximation, with the characteristics of the interaction intended to reproduce nucleon-nucleon elastic scattering cross sections, in the Born approximation, at a semi-quantitative level [18]. Currently we concentrate on constructing correlated initial states for reaction calculations in one dimension.

7. Conclusions

To sum up, we have obtained encouraging results in our study of slab collisions in one dimension. Only a limited range of arguments in the density matrix is of practical importance for evolution. Change of arguments in the functions facilitates dropping of far off-diagonal elements and decoupling of momentum and position resolution scales, important for application of the nonequilibrium Green's function to intermediate energy nuclear collisions. We found that initial states for reaction calculations can be arrived at, in practice through adiabatic changes in an interaction.

Acknowledgements

This work was partially supported by the U.S. National Science Foundation under Grants PHY-1068571 and PHY-1520971 by the U.K. Science and Technology Facilities Council under Grants ST/I005528/1 and J000051/1.

References

[1] Negele J 1982 *Rev. Mod. Phys.* **54** 913
[2] Golabek C and Simenel C 2009 *Phys. Rev. Lett.* **103** 042701
[3] Tohyama M 1987 *Phys. Rev. C* **36** 187
[4] Umar A and Vberacker 2006 *Phys. Rev. C* **73** 054607
[5] Bertsch G F and Das Gupta S 1988 *Phys. Rept.* **160** 189–233

[6] Danielewicz P 1984 *Annals of Physics* **152** 239
[7] Dickhoff W H 1998 *Phys. Rev. C* **58** 2807
[8] Nagamiya S *et al.* 1981 *Phys. Rev. C* **24** 971
[9] Danielewicz P and Bertsch G F 1991 *Nucl. Phys. A* **533**(4) 712–748
[10] Kadanoff L P and Baym G 1962 *Quantum Statistical Mechanics* (New York: Benjamin)
[11] Rios A, Barker B, Buchler M and Danielewicz P 2011 *Annals of Physics* **326** 1274–1319
[12] Tohyama M 1994 *Progress of Theoretical Physics* **92** 905–908
[13] Pfitzner A, Cassing W and Peter A 1994 *Nuclear Physics A* **577** 753–772
[14] Watanabe M and Reinhardt W P 1990 *Phys. Rev. Lett.* **65** 3301–3304 URL `http://link.aps.org/doi/10.1103/PhysRevLett.65.3301`
[15] Danielewicz P and Pratt S 2007 *Physical Review C* **75** 034907
[16] Shanker B and Huang H 2007 *Journal of Computational Physics* **226** 732–753
[17] Barker B 2013 Ph.D. thesis Michigan State University East Lansing, Michigan, USA
[18] Danielewicz P 1984 *Ann. Phys.* **152** 305
[19] Köhler H S, Kwong N H and Yousif H A 1999 *Comp. Phys. Comm.* **123** 123
[20] Stan A, Dahlen N E and Leeuwen R v 2009 *The Journal of Chemical Physics* **130** 224101
[21] Kremp D, Semkat D and Bonitz M 2005 *Journal of Physics: Conference Series* **11** 1

Journal of Physics: Conference Series **696** (2016) 012011 doi:10.1088/1742-6596/696/1/012011

Real-time Kadanoff-Baym approach to nuclear response functions

H S Köhler[1] and N H Kwong[2]

[1] Physics Department, University of Arizona, Tucson, Arizona 85721, USA [2] College of Optical Sciences, University of Arizona, Tucson, Arizona, 85721, USA

E-mail: kohler@physics.arizona.edu,kwong@optics.arizona.edu

Abstract. Linear density response functions are calculated for symmetric nuclear matter of normal density by time-evolving two-time Green's functions in real time. Of particular interest is the effect of correlations. The system is therefore initially time-evolved with a collision term calculated in a direct Born approximation until fully correlated. An external time-dependent potential is then applied. The ensuing density fluctuations are recorded to calculate the density response. This method was previously used by Kwong and Bonitz for studying plasma oscillations in a correlated electron gas. The energy-weighted sum-rule for the response function is guaranteed by using conserving self-energy insertions as the method then generates the full vertex-functions. These can alternatively be calculated by solving a Bethe -Salpeter equation as done in works by Bozek et al. The (first order) mean field is derived from a momentum-dependent (non-local) interaction while 2nd order self-energies are calculated using a particle-hole two-body effective (or residual) interaction given by a gaussian *local* potential.

We show results of calculations of the response function $S(\omega, q_0)$ for $q_0 = 0.2, 0.4$ and $0.8 fm^{-1}$. Comparison is made with the nucleons being un-correlated i.e. with only the first order mean field included.

We discuss the relation of our work with the Landau quasi-particle theory as applied to nuclear systems by Babu and Brown and followers.

1. Introduction

Response functions, the response of a many-body system to an external perturbation is instrumental in our understanding of the properties and interactions involved in the excitations of the system. In the study of nuclear systems these response functions are of particular interest when it comes to calculate the mean free path and absorpton of neutrinos in a neutron gas [1, 2], a subject of interest in astrophysical studies.[3, 4] It has been shown that the neutrino interaction in a nuclear medium is very sensitive to the afore-mentioned response function, with the mean free path λ given by:

$$\frac{1}{\lambda(k_\nu, T)} = \frac{G_F^2}{16\pi^2} \int d\mathbf{k}_3 [c_V^2(1 + cos\theta) S^{(0)}(q, T) + c_A^2(3 - cos\theta) S^{(1)}(q, T)] \quad (1)$$

with k_ν and k_3 being initial and final neutrino momenta, θ the scattering angle. $S^{(0,1)}$ are the spin 0 and 1 response functions. G_F is the Fermi constant and $c_{V,A}$ the coupling constants.

This report concerns the calculation of linear density response functions for symmetric nuclear matter. Of particular interest here is the effect of correlations i.e. dressed propagators and

Journal of Physics: Conference Series **696** (2016) 012011 doi:10.1088/1742-6596/696/1/012011

collision-terms. It was shown by Baym and Kadanoff [5] that, if one wishes to construct the linear response function with dressed equilibrium Green's functions, appropriate vertex corrections to the polarization bubble are necessary to guarantee the preservation of the local continuity equation for the particle density and current in the excited system, which in turn implies the satisfaction of the energy-weighted sum rule. This problem has traditionally involved solving a Bethe-Salpeter equation to calculate these vertex corrections. Such a calculation was done by Bozek for nuclear matter[6] and further applied in ref.[7]. Response functions have of course since long been the focus of intense studies for the electron gas with numerous contributions. Early works by Lundqvist and Hedin are noticeable leading to the often cited GW method. [8, 9] More recent works include those of Faleev and Stockman focussing on electrons in quantum wells.[10] An alternative method, seemingly simpler than the Bethe-Salpeter, is a real time solution of the Kadanoff Baym two-time equations using conserving approximations of the self-energies. This guarantees that the energy-weighted sum-rule (assuming an effective mass m^*) $\int_0^\infty \omega S(\omega, q_0) d\omega = \frac{q_0^2}{2m^*}$ is satisfied. This method was first used by Kwong and Bonitz for the plasma oscillations in an electron gas. [11] The details of this method were already presented in the paper by Kwong and Bonitz. We use it here to calculate response functions in symmetric nuclear matter, emphasizing the relative ease of these computations and also the relation of these calculations to the nuclear many-body problem in general.

The relevant KB-equations, already presented in ref. [11] are shown in the following section followed in section 2.1 by an exposé and explanation of the interactions used at the separate level of approximations. When the system initially correlates with the total energy being conserved, the potential energy of interactions decreases and the kinetic energy increases, i.e. the temperature increases. Section 2.2 shows how to calculate the resulting temperature at equilibrium. Results are shown in section 3. The relation of of our work to the Landau quasi-particle theory is addressed in section 4 and section 5 contains a summary of our results and some conclusions.

2. The two-time KB-equations

The Kadanoff-Baym equations in the two-time form for the specific problem at hand was already shown in previous work, where it was applied to the electron gas in a linear response calculation.[11] Some modifications (related to the difference in interactions and self-energies) are necessary for the present nuclear problem. Correlations are included by defining self-energies in a second order Born approximation with a residual interaction defined by eq.(16) below. This is in addition to a first order mean (Hartree-Fock) field. This then is a 'conserving approximation' [5] a necessary requirement when applying the two-time formalism consistently. We compare our results with calculations where correlations are neglected with only the mean field included, the 'HF+RPA' approximation. As made clear in ref. [11] there is no need to separately calculate vertex corrections when applying the two-time formalism. It is a major obstacle in using the Bethe-Salpeter formalism. The proper vertex corrections are in our method generated by the time-evolution of the Green's functions when subjected to the external disturbance. We separate the Green's function into a spatially homogeneous part G_{00} and a linear response part G_{10}.

Calculations proceed as follows: Equilibrium Green's functions are first constructed for an uncorrelated fermi distribution of specified density and temperature. The G_{00} functions are then time-evolved (for typically $10 fm/c$) with the chosen mean field and correlations until stationary. An external field $U(\mathbf{q}, t) = U_0(t)\delta_{\mathbf{q},\mathbf{q_0}}$ is then applied which generates particle-hole Green's functions $G_{10}^{\gtrless}$ that propagate in time according to equations already shown in ref. [11] and for completeness repeated here. (Summation over $m = 0, 1$ and integration over $\bar{t}$ from $-\infty$ to $+\infty$ is implied).

$$\left(i\hbar\frac{\partial}{\partial t} - \epsilon_{\mathbf{k}+\mathbf{q_0}}\right) G^{\gtrless}_{10}(\mathbf{k}tt') = U_0(t)G^{\gtrless}_{00}(\mathbf{k}tt') + \Sigma^{HF}_{1m}(\mathbf{k}t)G^{\gtrless}_{m0}(\mathbf{k}tt')$$
$$+\Sigma^{R}_{1m}(\mathbf{k}t\bar{t})G^{\gtrless}_{m0}(\mathbf{k}\bar{t}t') + \Sigma^{\gtrless}_{1m}(\mathbf{k}t\bar{t})G^{A}_{m0}(\mathbf{k}\bar{t}t') \tag{2}$$

and

$$\left(-i\hbar\frac{\partial}{\partial t'} - \epsilon_{\mathbf{k}}\right) G^{\gtrless}_{10}(\mathbf{k}tt') = U_0(t')G^{\gtrless}_{11}(\mathbf{k}tt') + G^{\gtrless}_{1m}(\mathbf{k}tt')\Sigma^{HF}_{m0}(\mathbf{k}t')$$
$$+G^{R}_{1m}(\mathbf{k}t\bar{t}')\Sigma^{\gtrless}_{m0}(\mathbf{k}\bar{t}t') + G^{\gtrless}_{1m}(\mathbf{k}t\bar{t}')\Sigma^{A}_{m0}(\mathbf{k}\bar{t}t') \tag{3}$$

Retarded and advanced parts are given by

$$\Sigma^{R/A}_{nm}(\mathbf{p},t,t') = \pm\theta(\pm(t-t')[\Sigma^{>}_{nm}(\mathbf{p},t,t') - \Sigma^{<}_{nm}(\mathbf{p},t,t')] \tag{4}$$

$$G^{R/A}_{nm}(\mathbf{p},t,t') = \pm\theta(\pm(t-t')[G^{>}_{nm}(\mathbf{p},t,t') - G^{<}_{nm}(\mathbf{p},t,t')] \tag{5}$$

For the The Hartree-Fock self-energy in the (00)-channel we choose a parametrisation shown in section 2.1. The Hartree-Fock self-energy in the (10)-channel is given by

$$\Sigma^{HF}_{10}(\mathbf{p},t) = -iV_a(q_0)\sum_{\mathbf{p}'} G^{<}_{10}(\mathbf{p}',t,t) + i\sum_{\mathbf{p}'} G^{<}_{10}(\mathbf{p}-\mathbf{p}',t,t)V_a(p) \tag{6}$$

with V_a specified below in section 2.1. Selfenergies $\Sigma^{\gtrless}_{00}$ are given by

$$\Sigma^{\gtrless}_{00}(\mathbf{k},t,t') = i\sum_{\mathbf{p}} G^{\gtrless}_{00}(\mathbf{k}-\mathbf{p},t,t')V^{\gtrless}_{s}(\mathbf{p},t,t'). \tag{7}$$

with

$$V^{\gtrless}_{s}(\mathbf{p},t,t') = V^2(\mathbf{p})\Pi^{\gtrless}_{00}(\mathbf{p},t,t') \tag{8}$$

where $V(\mathbf{p})$ is the momentum-representation of the residual potential, local in ccordinate space, eq. (16). The polarisation bubble Π_{00} is defined by

$$\Pi^{\gtrless}_{00}(\mathbf{p},t,t') = -i\sum_{\mathbf{p}'} G^{\gtrless}_{00}(\mathbf{p}',t,t')G^{\lessgtr}_{00}(\mathbf{p}'-\mathbf{p},t',t) \tag{9}$$

The selfenergies in the (10) channel are given by

$$\Sigma^{\gtrless}_{10}(\mathbf{k},t,t') = i\sum_{\mathbf{p}}[G^{\gtrless}_{10}(\mathbf{k}-\mathbf{p},t,t')V^{\gtrless}_{s}(\mathbf{p},t,t')$$
$$+G^{\gtrless}_{00}(\mathbf{k}-\mathbf{p},t,t')V^{\gtrless}_{s(10)}(\mathbf{p},t,t')] \tag{10}$$

where

$$V^{\gtrless}_{s(10)}(\mathbf{p},t,t') = V_a(\mathbf{p})V_a(\mathbf{p}+\mathbf{q_0})\Pi^{\gtrless}_{10}(\mathbf{p},t,t') \tag{11}$$

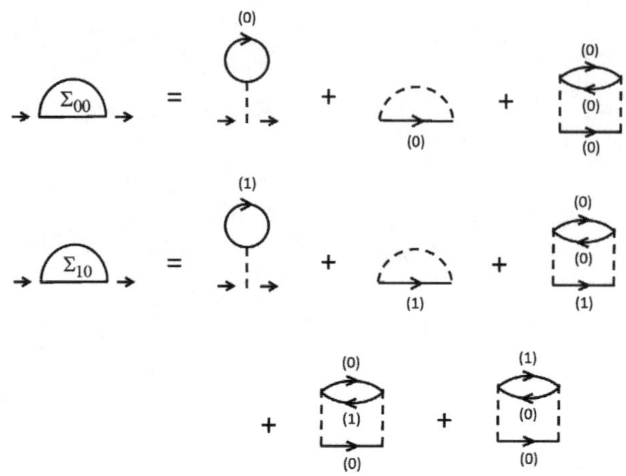

Figure 1. Diagrams representing contributions to the self energy in the Kadanoff-Baym equations to zeroth order (Σ_{00}) and first order (Σ_{10}) in the external perturbation U. The solid (dashed) lines are correlated Green's functions (NN interactions). The number near each Green's function line gives the order in U of that line.

and where the potential V_a is defined in section 2.1. The polarisation bubble in the (10)-channel is given by

$$\Pi^{\gtrless}_{10}(\mathbf{p},t,t') = -i\sum_{\mathbf{p}'}[G^{\gtrless}_{10}(\mathbf{p}',t,t')G^{\lessgtr}_{00}(\mathbf{p}'-\mathbf{p},t',t) + G^{\gtrless}_{00}(\mathbf{p}',t,t')G^{\lessgtr}_{10}(\mathbf{p}'-\mathbf{p}-\mathbf{q}_0,t',t)] \tag{12}$$

A diagrammatic representation of the self-energy is shown in Fig. 1.

The eqs. (2) and (3) are time-evolved and the response-function is, after a fourier-transformation of

$$\delta n(q_0,t) = -i\sum_{\mathbf{p}} G^{<}_{10}(\mathbf{p},t,t)$$

with respect to time t, obtained from

$$S(\omega,q_0) = \frac{\delta n(q_0,\omega)}{\pi n_0 U_0(\omega)} \tag{13}$$

2.1. Interactions

The nuclear two-body interaction in 'free space' involves' a combination of long-ranged one-pion exchange contributions and complicated strong short-ranged attractions and repulsions. This leads to even more complicated 'in-medium' interactions and strong correlations in the nuclear many-body environment. Simplifications are necessary to make the exploratory calculations presented here reasonably short. The nuclear model adopted in most similar works is that of nucleons moving in a mean (' Hartree-Fock self-consistent') field with mutual in-medium ('residual') two-body interaction (potential). A reasonable form of this potential could be a Skyrme- (as used in ref. [12]) or the Gogny-interaction (as used in ref. [13]) or some similar interaction. Apart from these in-medium effective potentials, a semi-realistic NN-interaction with a hard core was used in a Jastrow-correlated extension of the RPA.[14]. We are taking a somewhat different approach here adopting three different potentials (I,II and III) each representing different aspects (domains) of the interaction that for the purpose of presentation we assume to be a Brueckner state-independent reaction matrix $< \mathbf{k}|K|\mathbf{k}' >$ or $K(\mathbf{p},\mathbf{q})$ with $q = \mathbf{k} - \mathbf{k}'$ and $p = \frac{1}{2}(\mathbf{k} + \mathbf{k}')$.

I. An important feature of the Hartee-Fock mean field is that it is momentum-dependent and only the diagonal matrix-elements $K(p, 0)$ are needed. We adopt a parametrisation due to Welke et al [15] leading to:

$$\Sigma_{00}^{HF}(\mathbf{p},t) = A\frac{\rho}{\rho_o} + B\left(\frac{\rho}{\rho_0}\right)^{\sigma} + 2\frac{C}{\rho_o}\int \frac{n(\mathbf{p}',t)}{1+(\frac{\mathbf{p}-\mathbf{p}'}{\Lambda})^2}\frac{d\mathbf{p}'}{(2\pi)^3} \tag{14}$$

with $A = -110.44$ MeV, $B = 140.9$ MeV, C=-64.95 MeV, while ρ is the density of the symmetric nuclear matter under consideration and ρ_0 the saturation density. The momentum-distribution function $n(\mathbf{p}, t)$ is defined below.

We shall also, in separate calculations substitute this mean field by assuming an effective mass.

II. To calculate the second order Born diagrams we need off-diagonal matrix-elements of the interaction i.e. its q-dependence. We choose here an interaction often used in this context, first introduced by Danielewicz [16, 6, 7, 17] and given by:

$$V(r) = V_0 e^{-(\frac{r}{\eta})^2} \tag{15}$$

with $\eta = 0.57$fm and $V_0 = -453$ MeV.

In momentum-space it reads:

$$V(\mathbf{q}) = \pi^{3/2}\eta^3 V_0 e^{-\frac{1}{4}\eta^2 q^2} \tag{16}$$

We use this interaction to calculate self-energies $\Sigma_{00}^{<}$ and $\Sigma_{00}^{>}$.

III. The linear response calculation involves excitations near the fermi-surface i.e. matrix-elements $K(q, p \sim p_F)$ for which the p-independent interaction (16) is too strong, leading to divergences if used for calculating the (10) field components. This is related to a too large numerical value for the Landau parameter. Numerical results shown below are therefore made with an interaction $V_a(q) = a_{10}V(q)$ used in the calculation of Σ_{10}. The majority of the calculations presented below are with $a_{10} = 0.3$. Other values are chosen below together with a discussion of the Landau parameter. (See section 4.)

2.2. *Equilibrium Temperature*

In a KB-calculation with selfenergies defined in a conserving approximation the total energy is conserved; the potential energy decreases and kinetic energy increases with the same amount until internal equilibrium. The result is an increase in temperature from the initially set value. This situation can be moderated by an imaginary time-stepping method [16, 17, 18]. The final temperature T_f will be a function of the initial imaginary time τ (and initial temperature) with $T_f \to 0$ in the limit $\tau \to \infty$.

For finite values of τ this final equilibrium temperature $T_f = 1/\beta$ and chemical potential μ are related to the equilibrium self-energies $\Sigma^<$ and $\Sigma^>$ by [19]:

$$\Sigma^>(\mathbf{p},\omega) = -e^{\beta(\omega-\mu)}\Sigma^<(\mathbf{p},\omega). \tag{17}$$

(The ratio of the selfenergies is consequently independent of momentum $\mathbf{p}$, which serves as a check on numerical accuracy.)

Journal of Physics: Conference Series **696** (2016) 012011 doi:10.1088/1742-6596/696/1/012011

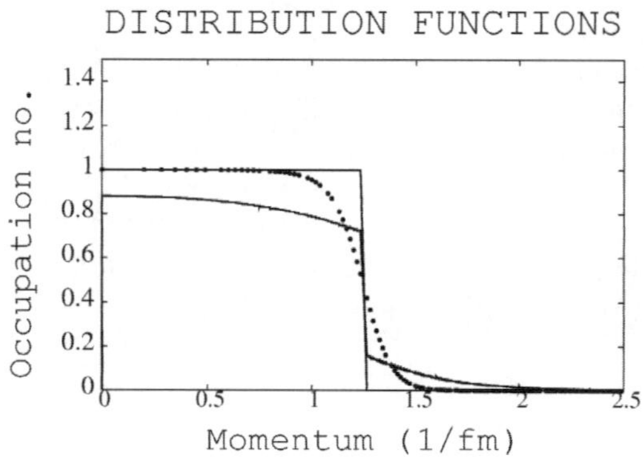

Figure 2. Solid lines show the zero temperature Fermi-distribution (top curve) and the correlated distribution with the second-order selfenergy. The dotted curve shows the $T = 5$ MeV Fermi-distribution referred to in the text.

With β and μ calculated from eq. (17) the equilibrium uncorrelated distribution function is given by

$$f(p) = 1/(1 + e^{\beta(\omega(p)-\mu)}) \tag{18}$$

with

$$\omega(p) = p^2/2m + Re\Sigma^+(\mathbf{p}, \omega) + \Sigma_{00}^{HF}((\mathbf{p})) \tag{19}$$

The real part of Σ^+ is calculated from the imaginary part by the dispersion relation, and using the relation

$$2Im\Sigma(\mathbf{p}, \omega) = i(\Sigma^<(\mathbf{p}, \omega) - \Sigma^>(\mathbf{p}, \omega)) \tag{20}$$

While the uncorrelated distribution is given by eq. (18) the correlated one is given by

$$n(\mathbf{p}, t) = -iG^<(\mathbf{p}, t, t). \tag{21}$$

3. Numerical results

Calculations are made for symmetric nuclear matter at normal nuclear matter density; $\rho = 0.13\text{fm}^{-3}$, i.e. fermi-momentum $k_F = 1.25\text{fm}^{-1}$. The external momentum transfer is chosen to be either $q_0 = 0.2, 0.4$ or 0.8fm^{-1}.The calculations follow essentially the computing-methods described in ref.[20] with the additional requirement to also time-evolve G_{10}.

All numerical results shown below are without the initial imaginary time-stepping. The initial (t=t'=0) distribution $n(\mathbf{p}, 0) = G_{00}^<(\mathbf{p}, 0, 0)$ is taken to be a zero temperature fermi-distribution except in the uncorrelated case where $T = 5$ MeV.(See below.) The system is then time-evolved and correlations are completed at $t \geq t_c$ after which time the distribution $n(p)$ is stationary. In the linear response evolution used here, separating the Green's function into G_{00} and G_{10} components, the collision term for $G_{00}(p, t, t)$ evolution will vanish for $t > t_c$. (See ref. [18] for a detailed discussion of correlation times t_c.)

Fig. 2 shows the relevant distribution-functions. Solid curves show the initial zero-temerature Fermi-distribution and the correlated distribution. The temperature obtained as described above was 5 MeV. The dotted curve in Fig. 2 shows the Fermi-distribution for this temperature. With the correlations completed in the (00) channel the external field $U_0(t)$ is now applied generating the G_{10} field as shown by eqs. (2) and (3). Note that with the conserving approximation satisfied the vertex-correction is 'automatically' included in the ensuing time-evolution.[11]

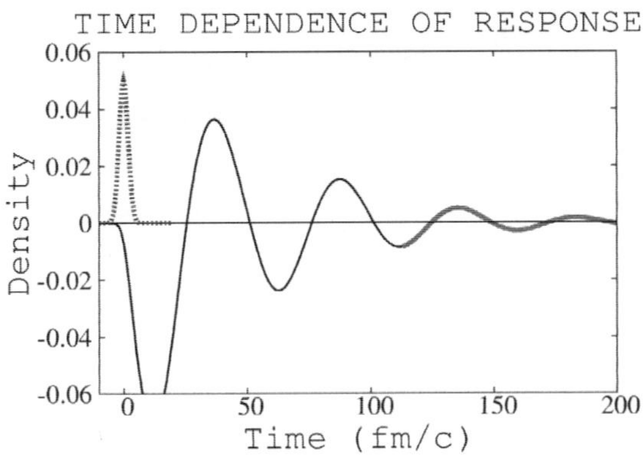

Figure 3. The solid curve shows the density $\delta n(q_0, t)$ with $q_0 = 0.4\text{fm}^{-1}$ calculated for $t < 120 fm/c$ and extrapolated for $t > 120 fm/c$. The broken curve shows the external field $U(q_0, t)$ in units of MeV/1000.

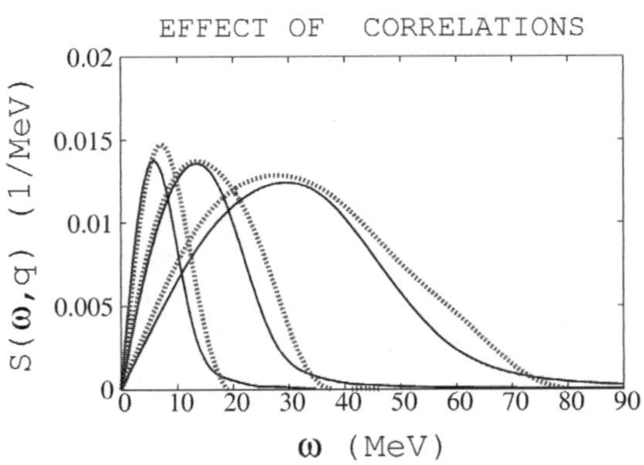

Figure 4. Solid curves show response functions $S(\omega, q_0)$ with 2nd order correlations included. External momentum transfers are (from left to right) $q_0 = 0.2, 0.4$ and 0.8fm^{-1} respectively. They are compared with 'HF+RPA' results at temperatures T=5 MeV: broken curves. See text for further explanations.

The resulting time dependent density

$$\delta n(q_0, t) = -i \sum_{\mathbf{p}} G^<_{10}(\mathbf{p}, t, t)$$

is recorded until it is fully damped. If the damping time is too large (number of time-steps $>\sim 150$) then $n(q_0, t)$ is extrapolated using the amplitudes and frequencies of the oscillations for lesser times. A typical result of the evolution in real time is shown in Fig. 3. This time-function is then fourier-transformed to ω-space to obtain linear response functions $S(\omega, q_0)$. Results are shown in Fig. 4 for three different external momenta tranfers q_0. The solid lines show the results including correlations while the broken lines show the uncorrelated ('HF+RPA') results. These latter results are obtained using a temperature $T = 5MeV$ obtained from eq. 17. They are in overall agreement with previous RPA calculations e.g. [12, 13, 7]. There are some differences to be expected because of the variety of two-body interactions and mean fields that have been used. The heating to $T = 5$ MeV causes only a very slight broadening in comparison with a $T = 0$ calculation.

As regards the correlated results (the solid lines) , the initial temperature at time $t = 0$ is, as already stated above, equal to zero; the momentum-distribution is that of a zero-temperature fermi-distribution. The correlations result in a heating to $T = 5$ MeV and an associated smoothing of the fermi-surface. An additional and more significant smoothing is due to the broadening of the static spectral-functions in the correlated medium. There is also a considerable depletion of the distribution at low momenta caused by the correlations.(See Fig. 2) This is the most probable cause for the difference between the correlated and uncorrelated cases in the

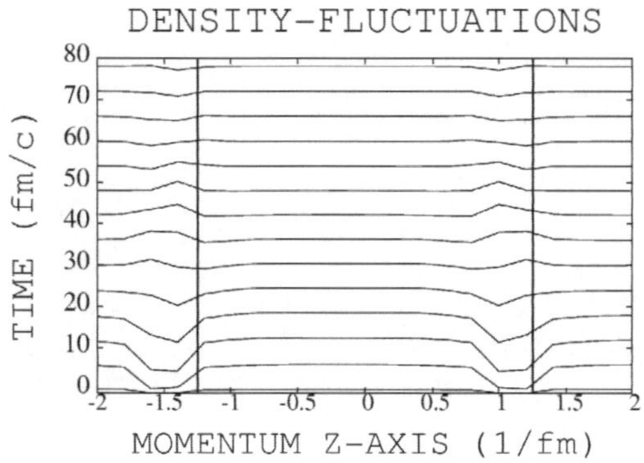

Figure 5. The perturbed density integrated over perpendicular momenta is shown as a function of momentum p_z (with $\mathbf{q_0}$ along the z-axis) and time t. The selfenergies are here calculated to second order. The vertical lines are at $p_z = \pm p_F$ (fermi-momentum). See text for further explanations.

region of the large values of frequencies ω. Other than that there is in fact no other major effect of the correlations.

In order to show the approach to equilibrium in some detail the density $iG^<_{10}(\mathbf{p}, t, t)$ is integrated over p_x and p_y , (p_z being the component along $\mathbf{q_0}$). The resulting t and p_z dependent function is shown in Fig. 5 for the correlated case.

It shows that the excitations take place close to the fermi-surface where the quasi-particle picture dominates and where spectral functions have a vanishing width. The correlations affect mainly the deeper lying states. This fact serves to explain the relatively small effect of the correlations, other than at the high frequency end of the response functions. This situation may however not be the case for tensor correlations [21], that have a longer range (in coordinate-space) and not included here.

The results in this section were all obtained with the factor $a_{10} = 0.3$. Fig. 6 below, also shows response functions for $a_{10} = 0.4, 0.3$ and 0.1 and with an effective mass $m^* = 0.7$ in lieu of the mean field eq. 14.

4. Landau parameters

Owing to the relatively low temperature and long fluctuation wavelenghths considered here, the excitations in our linear response calculations are concentrated around the fermi-surface. This is clearly evident from Fig. 5 which, as pointed out above, can explain why the NN-collisions have a smaller effect than might be expected. This is also why Landau-theory is applicable here. Another related reason why this theory is pertinent to our work is the well known relation [22] between the Landau-parameter

$$F_0 = \frac{2m^* k_F}{2\pi^2 \hbar} f_0 \qquad (22)$$

and the ring-diagram correlations, the collective excitation we are dealing with here. The factor f_0 is here the angular momentum ($l = 0$) quasi-particle (effective) interaction. The Landau parameter relates to the compressibility K by

$$K = 6e_F(1 + F_0) \qquad (23)$$

with e_F being the Fermi energy. The interaction f_0 is composed of a 'direct' part defined by the functional derivative

$$f_{\mathbf{k}\mathbf{k}'} = \frac{\partial^2 E}{\partial n_{\mathbf{k}} n_{\mathbf{k}'}} \qquad (24)$$

(with $\mathbf{k} \rightarrow \mathbf{k}'$) and an 'induced' part. The 'induced' interaction is essentially the effect of screening by rings and further defined below.

Journal of Physics: Conference Series **696** (2016) 012011 doi:10.1088/1742-6596/696/1/012011

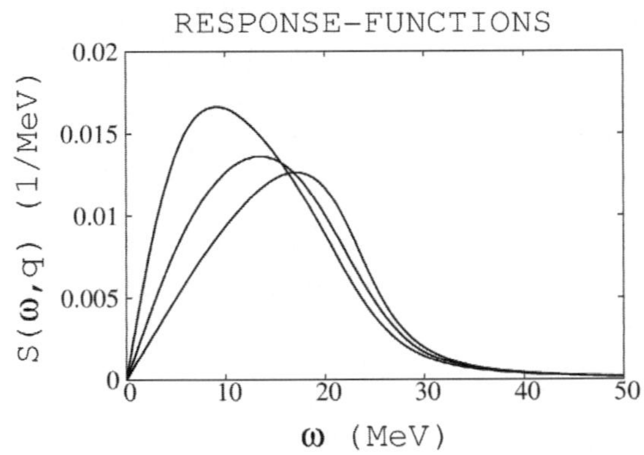

Figure 6. Solid curves show response functions $S(\omega, q_0)$ with second order self-energies for $q_0 = 0.4\text{fm}^{-1}$ and Landau parameters $F_0^{(d)} = -.799, -0.599$ and -0.200 ($a_{10} = .4, .3$ and 0.1 respectively), from left to right. An effective mass $m^* = 0.7$ was assumed. [15]

Guided by methods used by Babu and Brown[22], who considered interactions in liquid ^3He, Sjöberg [23] and later others (see e.g. ref. [26]) calculated Landau parameters for nuclear matter. They used Brueckner's K-matrix expression for the total energy E in their analysis. The 'direct' interaction is then just the Brueckner K-matrix while the 'induced', f_i is of higher orders of K. They find that it is defined by

$$f_i(\mathbf{k}, \mathbf{k}) = -\sum_{\mathbf{p},\mathbf{p}'} f(\mathbf{k}, \mathbf{p}) \left(\frac{\delta n_{\mathbf{p}}(\mathbf{q}, \omega)}{\delta U_{\mathbf{p}'}(\mathbf{q}, \omega)} \right)_{q,0} f(\mathbf{p}'\mathbf{k}) \tag{25}$$

Note that the perturbative potential $U_{\mathbf{p}'}$ is here assumed to act on particles of momenta $\mathbf{p}'$ only, to generate density fluctuation $\delta n_{\mathbf{p}}$. Our interactions are not 'realistic' enough to warrant a detailed comparison with Sjöberg's (or other's based on Brueckner theory). It is here made only to illustrate the connection of our work with that based on Brueckner theory.[22, 23] From the result of the calculation of response function shown in Fig. 4 (2nd order correlations) for $a_{10} = .3$ and for $q = 0.2\text{fm}^{-1}$ we find $\frac{\delta n}{\delta U} = .00126$ to get for the 'induced' interaction 24 MeV with the 'direct' being 140 MeV to get $F_0 = -0.495$ to compare with Sjöberg's $F_0 = -.373$.

Landau theory is a quasi-particle theory and applies to the limit $q = .0$ which is not attainable in our method of calculation. Furthermore, our calculation goes beyond the quasi-particle. The strength of the interaction, our factor a_{10}, does of course also affect the result. Fig. 6 shows the dependence of the the response on this strength-factor. The values of the Landau parameters $F_0^{(d)}$ are in this figure calculated using only the 'direct' interactions.

5. Summary and Conclusions

The main purpose of this presentation is to illustrate the usefulness of the two-time Green's function (Kadanoff-Baym) method for the study of nuclear response. It was previously applied to the electron gas to see the effect of correlations.[11] The time-evolution of Green's function in two-time space with conserving approximations for the self-energies guarantees the preservation of sum-rules. An alternative method, the solution of the Bethe-Salpeter (BS) equation (in ω-space) was been carried out by Bozek et al.[6, 7].

The rise in temperature associated with the onset of correlations when using the two-time method is sometimes cited as a drawback. It is however not difficult to remedy by allowing the correlations to initially form by propagating along the imaginary time-axis. It requires of course to have access to the proper computer-codes. It is our aim to utilise our existing codes in future work.

Our calculations show some differences between the two approximations (collisionless and

2nd-order) but mainly at the high frequency end of the response functions. This is accompanied by a corresponding difference in peak energies, but without any significant shift of these energies.

The results shown are for symmetric nuclear matter. Response calculations for neutron matter is of particular interest related to astrophysical problems, in particular neutrino-absorption.

One reason for choosing a local potential (eq. 15) as done here is that the 2-time computer-code ([20]) used here is restricted to this choice although in-medium effective interactions are known to be non-local. Separable interactions have been obtained from inverse scattering [25] and these can now be used in a new 2-time code developped specifically for these separable (i.e. non-local) potentials.

6. Acknowledgements

One of us (HSK) wishes to thank The University of Arizona and in particular the Department of Physics for providing office space and access to computer facilities. This work has not been supported by any external agency.

References

[1] J Margueron, J Navarro, P Blottiau, *Phys. Rev.* C **70** 028801 (2004).
[2] Naoki Iwamoto and C J Pethick, *Phys. Rev.* D **25** 313 (1982).
[3] S Reddy, M Prakash, J Lattimer and J Pons *Phys. Rev.* C **59** 2888 (1998).
[4] A Sedrakian and A Dieperink, *Phys. Rev.* D **62** 083002 (2000).
[5] Gordon Baym and Leo P Kadanoff, *Phys. Rev.* **124** 287 (1961)
Gordon Baym, *Phys. Rev.* **127** 1391 (1962).
[6] P Bozek, *Phys. Lett.* **B579** 309 (2004).
[7] P Bozek,J Margueron and H Müther, *Ann. Phys.* (N.Y.) **318** 245 (2005).
[8] L Hedin , *Phys. Rev.* **139** A796 (1965).
[9] L Hedin and S Lundqvist, *Solid State Physics* (Acadenmic, New York, 1969) Vol 23.
[10] Sergey V Faleev and Mark I Stockman *Phys. Rev.* B **63** 193302 (2001)
Phys. Rev. B **66** 085318 (2002).
[11] N H Kwong and M Bonitz, *Phys. Rev. Lett.* **84** 1768 (2000).
[12] C Garcia-Recio, J Navarro,Van Gai Nguyen and L L Salcedo, *Ann. Phys.* (N.Y.) **214** 340 (1992).
[13] D Gogny and R Padjen, *Nucl. Phys.* A **293** 365(1977)
[14] N H Kwong and D G Sandler, *Phys. Lett.* B **136** 1 (1984).
[15] G M Welke,M Prakash, T T S Kuo and S Das Gupta, *Phys. Rev.* C **38** 2101 (1988).
[16] P Danielewicz, *Ann. Phys.* (N.Y.) **152** 305 (1984).
[17] H S Köhler, *Phys. Rev.* C **51** 3232 (1995).
[18] H S Köhler and K Morawetz, *Phys. Rev.* C **64** 024613 (2001).
[19] L P Kadanoff and G Baym, *Quantum Statistical Mechanics* (Benjamin,New York, 1962).
[20] H S Köhler, N H Kwong and Hashim A Yousif, *Comp.Phys.Comm.* **123** 123 (1999)
[21] E Olsson, P Haensel and C J Pethick, *Preprint nucl-th/0403066*
G I Lykasov, E Olsson and C J Pethick, *Preprint nucl-th/0502026*
[22] S Babu and G E Brown, *Ann. Phys.* (N.Y.) **78** 1 (1973).
[23] O Sjöberg. *Ann. Phys.* (N.Y.) **78** 39 (1973).
[24] Armen Sedrakian and Jochen Keller, *Preprint nucl-th/1001.0395*
Jochen Keller and Armen Sedrakian *Preprint nucl-th/1205.6902*
[25] N H Kwong and H S Köhler, *Phys. Rev.* C **55** 1650 (1997).
[26] D Gambacurta, U Lombardo and W Zuo, *Physics of Atomic Nuclei* **74** 1424 (2011).

Journal of Physics: Conference Series **696** (2016) 012012 doi:10.1088/1742-6596/696/1/012012

Time Dependent Study of Multiple Exciton Generation in Nanocrystal Quantum Dots

Fikeraddis A. Damtie
Mathematical Physics, Lund University
E-mail: Fikeraddis.Damtie@teorfys.lu.se

Andreas Wacker
Mathematical physics, Lund University.
E-mail: Andreas.Wacker@fysik.lu.se

Abstract. We study the exciton dynamics in an optically excited nanocrystal quantum dot. Multiple exciton formation is more efficient in nanocrystal quantum dots compared to bulk semiconductors due to enhanced Coulomb interactions and the absence of conservation of momentum. The formation of multiple excitons is dependent on different excitation parameters and the dissipation. We study this process within a Lindblad quantum rate equation using the full many-particle states. We optically excite the system by creating a single high energy exciton E_{SX} in resonance to a double exciton E_{DX}. With Coulomb electron-electron interaction, the population can be transferred from the single exciton to the double exciton state by impact ionisation (inverse Auger process). The ratio between the recombination processes and the absorbed photons provide the yield of the structure. We observe a quantum yield of comparable value to experiment assuming typical experimental conditions for a 4 nm PbS quantum dot.

1. Introduction

Multiple exciton generation (MEG) is the process by which a single absorbed photon with high energy creates more than one electron hole pair [1, 2, 3, 4]. It allows for converting the excess energy of absorbed photon into an additional photocurrent. In conventional solar cells based on bulk semiconductors, the conversion efficiency is limited and most of the excess energy is lost as heat within picosecond time scale after absorption. Due to quantum confinement effects, such as, relaxed momentum conservation, modified carrier-cooling rates, reduced dielectric screening of the quantum dot surface and enhanced Auger processes, an enhancement of MEG is expected in quantum dots compared to their bulk counterparts [5, 6]. Lead Chalcogenide QDs have been intensively studied in relation to MEG since the first demonstration of MEG in PbSe quantum dots [4, 7]. These quantum dots are known to exhibit high confinement effects combined with a small band gap of bulk materials, e.g. 0.41 eV for PbS and 0.28 eV for PbSe at room temperature [8]. For a small lead based quantum dot, a quantum yield of 1.21 ± 0.05 has been shown in a recent paper [9] which means that the number of induced excitons surpassed the absorbed photons by about 20%. Apart from lead based dots, Cadmium Chalcogenide quantum dots, indium based quantum dots (InAs and InP) and Silicon quantum dots have been studied for exploring the efficiency of MEG in quantum dots [10, 11, 12, 13, 14]. From an experimental

Journal of Physics: Conference Series **696** (2016) 012012 doi:10.1088/1742-6596/696/1/012012

perspective, very short time resolution and very low excitation fluence are two requirements that needs to be fulfilled when measuring MEG quantum yield [15]. The first requirement is due to the rapid Auger recombination ($\approx 10-100$ ps) [16] for multi-excitons in quantum dots in comparison to lifetime of single excitons ($\approx 10-100$ ns) [17]. To be able to capture these fast processes, techniques with time resolution much faster than the recombination times is needed. Ultrafast lasers can produce pulses as short as 100 fs or less in duration which offer adequate time resolution needed [18]. The second requirement is to avoid multi-exciton formation due to successive excitation of quantum dots with two or more photons within the life time of a single exciton. The observation of multi-exciton recombination at low pump fluence can then be considered as the signature for MEG. Ultrafast transient absorption spectroscopy, which is a pump probe technique, is one of the mostly used experimental techniques for measuring MEG [19, 9, 20]. MEG in quantum dots has been simulated by several groups, with a strong focus on the impact ionisation process. Shabaev *et al.* [3] used a time dependent density matrix approach to study efficient MEG in nanocrystals as a coherent superposition of single and multi-exciton states and showed that efficient MEG is a consequence of suppressed thermalization rate of the initially excited single exciton. Allan *et al.* [21] showed that experimentally observed MEG yield in PbSe nanocrystals can be explained by impact ionisation process which is very efficient in PbSe and PbS nanocrystals with very fast ($\sim$ fs) lifetime for the single exciton to relax into the double exciton. Beard *et al.* [5] compared MEG in nanocrystal with impact ionisation in bulk and showed that the calculated quantum yield for nanocrystals is much larger than its bulk counterpart. The effect of relaxation has been studied by Schulze *et al.* [22]. In the same spirit, Azizi and Machnikowski [23] showed that relaxation can even increase the yield under certain conditions in a phenomenological three-state model.

In this work, we perform a quantum kinetic description based on a selected set of microscopic single particle states similar to [22]. However, we work in the corresponding many-body basis, which allows for a better identification of the exciton states. We use specific parameters for the recent experiment [9] with 4 nm PbS dots, where MEG is studied after the excitation by a resonant pulse. This situation is schematically shown in Fig. 1(a). Within this method, the transient changes to the single exciton and double exciton states by the light pulse can be examined to see the various intermediate processes before the system relaxes to its ground state. Since the transient properties depend highly on the different parameters which can be controlled by suitable experimental setup, understanding these processes at the microscope level can greatly benefit for optimizing MEG experiments. The method can be used for similar quantum dot systems for interpreting experimental measurements on MEG yield and transient absorption experiments.

2. Method and Model

In order to describe the quantum kinetics in a PbS quantum dot under irradiation by a short laser pulse, we construct a set of many-body basis states which are diagonal with respect to the electron-electron interaction. Within this basis we formulate a Lindblad kinetics to include relaxation, recombination and dephasing. The different steps in the setup are addressed below.

2.1. Single particle levels

The energies of the single particle states for the spherical PbS quantum dot are evaluated using a K·P method neglecting the multiplicity of valleys [8]. As an input for the K·P calculations, the material parameters for PbS quantum dot were adopted from Table 1 of the paper [8]. $m^+ = m_e/3$ and $m^- = m_e/2.5$ are the conduction and valence band-edge effective masses respectively. The momentum-matrix element taken between the band-edge states of the conduction and valence band states is given by $P = \sqrt{2.5(eV)m_e/2}$. Here m_e is the free electron mass. Furthermore, an energy gap $E_g = 0.41$ eV at room temperature was also taken from the

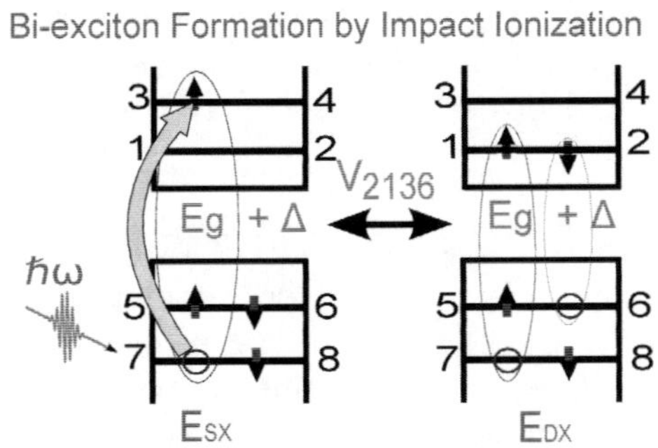

(a) Schematic single particle energies for the quantum dot considered showing the single exciton and double exciton configurations

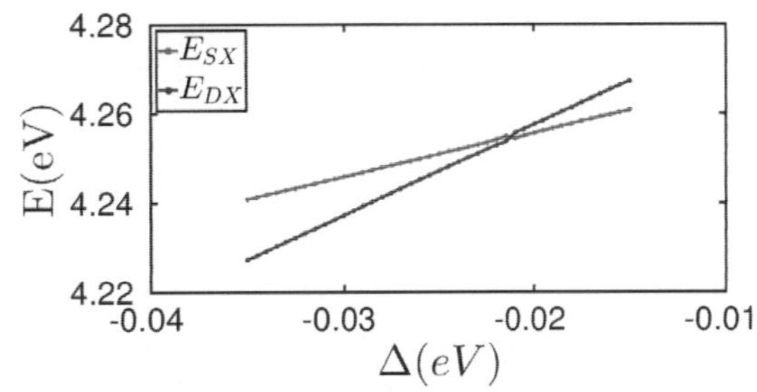

(b) Calculated energies of the high energy exciton E_{SX} and double exciton E_{DX} as a function of the parameter Δ taking into account all interactions.

Figure 1. Model studied and many particle energies of two important exciton states for the study

Orbital	n	l	$E^c_{nl}(eV)$	$E^v_{nl}(eV)$
1s	1	0	$E_{1/2} = 0.650280$	$E_{5/6} = -0.6972889$
1p	1	1	0.964789	-1.0609576
1d	1	2	1.319638	-1.4778528
2s	2	0	1.482883	-1.6709188
1f	1	3	1.721845	-1.954428
2p	2	1	$E_{3/4} = 1.99421$	$E_{7/8} = -2.27846$
1g	1	4	2.174896	-2.493800
2d	2	2	2.562865	-2.956857

Table 1. The lowest single-particle levels for the quantum dot considered in the study. The 1s and 2p levels are identified as most relevant and used in the calculations with the notation given in Fig. 1(a)

table. This gap is very narrow compared to most semiconductor materials which makes PbS quantum dots an excellent platform for studying MEG.

In table 1 the energies of the lowest single-particle levels of a 4 nm PbS quantum dot are listed. The most relevant levels are the ground states (1s) of the conduction and valence band, which are the levels 1,2 and 5,6 in Fig. 1(a), respectively. Here the different spin orientations ($\uparrow$ for odd numbers and $\downarrow$ for even numbers) have been taken into account. In a single particle picture, MEG requires the energy difference between the excited and the ground state matching the gap $E_1 - E_5$. The 2p conduction band level is the first to be close to this resonance. Thus this level, as well as it's corresponding valence band state (where optical transitions are in particular strong) are selected for the excited states 3,4 and 7,8. The precise shape of the quantum dots is not known and the K·P modeling is not perfect for energies far beyond the gap. In order to mimic the the impact of such modifications, we consider a phenomenologically modified band gap $E_g + \Delta$ in the following. The detuning parameter Δ shifts all the conduction band levels E^c_{nl} as listed in Table 1. By varying the Δ, the resonance between the $|SX\rangle$ and $|DX\rangle$ states can be controlled which determine the probability of the population transfer between these two exciton states. In experiment it can be achieved by changing the geometry of the quantum dots. Fig. 1b shows the many particle energies for different values of the detuning parameter. As we go along the horizontal axis, the detuning parameter is increasing which has the effect of increasing

Journal of Physics: Conference Series **696** (2016) 012012 doi:10.1088/1742-6596/696/1/012012

the band gap. As a result, the many particle energies rearrange themselves to this new band gap energy. In the specific spectrum considered, for $\Delta = 0$ the $|SX\rangle$ is higher in energy by $\approx 0.0215eV$ than the $|DX\rangle$ state. Thus, the resonance can be achieved at $\Delta = -0.0215eV$, which is an avoided crossing as shown in Fig. 1b.

2.2. Coulomb matrix elements

MEG by impact ionisation is resulting from the Coulomb electron-electron interaction which is highly enhanced in quantum dots due to the high overlap of states. Here we calculate the corresponding matrix elements for a cuboid geometry which has a volume equivalent to the volume of a 4 nm spherical quantum dot. This simplifies the calculations with wave functions $\sin(n_x\pi x/L_x)\sin(n_y\pi y/L_y)\sin(n_z\pi z/L_z)$. Here we identify the 1s state with $(n_x, n_y, n_z) = (1, 1, 1)$ and the 2p state with $(1, 1, 4)$, on the basis of their parity and the level sequence for the respective geometries. We estimate that the corresponding error is comparable to others resulting from the simplified screening with constant ϵ and the use of envelope functions. The general expression for the Coulomb matrix element is given by the integral:

$$V_{n_1n_2m_1m_2} = \frac{e^2}{4\pi\epsilon\epsilon_0}\int\int d^3\mathbf{r}d^3\mathbf{r'}\phi^*_{n_1}(\mathbf{r})\phi^*_{n_2}(\mathbf{r'})\frac{e^{-\lambda|\mathbf{r}-\mathbf{r'}|}}{|\mathbf{r}-\mathbf{r'}|}\phi_{m_1}(\mathbf{r'})\phi_{m_2}(\mathbf{r}) \tag{1}$$

which enter the Hamiltonian in occupation number representation as

$$\hat{H}_{ee} = \sum_{n_2<n_1,m_1<m_2} V^{\text{eff}}_{n_1n_2m_1m_2}\hat{a}^\dagger_{n_1}\hat{a}^\dagger_{n_2}\hat{a}_{m_1}\hat{a}_{m_2} \quad \text{with } V^{\text{eff}}_{n_1n_2m_1m_2} = V_{n_1n_2m_1m_2} - V_{n_1n_2m_2m_1} \tag{2}$$

In Eq.(1) a screening length $\lambda^{-1} \approx 5L_z$ was assumed which is a length much larger than the dimension of the dot. This has no significant effect on the overall result of the integral. As it will be described below, the evaluation of the Coulomb matrix element involve an integral in Fourier space where λ stabilizes the numerics. In addition, a relative permittivity of $\epsilon = 17.2$ was used [24]. The integral in Eq.(1) is solved by splitting it into parts which involve $\mathbf{r}$ and $\mathbf{r'}$ and taking the Fourier transform of the screened Coulomb potential. This gives pairs of integrals in Fourier space of the form

$$\begin{aligned} A(\mathbf{q})_{n_1,m_2} &= \int d^3\mathbf{r}\phi^*_{n_1}(\mathbf{r})e^{i\mathbf{q}\cdot\mathbf{r}}\phi_{m_2}(\mathbf{r}) \\ A^*(\mathbf{q})_{m_1,n_2} &= \int d^3\mathbf{r'}\phi^*_{n_2}(\mathbf{r'})e^{-i\mathbf{q}\cdot\mathbf{r'}}\phi_{m_1}(\mathbf{r'}) \end{aligned} \tag{3}$$

which contain pairs of indices (n_1, m_2) and (n_2, m_1) and $\mathbf{q}$ is the wave vector. Generally the Coulomb matrix elements can involve envelope functions ϕ_i either all from the same band or at least one is from the different band. If the pairs, (n_1, m_2) originate from the same band, the the integrals in Eq.(3) can be evaluated directly by using the envelope functions. However if they originate from different bands, the evaluation is more complicated. In this case, one has to consider a multi-band model. In this work, a two band model [25] was used for evaluating the integral which involve the conduction and valence bands. The result gives a prefactor times the integral of the corresponding intraband pair with this prefactor being a constant proportional to the interband dipole matrix element times the wave vector $\mathbf{q}$. Two types of electron-electron interactions dominate. Direct terms of the type $V^{\text{eff}}_{n_1n_2n_2n_1}$ which account for direct electron-electron interaction within the same band (intraband) e.g V^{eff}_{2112} or with different bands (interband) e.g. V^{eff}_{5115}. These direct terms are the strongest ones with calculated values of the order of ≈ 100 meV. The other types of elements which are of interest are the scattering terms $V^{\text{eff}}_{n_1n_2m_1m_2}$ which account for several different types of interactions. The matrix element V^{eff}_{1263}, which couples the single exciton and the double exciton for MEG (Auger term), is of the order of ≈ -0.6 meV.

Journal of Physics: Conference Series **696** (2016) 012012 doi:10.1088/1742-6596/696/1/012012

2.3. Interaction with light and dipole matrix elements

The coupling to the laser field is described by the Hamiltonian

$$\hat{H}_I(t) = \sum_{n,m} eE(t)\langle n|\hat{z}|m\rangle \hat{a}_n^\dagger \hat{a}_m \tag{4}$$

where e is the elementary charge, and $E(t)$ the electrical component of the laser field, which is assumed to be z-polarised.

The dipole matrix elements $\langle n|\hat{z}|m\rangle$ are divided into two groups: interband and intraband matrix elements. The interband dipole matrix elements are essentially determined by the z-matrix element of the lattice periodic functions and diagonal in the spatial structure of the envelope functions [26]. On the basis of the p-matrix element between the valence and conduction band states from [8] we find

$$\langle 1|\hat{z}|5\rangle = \langle 3|\hat{z}|7\rangle = \langle 2|\hat{z}|6\rangle = \langle 4|\hat{z}|8\rangle = 0.7371 \text{ nm} \tag{5}$$

For the intraband case, the matrix element is evaluated by the envelope functions. Using the single particle wavefunctions of cuboid quantum dots we find

$$\langle 3|\hat{z}|1\rangle = \langle 7|\hat{z}|5\rangle = \langle 4|\hat{z}|2\rangle = \langle 8|\hat{z}|6\rangle = -0.05764 \text{ nm} \tag{6}$$

A similar result (0.045 nm) is obtained from the 1s and 2p eigenstates of the spherical dot in a single band approximation, which shows that the cuboid states are a reasonable approximation. For PbSe quantum dots, An *et al.* [31] calculated the electronic band structure using an atomistic pseudopotential method which indicated that care has to be taken when using K·P method as it misses some optical selection rules.

The oscillating electric field is assumed to have a Gaussian envelope

$$E(t) = E_0 \exp\left(-\frac{(t-t_0)^2}{\tau^2}\right) \sin(\omega(t-t_0)) \tag{7}$$

where in the above expression ω is the central frequency for the pulse which can be tuned to the resonance frequencies. τ describes the width of the pulse and E_0 is the electric field amplitude. Here we parametrise E_0 by the pulse area

$$\Theta = \frac{e}{\hbar} E_0 \langle 1|\hat{z}|5\rangle \int_{-\infty}^{\infty} dt \exp\left(-\frac{(t-t_0)^2}{\tau^2}\right) \tag{8}$$

In a two two-level system, a $\Theta = \pi$ pulse flips the ground state to the excited [27]. From the experimental pulse parameters given in the supplementary information of [9] a pulse strength of the order of $\Theta \approx 1$ is possible. However for the calculations in this work, a value of Θ smaller than one is used to be able to achieve the low excitation fluence requirement [15] which is about $1\mu J$ per pulse for the sample assumed [9].

2.4. Many particle states

The total Hamiltonian for the system can be divided into two main parts, time independent $\hat{H}_0$ and time dependent interaction part $\hat{H}_I(t)$.

$$\hat{H}_{\text{eff}}(t) = \underbrace{\sum_i E_i \hat{a}_i^\dagger \hat{a}_i + \hat{H}_{ee}}_{\hat{H}_0} + \hat{H}_I(t) \tag{9}$$

The many body states are obtained by exact diagonalisation of $\hat{H}_0$. Fixing the total number of electrons to 4, the 4 particle many-body ground state is where all the levels in the conduction band are empty and all the levels in the valence band are occupied. In Fock state representation it is given by

$$|GS\rangle \approx |\underbrace{0000}_{c}\underbrace{1111}_{v}\rangle \tag{10}$$

where the occupation numbers $|n_1, n_2, n_3, n_4, n_5, n_6, n_7, n_8\rangle$ relate to the 8 states in Fig. 1(a). The lowest energy excitation $|ex\rangle$ is due to an electronic transition across the band gap from the highest level in the valence band to the lowest level in the conduction band, which is called exciton. For the system considered, the energy of the lowest singlet excitation is E_{ex} = 1.334 eV above the ground state. The corresponding many body eigenstate is $|ex\rangle \approx 1/\sqrt{2}(|10000111\rangle - |01001011\rangle)$. The energy E_{ex} is slightly smaller than the difference between the single-particle levels 1 and 5 (1.347 eV from table 1) due to small admixtures of other states. (The experimentally measured gap of 1.07 meV is even smaller. This might be due a different shape, but the admixture from lower lying levels (e.g. 1p) neglected in our model will also play a role.) Two other many-body states play a very important role in the description of MEG: The high energy single exciton $|SX\rangle \approx 1/\sqrt{2}(|00101101\rangle - |00011110\rangle)$ and the double exciton $|DX\rangle \approx 1/\sqrt{2}(|11001001\rangle - |11000110\rangle)$ which are both singlets and close in energy. The detuning between these two states can be controlled by the parameter Δ (which modifies the band gap, see section 2.1) as shown in Fig. 1(b). At $\Delta \approx -0.0215$ eV the two states show an anticrossing with the difference $|E_{DX} - E_{SX}| = 2V_{DX,SX}$, where $|V_{DX,SX}| \approx 0.6$ meV is the Coulomb matrix element between the states $|DX\rangle$ and $|SX\rangle$.

2.5. Equation of motion for the density matrix

The system consists of a classical light field and a quantum system with discrete energies. The change to the system induced by the light field can be best described by the time evolution of the reduced density operator for the system given by the Lindblad equation [28].

$$\hbar\frac{d}{dt}\hat{\rho}_S(t) = i[\hat{\rho}_S(t), \hat{H}_{\text{eff}}(t)] + \sum_{j=1}^{N_{\text{jump}}} \Gamma_j \left[\hat{L}_j\hat{\rho}_S(t)\hat{L}_j^\dagger - \frac{1}{2}(\hat{\rho}_S(t)\hat{L}_j^\dagger\hat{L}_j + \hat{L}_j^\dagger\hat{L}_j\hat{\rho}_S(t))\right] \tag{11}$$

The time evolution in Eq. (11) was solved in the many particle basis with a 4th order Runge-Kutta method. The jump operators $\hat{L}_j$ describe different dissipation processes (with rate $\Gamma_j/\hbar$) which are restoring the ground state at sufficiently long time after the excitation. Here we use: Relaxation in the conduction band

$$\hat{L}_{\text{rel}} = \hat{a}_{1\uparrow}^\dagger\hat{a}_{3\uparrow} + \hat{a}_{2\downarrow}^\dagger\hat{a}_{4\downarrow} \quad \text{with strength} \quad \Gamma_{\text{Relaxation}} \tag{12}$$

Relaxation in the valence band

$$\hat{L}_{\text{rel}} = \hat{a}_{7\uparrow}^\dagger\hat{a}_{5\uparrow} + \hat{a}_{4\downarrow}^\dagger\hat{a}_{8\downarrow} \quad \text{with strength} \quad \Gamma_{\text{Relaxation}} \tag{13}$$

Recombination across the band gap

$$\hat{L}_{\text{rec}} = \hat{a}_{5\uparrow}^\dagger\hat{a}_{1\uparrow} + \hat{a}_{6\downarrow}^\dagger\hat{a}_{2\downarrow} \quad \text{with strength} \quad \Gamma_{\text{Recombination}} \tag{14}$$

Dephasing in the conduction band

$$\begin{aligned}\hat{L}_{\text{deph}} &= \hat{a}_{1\uparrow}^\dagger\hat{a}_{1\uparrow} + \hat{a}_{2\downarrow}^\dagger\hat{a}_{2\downarrow} \quad \text{with strength} \quad \Gamma_{\text{Dephasing}} \\ \hat{L}_{\text{deph}} &= \hat{a}_{3\uparrow}^\dagger\hat{a}_{3\uparrow} + \hat{a}_{4\downarrow}^\dagger\hat{a}_{4\downarrow} \quad \text{with strength} \quad \Gamma_{\text{Dephasing}}\end{aligned} \tag{15}$$

Journal of Physics: Conference Series **696** (2016) 012012 doi:10.1088/1742-6596/696/1/012012

Dephasing in the valence band

$$\begin{aligned} \hat{L}_{\text{deph}} &= \hat{a}^\dagger_{5\uparrow}\hat{a}_{5\uparrow} + \hat{a}^\dagger_{6\downarrow}\hat{a}_{6\downarrow} \quad \text{with strength} \quad \Gamma_{\text{Dephasing}} \\ \hat{L}_{\text{deph}} &= \hat{a}^\dagger_{7\uparrow}\hat{a}_{7\uparrow} + \hat{a}^\dagger_{8\downarrow}\hat{a}_{8\downarrow} \quad \text{with strength} \quad \Gamma_{\text{Dephasing}} \end{aligned} \tag{16}$$

These jump operators are defined in such a way that they all conserve the total spin indicated by the spin indices. The different decoherence mechanism which were assumed phenomenologically in Eq. (11) can be associated to all forms of intrinsic scattering mechanisms other than electron-electron scattering, which has already been included in the effective Hamiltonian.

3. Results

We simulate Eq. (11) in order to determine the density matrix as a function of time. In all calculations we use $\tau = 150$ fs and $t_0 = 200$ fs together with the initial condition $\hat{\rho} = |GS\rangle\langle GS|$ at $t = 0$. Regarding the dissipation mechanisms, dephasing is the fastest scale and we use $\Gamma_{\text{Dephasing}} = 6$ meV, which corresponds to a $\tau_{\text{Dephasing}} = \hbar/\Gamma_{\text{Dephasing}} \approx 100$ fs. Relaxation is typically slightly slower, and we use $\Gamma_{\text{Relaxation}} = 3.3$ meV. These are typical values in solids. Recombination is typically on the ns time scale. In order to save simulation time, we use a larger value of $\Gamma_{\text{Recombination}} = 0.33$ meV (corresponding to $\tau_{\text{Recombination}} = \hbar/\Gamma_{\text{Recombination}} \approx 2$ ps), which is still a factor of 10 slower than the other processes.

Our main observable is the recombination rate

$$\text{Rec(t)} = \Gamma_{\text{Recombination}} Tr\{\hat{L}_{\text{rec}}\hat{\rho}\hat{L}^\dagger_{\text{rec}}\} \tag{17}$$

Integrating over time provides the number of recombined electron-hole pairs generated by the pulse. Fig. 2(a) shows the calculated number of recombinations for excitations in the vicinity of the $2p$ resonance for different values of Δ. The average full width at half maximum (FWHM) of the peaks is about ≈ 22 meV which is mainly due to the pulse width $\tau = 150$ fs. It can be seen that recombination is enhanced when the two excitons are close in energy near $\Delta = -0.0215$ eV. The other ingredient needed for calculating the yield is the power transferred to the system by the oscillating field. It can be defined as the absorbed energy per unit time and can be calculated as

$$\text{P(t)} = \frac{d}{dt}\langle H_0\rangle = \frac{i}{\hbar}\langle[H_{\text{eff}}(t), H_0]\rangle = \frac{i}{\hbar}\langle[H_I(t), H_0]\rangle \tag{18}$$

Integrating over time provides the absorbed energy, which is displayed in Fig. 2(b). This has a similar shape, except for a small dip near $\Delta = -0.0215$ eV.

The yield is then defined as the ratio between the number of recombinations and the absorbed photons. The latter is given by the absorbed energy divided by the photon energy. Due to the short pulse width the photon energy is not well defined. Thus we take the energy $\hbar\omega_{\text{pulse}}$, where the number of recombinations peaks for a given detuning Δ. The corresponding result is plotted in Fig. 3

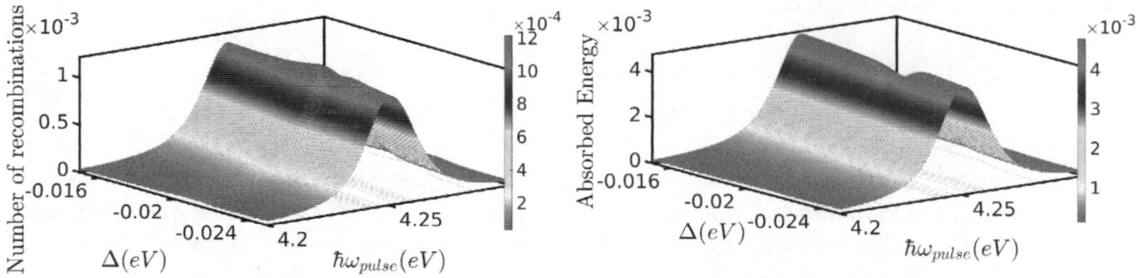

(a) Calculated number of recombinations (b) Calculated absorbed energy

Figure 2. Calculated number of recombinations (a) and absorbed energy (b) as a function of Δ and pulse frequency close to the transition between the 2p states. The pulse area is $\Theta = 0.035\pi$.

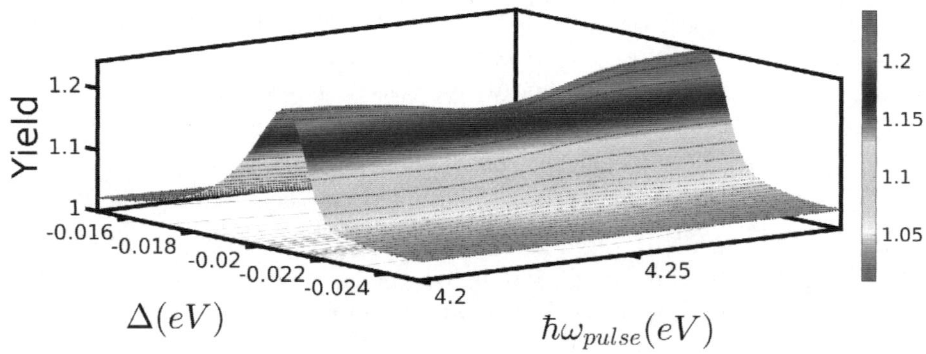

Figure 3. Quantum yield combining the data of Figs. 2(a,b)

We observe a quantum yield of ≈ 1.2 close to resonance $\Delta \approx -0.0215$ eV. This yield agrees well with the experimentally observed values in these dots reported by Karki *et al.*[9]. However we obtain this number only for a specific type of dots, characterised by our phenomenological parameter Δ.

Now we want to study the kinetics in detail for $\Delta = -0.0215$ eV, where the single exciton and double exciton states are very close in energy. Thus, the states $|SX\rangle$ and $|DX\rangle$ mix significantly, which results in the singlet states $|22\rangle$ and $|26\rangle$ in our many-body calculation. Analysing these states they can be decomposed as

$$|22\rangle \approx \alpha_1|SX\rangle + \beta_1|DX\rangle \tag{19}$$

and

$$|26\rangle \approx \alpha_2|SX\rangle + \beta_2|DX\rangle \tag{20}$$

where further contributions are neglected. In terms of these coefficients the probability for the double exciton and single exciton is given by

$$\rho_{\mathrm{DX,DX}} = \langle DX|\rho|DX\rangle = \alpha_1^2\rho_{26,26} + \alpha_2^2\rho_{22,22} - 2\alpha_1\alpha_2\Re\{\rho_{22,26}\} \tag{21}$$

$$\rho_{SX,SX} = \langle SE|\rho|SX\rangle = \beta_1^2\rho_{26,26} + \beta_2^2\rho_{22,22} - 2\beta_1\beta_2\Re\{\rho_{22,26}\} \tag{22}$$

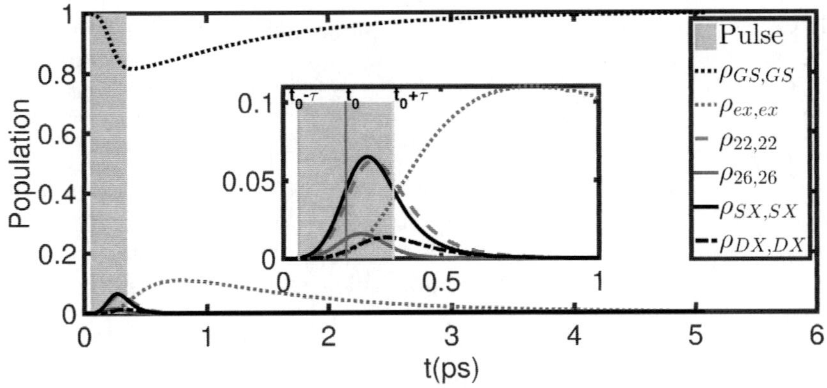

Figure 4. Time evolution for selected many particle states for $\Delta = -0.0215$ eV , $\hbar\omega_{\text{pulse}} = 4.255$ eV. An increased pulse area of $\Theta = \pi/2$ is used in order to emphasise the effect. The grey area depicts the duration of the pulse. The inset is an enlargement for small times.

respectively where $\Re$ denote the real part of the coherences between the states $|22\rangle$ and $|26\rangle$.

Fig. 4 shows the time evolution of some selected many-particle states when the two excitons are close and a pulse in resonance with the state $|SX\rangle$ is used. In the region where the pulse is active, coherent population transfer is dominating mechanism. It can be seen that for a strong pulse $\Theta = \pi/2$ the system is excited significantly. The $|SX\rangle$ state is the first to be excited as the pulse is tuned to be in resonance with this state and the transition is optically allowed. Subsequently (around t_0) the state $|DX\rangle$ starts to grow, which can be understood as impact ionisation due to the Coulomb interaction sketched in Fig. 1(a). Almost at the same time, the single exciton ground states $|ex\rangle$ starts growing as it is fed by relaxation processes from the excited exciton $|SX\rangle$ via intermediate states not plotted here. Finally, the $|SX\rangle$ and $|DX\rangle$ states lose their population after the pulse due to the slow recombination rate. In addition there is an Auger contribution, where $|DX\rangle$ is transferred back to $|SX\rangle$.

Within the many-body eigenstates, the same story is told differently: The optically excited state $|SX\rangle$ is a linear combination of $|22\rangle$ and $|26\rangle$. Thus the population of these two states is evolving in parallel. However, due to the energy splitting $\Delta \geq 2V_{DX,SX}$, the coherence $\rho_{22,26}$ is oscillating in time with the angular frequency $\Delta/\hbar$. Eqs. (21,22) show that this oscillation implies population changes in $|SX\rangle$ and $|DX\rangle$. Relaxation processes reduce the populations of $|SX\rangle$ and $|DX\rangle$ slightly after the peak of the pulse at $t_0 = 0.2$ ps was reached and the generation becomes weaker. Due to this relaxation and, equally important, dephasing, the oscillation period is never completed for the parameters used.

4. Conclusion and outlook

In this work, we have studied MEG due to optical excitations in 4 nm PbS quantum dots. We took into account the electron-electron interaction by diagonalising the Hamiltonian. In this basis we solved the equation of motion for the density matrix in the Lindblad form. We find a yield of 1.2 for a $\hbar\omega_{\text{pulse}} \approx 4.2$ eV. This goes well with the experimental observation of a yield of 1.2 for photon energies surpassing 3 times the lowest exciton absorption [9]. Furthermore, we clarified the relation between the picture of impact-ionisation and a description with diagonal many-particle states.

While the values of the yield are promising, it is not clear, why these values for a particular range of parameters Δ are representative for the actual experiment. One might argue that the multitude of many-particle states arising if further single-particle levels are taken into account for, offers a plentitude of such resonances, so that they become generic similar to the scenario discussed for tunneling in [29]. This has, however, to be studied further in detail. Another interesting point for further study is the optimisation of the yield and the harvesting of multiple excitons by appropriate contacts. To be able to benefit from MEG, an efficient extraction mechanism is a key. One promising demonstration of efficiently collecting the multiple excitons produced is by Sambur *et al.* [30] in which they attached a single layer of PbS quantum dot to a thin oxide (TiO_2). They claim that the strong electronic coupling and favorable energy level alignment between the quantum dot and the oxide facilitates a quick extraction of multiple excitons before they recombine. This can actually be quantified using an extension of the model used here. Corresponding work is ongoing.

5. Acknowledgements
We thank Tönu Pullerits and Khadga Karki for helpful discussions and the Knut and Alice Wallenberg foundation as well as NanoLund for financial support.

6. References
[1] Beard M C 2011 *J. Phys. Chem. Lett.* **2** 1282
[2] Nozik A J 2008 *Chemical Physics Letters* **457** 3
[3] Shabaev A, Efros A L and Nozik A J 2006 *Nano Letters* **6** 2856
[4] Schaller R D and Klimov V I 2004 *Phys. Rev. Lett.* **92** 186601
[5] Beard M C, Midgett A G, Hanna M C, Luther J M, Hughes B K and Nozik A J 2010 *Nano Letters* **10** 3019
[6] Klimov V I 2007 *Annu. Rev. Phys. Chem.* **58** 635
[7] Wise F W 2000 *Accounts of Chemical Research* **33** 773–780
[8] Kang I and Wise F W 1997 *J. Opt. Soc. Am. B* **14** 1632
[9] Karki K J, Ma F, Zheng K, Zidek K, Mousa A, Abdellah M A, Messing M E, Wallenberg L R, Yartsev A and Pullerits T 2013 *Scientific Reports* **3** 2287
[10] Schaller R D, Petruska M A and Klimov V I 2005 *Applied Physics Letters* **87** 253102
[11] Gachet D, Avidan A, Pinkas I and Oron D 2010 *Nano Letters* **10** 164
[12] J J H Pijpers, E Hendry, M T W Milder, R Fanciulli, J Savolainen, J L Herek, D Vanmaekelbergh, S Ruhman, D Mocatta, D Oron, A Aharoni , U Banin, and Bonn M 2007 *The Journal of Physical Chemistry C* **111** 4146
[13] Stubbs S K, Hardman S J O, Graham D M, Spencer B F, Flavell W R, Glarvey P, Masala O, Pickett N L and Binks D J 2010 *Phys. Rev. B* **81** 081303
[14] Matthew C Beard, Kelly P Knutsen, Pingrong Yu, Joseph M Luther, Qing Song, Wyatt K Metzger, Randy J Ellingson, and Arthur J Nozik 2007 *Nano Letters* **7** 2506
[15] Smith C and Binks D 2013 *Nanomaterials* **4** 19
[16] Klimov V I, Mikhailovsky A A, McBranch D W, Leatherdale C A and Bawendi M G 2000 *Science* **287** 1011
[17] Crooker S A, Barrick T, Hollingsworth J A and Klimov V I 2003 *Applied Physics Letters* **82**
[18] Träger F 2007 *Springer handbook of lasers and optics* (Springer Science & Business Media)
[19] Choi Y, Sim S, Lim S C, Lee Y H and Choi H 2013 *Scientific Reports* **3** 3206
[20] Semonin O E, Luther J M, Choi S, Chen H Y, Gao J, Nozik A J and Beard M C 2011 *Science* **334** 1530
[21] Allan G and Delerue C 2006 *Phys. Rev. B* **73** 205423
[22] Schulze F, Schoth M, Woggon U, Knorr A and Weber C 2011 *Phys. Rev. B* **84** 125318
[23] Azizi M and Machnikowski P 2013 *Phys. Rev. B* **88** 115303
[24] O Madelung U Rössler M S 1998 Lead sulfide (pbs) optical properties, dielectric constants *Landolt-Börnstein, Group III Condensed Matter* vol 41C (Springer-Verlag)
[25] Sirtori C, Capasso F, Faist J and Scandolo S 1994 *Phys. Rev. B* **50** 8663
[26] Davies J H 1997 *The physics of low-dimensional semiconductors: an introduction* (Cambridge university press)
[27] Mukamel S 1995 *Principles of nonlinear optical spectroscopy* (Oxford University Press)
[28] Lindblad G 1976 *Comm. Math. Phys.* **48** 119
[29] Goldozian B, Damtie F A and Wacker A 2015 *preprint arXiv:1501.02974*
[30] J B Sambur, T Novet and B A Parkinson 2010 *Sciences* **330** 63
[31] J M An, A Franceschetti, S V Dudiy and A Zunger 2006 *Nano Letters* **6** 2728

Journal of Physics: Conference Series **696** (2016) 012013 doi:10.1088/1742-6596/696/1/012013

Simple electron-electron scattering in non-equilibrium Green's function simulations

David O. Winge[1], Martin Franckié[1], Claudio Verdozzi[1], Andreas Wacker[1] and Mauro F. Pereira[2]

[1] Division of Mathematical Physics, Department of Physics, Lund University, Box 118, SE-221 00 Lund, Sweden

[2] Materials and Engineering Research Institute, Sheffield Hallam University, City Campus, Howard Street, Sheffield S1 1WB, UK

E-mail: david.winge@teorfys.lu.se

Abstract. In this work we include electron-electron interaction beyond Hartree-Fock level in our non-equilibrium Green's function approach by a crude form of GW through the Single Plasmon Pole Approximation. This is achieved by treating all conduction band electrons as a single effective band screening the Coulomb potential. We describe the corresponding self-energies in this scheme for a multi-subband system. In order to apply the formalism to heterostructures we discuss the screening and plasmon dispersion in both 2D and 3D systems. Results are shown for a four well quantum cascade laser with different doping concentration where comparisons to experimental findings can be made.

1. Introduction

Since the quantum cascade laser (QCL) was introduced more than 20 years ago [1] it has continuously been improved and redesigned to operate from the mid-infrared all the way down to the Terahertz (THz) range. Compact laser sources at these wavelengths are valuable for spectroscopic applications [2] but a major problem is that THz QCLs do not operate at room temperature. This can be overcome via difference frequency generation which recently has been demonstrated at powers in the milli-watt range [3]. Direct THz-QCLs still generate a lot of interest in the research community due to the promise of higher wall-plug efficiency and the prospects of miniaturization of cryo-coolers.

The main temperature degrading mechanism of the THz QCLs is currently not fully understood and is still debated by the community [4]. This is a challenge to theory and there is a great need of realistic modeling tools that are able to treat all important quantum effects on the same footing. A summary of different methods for simulating these structures is found in [5]. Monte Carlo simulations have shown that electron-electron interaction, beyond the meanfield or Hartree approximation can influence the dynamics of THz QCLs [6, 7, 8].

In this work we include and study the effects of a simple electron-electron scattering via the Single Plasmon Pole Approximation (SPPA) [9, 10]. In this approximation we capture both the static limit as well as dynamic effects. This gives an energy dependent (non-local in time) interaction beyond the Hartree-Fock approximation. This has been studied in a similar model with promising results [11], and with this work we want to adapt the idea into our model described in Ref. [12]. In other methods based on Non-Equilibrium Green's Functions (NEGF) and applied to QCLs, electron-electron scattering has previously been implemented in a low order GW_0 approximation employing a purely static screening model [13].

To simulate the periodic structure of the QCL, we use a product space of plane waves and confined states. The Green's function evaluation provides us with the energy resolution of the states, labeled by state and k index, at the cost of inverting the self-energy matrices at each k and energy grid point. A number of elastic scattering processes as well as longitudinal optical phonon scattering is treated in this formalism at a high computational cost. Adding a many-body term fully dependent on the polarization would increase the complexity by one order, a formidable task in this model, leading us to consider simpler schemes as done in this work.

This paper is organized as follows: first the theory of the SPPA is discussed and we show how it is implemented in the context of our formalism. Then the dispersion of the plasmons is discussed as well as the exchange shift. In the last part a test structure is used to evaluate the effect of the inclusion of the new scattering mechanism. Finally, we present the conclusions of our study.

2. Theory

In THz quantum cascade simulations, the relevant sources of scattering, beside electrons, are interface roughness, impurities from donor atoms, acoustic and optical phonons as well as alloy composition. All these effects have already been implemented and details can be found in Ref. [12].

The model discussed here employs a non-equilibrium Green's function formalism where the system observables are extracted from the retarded and lesser parts of the Green's function in the formalism of [14, 15]. These are solved using self-energies for the part of the Hamiltonian that is non-diagonal in $\mathbf{k}$ and solved through the Dyson equation and the Keldysh relation.

In this section we will describe how we formulate the self-energy expressions for the screened interaction in the single plasmon pole approximation (SPPA). First we will discuss how the conduction band is treated as an effective band in order to apply the named approximation. Then we will describe how we formulate and also evaluate the Coulomb matrix elements. After this we acquire the lesser and retarded part of the screened interaction before moving on to the actual construction of the self-energies. This is done in close correspondence to the treatment of phonons in Ref. [16] in order to keep the length of this section reasonable. To this treatment we will here add the necessary exchange part. Before we move on to the next section where we test the model, we will also describe how the plasmon dispersion is treated.

2.1. Screening from one effective band

In frequency space the retarded screened potential is given by its own Dyson equation, given here as a matrix equation:

$$\mathbf{W}(\mathbf{q},E) = \mathbf{V}(\mathbf{q}) + \mathbf{V}(\mathbf{q})\mathbf{P}(\mathbf{q},E)\mathbf{W}(\mathbf{q},E) \tag{1}$$

$$\Longleftrightarrow \mathbf{W}(\mathbf{q},E) = \underbrace{[1-\mathbf{V}(\mathbf{q})\mathbf{P}(\mathbf{q},E)]^{-1}}_{\epsilon(\mathbf{q},E)^{-1}}\mathbf{V}(\mathbf{q}). \tag{2}$$

where the $\mathbf{V}(\mathbf{q})$ is the bare interaction as a function of the momentum wave vector, described in more detail below, and $\mathbf{P}(\mathbf{q},E)$ is the polarization as a function of momentum and energy E. If we now assume an isotropic screening medium and that the dielectric function $\epsilon(\mathbf{q},E)$ is not level dependent, we can write the approximated retarded screened potential as

$$W^{\mathrm{ret}}_{ijkl}(\mathbf{q},E) = \frac{V_{ijkl}(q)}{\epsilon(\mathbf{q},E)} \tag{3}$$

where $V_{ijkl}(q)$ now holds the matrix structure. We consider the dielectric function $\epsilon(\mathbf{q}, E)$ given by the Lindhard formula [17] which is retarded due to poles in the lower half of the complex plane. While we assume an effective band for the screening, all other physical mechanisms are treated with full state dependence.

In this work we will assume a plasmon distribution in equilibrium with respect to the electron temperature. This approximation will ensure that energy is not dissipated from the distribution of non-equilibrium electrons. In addition, this assumption will allow us to link the lesser part of the screened interaction to the retarded through the fluctuation dissipation theorem [15] as

$$\mathbf{W}^{<}(\mathbf{q}, E) = \mathrm{i}f(E)A(\mathbf{q}, E) = \mathrm{i}f(E)\left[-2\Im\{\mathbf{W}^{\mathrm{ret}}\}\right] \tag{4}$$

where $f(E)$ is the (bosonic) plasmonic distribution function and $A(\mathbf{q}, E)$ is the spectral function, which can be written in terms of the retarded Green's function leading to the rightmost expression. The electron temperature will be represented as the expectation value $\langle E_k \rangle$ for the effective conduction band. This is with respect to the discrete energy for each state α as the single particle energies are $E_\alpha + E_k$. We calculate $\langle E_k \rangle$ self-consistently in the model. The lesser and retarded part are all that is needed to formulate the self-energy which will be described in detail below.

2.2. Coulomb matrix elements

The Coulomb matrix elements are given on the quasi-2D form following [17]

$$V_{ijkl}(q) = \frac{e^2}{2A\epsilon_0\epsilon_r}\frac{F_{ijkl}(q)}{q} \tag{5}$$

with ϵ_r being the dielectric constant and A is the system area while the other constants have their usual meaning. The relative constant ϵ_r is the equilibrium value for the main material system with no dopant atoms present. To this we will add the non-equilibrium contribution as shown in Eq. (3), making up the total dielectric constant. The *form factors* are expressed as

$$F_{ijkl}(q) = \int dz \int dz' \varphi_i^*(z)\varphi_j(z)\mathrm{e}^{-q|z-z'|}\varphi_k^*(z')\varphi_l(z'). \tag{6}$$

where the integral is taken along the growth direction, which defines the quantization direction of the quasi-2D system and $\varphi_i(z)$ are the quantized states in the z-direction used as a basis for the simulation. The numerical integration is greatly simplified by expressing the wave functions as a truncated Fourier series [18]. All form factors are then easily calculated and are thus available at a reasonable computational cost for all q. These form factors will also be close to exact as the number of Fourier components needed for the well behaved wave functions of the heterostructures is limited.

2.3. Lesser and retarded expressions for the screened interaction

We define the SPPA following [17] as

$$\frac{1}{\epsilon(\mathbf{q}, E)} \approx 1 + \frac{E_{pl}^2(q)}{(E + i\delta)^2 - E_{eff}^2} \tag{7}$$

where the effective plasmon pole is chosen to ensure that the approximation is correct both in the long wavelength *and* the static limit, as thoroughly discussed in [17], and given in the two cases of screening dimensions as

$$\text{2D:} \qquad E_{eff}^2(q) = E_{pl}^2(q)(1 + \frac{q}{\kappa}) + \nu_q^2 \tag{8}$$

Journal of Physics: Conference Series **696** (2016) 012013 doi:10.1088/1742-6596/696/1/012013

and

$$\text{3D:} \qquad E_{eff}^2(q) = E_{pl}^2(1 + \frac{q^2}{\kappa^2}) + \nu_q^2 \tag{9}$$

where the plasma frequency is defined as

$$\text{2D:} \qquad E_{pl}(q) = \sqrt{\frac{\hbar^2 e^2 n^{2D}}{2\epsilon_r \epsilon_0 m} q}$$

$$\text{3D:} \qquad E_{pl} = \sqrt{\frac{\hbar^2 e^2 n^{3D}}{\epsilon_r \epsilon_0 m}}$$

where n^{2D} and n^{3D} are, respectively, the effective 2D and 3D carrier densities. m is the effective mass of the electron, ν_q^2 is a term proportional to q^4 which should take into account the *pair continuum* [17] and κ is the inverse of the static screening length given in the two cases as

$$\text{2D:} \qquad \kappa = \frac{me^2}{2\pi\epsilon_0\epsilon_r\hbar^2} f_{k=0}$$

$$\text{3D:} \qquad \kappa = \sqrt{\frac{e^2 n^{3\mathrm{D}}}{\epsilon_0 \epsilon_r k_B T}}.$$

In 3D we use the Debye limit of screening and for the 2D case the screening becomes a function of the occupation probability f_k at $k = 0$ for the effective conduction band.

Using the SPPA and the Dirac identity we find for the retarded part of the screened interaction

$$\mathbf{W}^{\mathrm{ret}}(\mathbf{q}, E) = \mathbf{V}(\mathbf{q}) - \mathrm{i}\mathbf{V}(\mathbf{q})\frac{\pi}{2}\frac{E_{pl}^2}{\hbar E_q}\left[\delta(E - E_q) - \delta(E + E_q)\right]. \tag{10}$$

This can be linked to the lesser propagator using the fluctuation-dissipation theorem as discussed above and it is found to be

$$\begin{aligned} \mathbf{W}^{<}(\mathbf{q}, E) &= \mathrm{i}f(E)\left[-2\Im\{\mathbf{W}^{\mathrm{ret}}\}\right] \\ &= -\mathrm{i}\pi\mathbf{V}(\mathbf{q})\frac{E_{pl}^2}{\hbar E_q}\left[f(E_q)\delta(E - E_q) + (f(E_q) + 1)\delta(E + E_q)\right]. \end{aligned} \tag{11}$$

These are the standard non-interacting bosonic propagators, which are the same that we have used previously for the optical phonons in our model. Following the same treatment applied to phonons one can formulate expressions for the retarded and lesser self-energies. However the static term in the retarded interaction has to be taken into account. This is the exchange shift and it is crucial to preserve the limits of the SPPA, static and long wavelength cases.

According to standard Feynmann rules we express the diagram for the screened interaction self-energy in the GW approximation [19] as

$$\Sigma_{\alpha\alpha'}(\mathbf{k}, t, t') = \mathrm{i}\hbar \sum_{\beta\beta'} \sum_{\mathbf{k}'} G_{\beta\beta'}(\mathbf{k}', t, t') W_{\alpha\beta\beta'\alpha'}(\mathbf{k} - \mathbf{k}', t, t'), \tag{12}$$

where the times are now on the Keldysh contour. Using the Langreth rules [20] we can transform them to the real time axis. Schematically we have for products structured as above the following rules:

$$\Sigma^{<} = G^{<}W^{<} \tag{13}$$

$$\Sigma^{\mathrm{ret}} = G^{<}W^{\mathrm{ret}} + G^{\mathrm{ret}}W^{<} + G^{\mathrm{ret}}W^{\mathrm{ret}} \tag{14}$$

Journal of Physics: Conference Series **696** (2016) 012013 doi:10.1088/1742-6596/696/1/012013

where the time arguments are implicit and are ordered as in the product in Eq. (12). At this point we make a rather drastic approximation and express the screened interaction in the SPPA. The static term in the retarded screened interaction is present in the first and third term. In the first it gives rise to the exchange described below, however in the third, it does not contribute as the retarded Green's function, considered as a function of energy, only has poles in the lower imaginary half plane. Closing the contour in the upper plane prevents contributions from this term.

We can now proceed and formulate the self-energies for the screened interaction, and the treatment is analogous to that presented in Ref. [16] for optical phonons, with the addition of the exchange term treated below. Here we make the crucial assumption that the self-energies are functions of energy only, in contrast with the more computationally intensive approach of Ref. [11] that uses full energy and k-dependent terms. Here the full k-dependence is replaced by an average contribution, by evaluating the scattering matrix elements for a set of typical values for E_k and $E_{k'}$, as explained in detail in [12]. This brings the scattering matrix elements out of the integral over $E_{k'}$ and we write the lesser and retarded self-energy as

$$\begin{aligned}\Sigma^{<}_{\alpha\alpha'}(E) = &\sum_{\beta\beta'}\int\frac{\mathrm{d}\theta}{2\pi}X^{sppa-}_{\alpha\alpha'\beta\beta'}(\theta)f_{\mathrm{Bose}}(E_{eff}(\theta))\int_0^\infty \mathrm{d}E_{k'}G^{<}_{\beta\beta'}(k',E-E_{eff}(\theta))\\ &+\sum_{\beta\beta'}\int\frac{\mathrm{d}\theta}{2\pi}X^{sppa+}_{\alpha\alpha'\beta\beta'}(\theta)(f_{\mathrm{Bose}}(E_{eff}(\theta))+1)\int_0^\infty \mathrm{d}E_{k'}G^{<}_{\beta\beta'}(k',E+E_{eff}(\theta))\end{aligned} \tag{15}$$

and

$$\begin{aligned}\Sigma^{\mathrm{ret}}_{\alpha\alpha'}(E) = &\sum_{\beta\beta'}\int\frac{\mathrm{d}\theta}{2\pi}X^{sppa-}_{\alpha\alpha'\beta\beta'}(\theta)(f_{\mathrm{Bose}}(E_{eff}(\theta))+1)\int_0^\infty \mathrm{d}E_{k'}G^{\mathrm{ret}}_{\beta\beta'}(k',E-E_{eff}(\theta))\\ &+\sum_{\beta\beta'}\int\frac{\mathrm{d}\theta}{2\pi}X^{sppa+}_{\alpha\alpha'\beta\beta'}(\theta)f_{\mathrm{Bose}}(E_{eff}(\theta))\int_0^\infty \mathrm{d}E_{k'}G^{\mathrm{ret}}_{\beta\beta'}(k',E+E_{eff}(\theta))\\ &+\frac{1}{2}\sum_{\beta\beta'}\int\frac{\mathrm{d}\theta}{2\pi}X^{sppa-}_{\alpha\alpha'\beta\beta'}(\theta)\int_0^\infty \mathrm{d}E_{k'}G^{<}_{\beta\beta'}(k',E-E_{eff}(\theta))\\ &-\frac{1}{2}\sum_{\beta\beta'}\int\frac{\mathrm{d}\theta}{2\pi}X^{sppa+}_{\alpha\alpha'\beta\beta'}(\theta)\int_0^\infty \mathrm{d}E_{k'}G^{<}_{\beta\beta'}(k',E+E_{eff}(\theta))\\ &+\mathrm{i}\int\frac{\mathrm{d}E''}{2\pi}\mathcal{P}\left\{\frac{1}{E''}\right\}\Big[\sum_{\beta\beta'}\int\frac{\mathrm{d}\theta}{2\pi}X^{sppa-}_{\alpha\alpha'\beta\beta'}(\theta)\int_0^\infty \mathrm{d}E_{k'}G^{<}_{\beta\beta'}(\mathbf{k}',E-E_{eff}(\theta)-E'')\\ &\qquad-\sum_{\beta\beta'}X^{sppa+}_{\alpha\alpha'\beta\beta'}(\theta)\int_0^\infty \mathrm{d}E_{k'}G^{<}_{\beta\beta'}(\mathbf{k}',E+E_{eff}(\theta)-E'')\Big]\end{aligned} \tag{16}$$

where the last term contains an principal value integral. As this term involves yet another integral in energy space, it is of higher numerical complexity than the other terms and currently neglected to expedite the numerical calculations. The factor $f_{\mathrm{Bose}}(E_{eff}) = 1/(\exp(E_{eff}/k_BT)-1)$ is again the Bose distribution where the electron temperature enters. In each term there is an average over all possible scattering angles. The scattering matrix elements are given as

$$X^{sppa}_{\alpha\alpha'\beta\beta'}(\theta) = \frac{1}{2\pi}\frac{me^2}{2\epsilon_0\epsilon_r\hbar^2}\frac{F_{\alpha\alpha'\beta\beta'}(\theta)}{q(\theta)}\frac{E_{pl}(q(\theta))}{2E_{eff}(q(\theta))} \tag{17}$$

and the plus and minus variants used above comes from the choice of typical values for the evaluation. In our k-independent approximation we put

$$X^{\text{sppa}\pm}_{\alpha\alpha',\beta\beta'} = X^{\text{sppa}}_{\alpha\alpha',\beta\beta'}(E_{typ}, E_{typ} + |\Delta E \mp E_{\text{eff}}|) \tag{18}$$

with ΔE takes the level difference properly into account for all combinations of indices. E_{typ} is a representative value chosen to give a similar scattering rate compared to the case of thermalised subbands at the given lattice temperature.

2.4. Exchange shift

As the Hartree-Fock self-energies are local in time they depend only on the difference in coordinates, and the exchange term is normally evaluated directly from the diagrammatic rules to be

$$\Sigma^{x}_{\alpha\alpha'}(\mathbf{k}, t_1, t_2) = \hbar \sum_{\beta\beta'} \sum_{\mathbf{q}} V_{\alpha\beta\beta'\alpha'}(|\mathbf{q}|)\delta(t_1, t_2)\mathrm{i}G_{\beta\beta'}(\mathbf{k}-\mathbf{q}, t_1, t_2) \tag{19}$$

which simplifies if only the difference in time matters and we find

$$\begin{aligned}
\Sigma^{x}_{\alpha\alpha'}(\mathbf{k}, E) &= -\sum_{\beta\beta'} \sum_{\mathbf{q}} \int \mathrm{d}\tau \mathrm{e}^{\mathrm{i}E/\hbar\tau} V_{\alpha\beta\beta'\alpha'}(|\mathbf{q}|)\delta(\tau) \int \frac{\mathrm{d}E'}{2\pi\mathrm{i}} \mathrm{e}^{-\mathrm{i}E'/\hbar\tau} G^{<}_{\beta\beta'}(\mathbf{k}-\mathbf{q}, E') \\
&= -\sum_{\beta\beta'} \sum_{\mathbf{q}} V_{\alpha\beta\beta'\alpha'}(|\mathbf{q}|) \int \frac{\mathrm{d}E'}{2\pi\mathrm{i}} G^{<}_{\beta\beta'}(\mathbf{k}-\mathbf{q}, E') \\
&= -\sum_{\beta\beta'} \sum_{\mathbf{k}'} V_{\alpha\beta\beta'\alpha'}(|\mathbf{k}-\mathbf{k}'|)\rho_{\beta\beta'}(\mathbf{k}') \\
&\approx -\sum_{\beta\beta'} \bar{V}_{\alpha\beta\beta'\alpha'}(E_k, E_{k'}) \sum_{\mathbf{k}'} \rho_{\beta\beta'}(\mathbf{k}') \\
&= -\sum_{\beta\beta'} \bar{V}_{\alpha\beta\beta'\alpha'}(E_k, E_{k'}) \frac{A}{2} \bar{\rho}_{\beta\beta'}
\end{aligned} \tag{20}$$

where we moved the Coulomb matrix elements out of the sum over $\mathbf{k}$ in order to do the sum and reach the final 2D sheet densities $\bar{\rho}_{\beta\beta'}$. The factor of $1/2$ appears as we do not sum over spin for the exchange self-energy. The $\bar{V}(E_k, E_{k'})$ is now an angle averaged quantity over the possible angles between $\mathbf{k}$ and $\mathbf{k}'$, where we again use typical values for E_k and $E_{k'}$.

3. Results

Our model, without electron-electron scattering has been previously tested successfully on superlattices and QCL heterostructures [12, 21, 22]. The case considered here is a QCL design from 2009 [23]. The periodic structure is displayed in Fig. 1 together with the probability density for the subbands relevant for transport. These type of four-well THz QCL structures have previously been troublesome to simulate, see for example [24], due to a pre-peak feature which is not recovered experimentally, at least not to the same extent as in the simulations. In theory, a lot more weight is put on the pre-peak at the cost of a lower main peak, compared to the opposite situation in experiments.

This design was also tested experimentally for the different doping concentrations, which makes it ideal as we want to study the impact of electron-electron interaction. The experimental findings are summarized in Table 1. Sample A is doped to acquire a sheet density normal for THz QCLs while sample B has about a factor three higher doping concentration. For the SPPA,

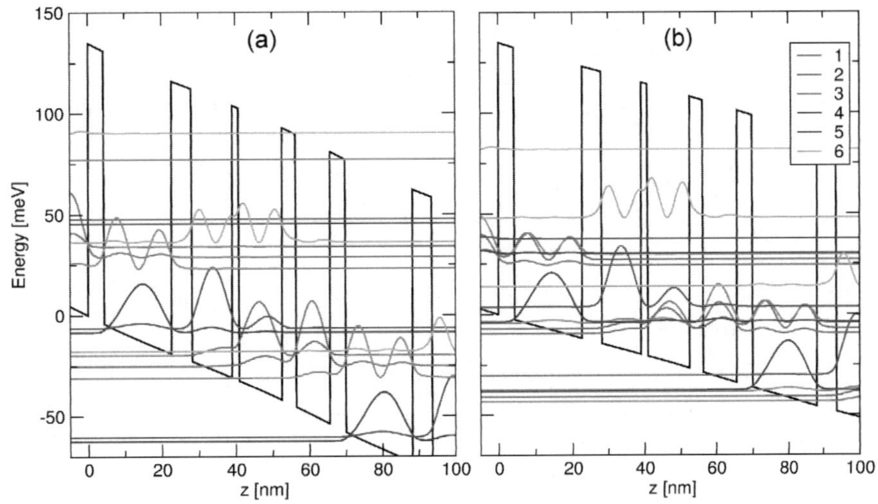

Figure 1. Band diagram of a four-well QCL from Ref. [23]. Here the conduction band edge is shown as solid black lines spanning a periodic sequence of quantum wells typical for the QCL. The barrier material is $Al_{0.15}Ga_{0.85}As$ while the wells are pure GaAs. In the widest well dopant atoms are placed in order to n-dope the structure, giving rise to impurity scattering, whereas interface roughness scattering originates from imperfect material interfaces. Probability densities are shown for the most relevant states. This design employs resonant tunneling injection from 4 to 5, building up inversion between 5 and 2. The lower laser state 2 is efficiently depopulated via resonant tunneling out to 3 via 1, combined with phonon extraction to the starting point; state 4, shifted by one period. In (a) the alignment at design bias is shown, at a bias of 54 mV per period. In (b) the alignment at the parasitic resonance at 34 mV is shown, giving rise to the pre-peak in Fig. 2.

there is clearly an issue with vanishing carrier sheet density. In such a situation the physical picture of a continuous spectra of modes of collective oscillations is questionable as a valid representation of the actual physics.

The results of adding the new self-energy to the model is displayed for both sample A and B in Fig. 2. As a reference we show currents calculated without the SPPA self-energy for both 100 and 200 K lattice temperature. For high temperature the current increases almost linearly up to design bias, around 54 mV as depicted in Fig. 1(a), however for low temperature the current enters a region of Negative Differential Resistance (NDR) after $\sim$ 38 mV for both samples. After entering this region it is no longer possible to converge the simulations; that is why the black line abruptly stops. The lack of convergence is attributed to artificially low lifetimes of the states due to the lack of important scattering mechanisms. In order to reach convergence in the self-consistent process, the energy resolution would have to be increased significantly. This is currently being investigated but the higher numerical complexity prevents us from showing any results here.

The pre-peak feature occurs at a low bias, before there is sufficient gain to start laser operation and it indicates that some important scattering is missing in the model as this is not the behavior seen in experiments. For sample A, a threshold current of 175 A/cm^2 is observed for 10 K heatsink temperature and this gradually increases to 430 A/cm^2 for the maximum operating temperature. The structure should thus be able to reach design bias also for low temperature. The NDR in both samples occurs close to the bias matching a potential drop of the energy of one optical phonon, which is about 36 meV for AlGaAs/GaAs systems. This strongly favors the

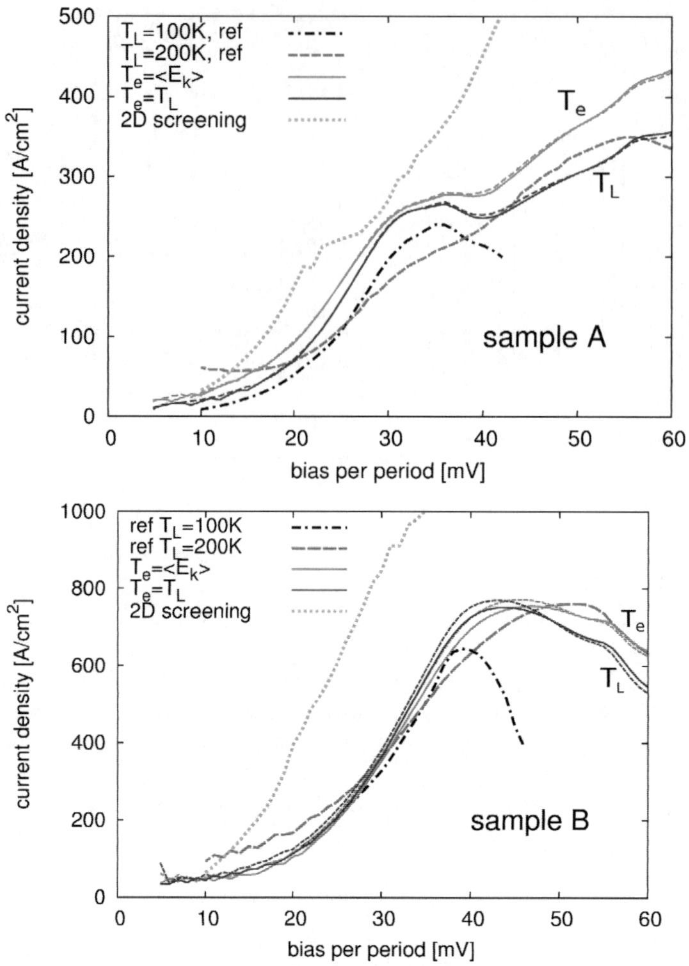

Figure 2. Current simulations for sample A and B in the respective plots. Reference simulations are shown as black (dot-dashed) and red (long-dashed) lines for 100 K and 200 K lattice temperature, respectively. Simulations with the SPPA self-energy at 100 K, are shown as blue and magenta lines for 3D screening using Eq. (9) and orange (square-dotted) lines for 2D screening using Eq. (8). Simulations without the exchange shift is indicated by short-dashed lines.

parasitic current channel shown in Fig. 1(b).

When the SPPA self-energy is included in the low temperature simulations the current changes drastically, and employing a 2D or 3D screening model makes a big difference. The orange lines in Fig. 2 shows simulations with 2D screening at a lattice temperature of 100K, and the current continues off the scale and is much too large compared to the experimental data. The reason for this is that the plasmon dispersion has a finite value for the $q \to 0$ limit for 3D but tends to zero in the 2D case, as shown in Eqs. (8-9). This effectively gives additional screening for the 3D case as the plasmon quanta, limited downwards by the plasma energy, always have to be exchanged with the collective modes that are assumed to be in thermal equilibrium. For the 2D case, there is nothing limiting the system from exchanging plasmons with vanishing energy, making both the occupation factor in Eqs. (15-16) and the Coulomb matrix element tend to large values. In the long wavelength ($q \to 0$) limit the system should tend to a 3D electron

	Sample	ρ^{2D}	J_{NDR}(10 K)	J_{max}($\sim$ 150 K)
Exp.	A(EV1157)	3.7×10^{10} cm^{-2}	225 A cm^{-2}	430 A cm^{-2}
	B(N907)	1.1×10^{11} cm^{-2}	810 A cm^{-2}	920 A cm^{-2}
Theory	A	3.7×10^{10} cm^{-2}	465 A cm^{-2}	460 A cm^{-2}
	B	1.1×10^{11} cm^{-2}	710 A cm^{-2}	800 A cm^{-2}

Table 1. Experimental results from Ref. [23] compared to simulations. Here we list, from left to right, sample name, sheet density, current at the NDR feature where lasing stops at low temperature and lastly the maximum current at the temperature, observed in experiments, where the laser ceases to operate.

gas, which is respected by the 3D but not the 2D screening model. This, together with the observation that the 2D screening model gives an unrealistic current contribution, is the reason for focusing the rest of our discussion on the 3D screening model.

A few different cases are displayed in Fig. 2 for the simulations with the SPPA self-energy with 3D screening at $T_L = 100$ K. Qualitatively we observe better agreement as the simulations are similar to the reference case at 200 K, which is expected. The quantative agreement is shown in Table 1. Here we use a lattice temperature 30 K higher than the *heatsink* temperature reported in the experiment. No laser field is included, only off-state simulations are shown. The agreement is reasonable except for the J_{NDR}(10K) for sample A. For sample B the agreement appears to be better but as we consider off-state current and on-state current probably is higher, this might be too large as well. The experimental decrease in current with lattice temperature is thus not seen in the simulations. The way we interpret this is that the possibility of exchanging plasmons with the environment opens up new transport channels for the electrons, and that this effect actually is larger than the hypothesized effect of increased dephasing which would decrease the tunneling currents, such as the parasitic resonance at the NDR peak. This effect plays a larger role at low temperature where phonon scattering is weak.

In Fig. 2 different limits of plasmon-phonon coupling are shown, indicated by T_L and T_e. In case of a strong coupling of the plasmon and phonon baths the plasmons would be quickly cooled by the phonons. This is shown as blue lines where the electron temperature has been set equal to the lattice temperature T_L. The opposite would be a weak coupling, and here the plasmons are only coupled indirectly through the mean electron temperature calculated as $T_e = \langle E_k \rangle / k_B$ with respect to the bottom of each subband. This is shown as magenta lines in Fig. 2. Due to the occupation factors in Eqs. (15-16) a higher electron temperature increases the strength of the SPPA self-energy. The impact of switching the exchange self-energy off is shown by the dashed line of the respective color.

In Fig. 3 the imaginary part of the retarded self-energy is plotted as a function of energy. This is directly related to the linewidths of the single particle states. One finds that the effect of this self-energy on the linewidth is below one meV, for all the states.

4. Conclusions

The screened electron-electron interaction has been included in a crude GW approximation via the SPPA, including dynamical screening. This leads to increased scattering in our model and a behavior that better represents the data from experimental studies, showing that these scattering mechanisms are indeed an important source of inelastic scattering when the other mechanisms of this type, such as acoustic phonon scattering, are weak.

Exchange effects are shown to play a minor role only at higher doping concentrations, while the choice of coupling of the plasmon and phonon baths has a larger impact. We observe that a 3D screening model gives reasonable results due to the finite value of the plasmon energy in

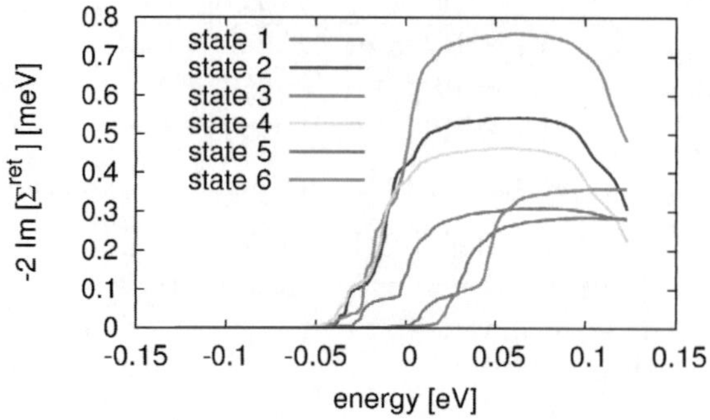

Figure 3. Here the imaginary parts of the diagonal elements of the retarded SPPA self-energy is shown for simulations at design bias. The state indexing follows the one in Fig. 1. Compared to other scattering mechanisms, the SPPA gives a small contribution to the linewidth: about 7% of the total at this bias. The lattice temperature is 100 K.

the low q limit, and we conclude that it also preserves the long wavelength limit. In contrast, the 2D screening model largely overestimates current and does not respect the long wavelength limit.

Acknowledgments

Financial support from the COST Action MP1204 in order to execute Short Term Scientific missions is thankfully acknowledged, as is support from the Swedish Research Council (VR).

References

[1] Faist J, Capasso F, Sivco D L, Sirtori C, Hutchinson A L and Cho A Y 1994 *Science* **264** 553–556
[2] Williams B S 2007 *Nat. Photonics* **1** 517
[3] Belkin M A and Capasso F 2015 *Phys. Scripta* **90** 118002
[4] Khanal S, Zhao L, Reno J L and Kumar S 2014 *J. Opt.* **16** 094001
[5] Jirauschek C and Kubis T 2014 *Appl. Phys. Rev.* **1** 011307
[6] Callebaut H, Kumar S, Williams B S, Hu Q and Reno J L 2004 *Appl. Phys. Lett.* **84** 645
[7] Manenti M, Compagnone F, Di Carlo A and Lugli P 2003 *J. Comput. Electron.* **2** 433–437
[8] Jirauschek C and Lugli P 2008 *phys. stat. sol. (c)* **5** 221–224
[9] Lundqvist B 1967 *Phys. Kondens. Mater.* **6** 193–205
[10] Ando T, Fowler A B and Stern F 1982 *Rev. Mod. Phys.* **54** 437–672
[11] Schmielau T and Pereira M 2009 *Appl. Phys. Lett.* **95** 231111
[12] Wacker A, Lindskog M and Winge D 2013 *Sel. Top. in Quantum Electron., IEEE Journal of* **19** 1200611
[13] Kubis T, Yeh C, Vogl P, Benz A, Fasching G and Deutsch C 2009 *Phys. Rev. B* **79** 195323
[14] Keldysh L V 1965 *Sov. Phys. JETP* **20** 1018 [Zh. Eksp. Theor. Fiz. **47**, 1515 (1964)]
[15] Kadanoff L P and Baym G 1962 *Quantum Statistical Mechanics* (New York: Benjamin)
[16] Wacker A 2002 *Phys. Rep.* **357** 1
[17] Haug H and Koch S 2004 *Quantum theory of the optical and electronic properties of semiconductors* (Singapore: World Scientific)
[18] Bonno O, Thobel J and Dessenne F 2005 *J. Appl. Phys.* **97** 043702
[19] Hedin L 1965 *Phys. Rev.* **139** A796
[20] Langreth D C 1976 *Linear and Nonlinear Electron Transport in Solids* ed Devreese J T and van Doren V E (New York: Plenum Press)
[21] Franckié M, Winge D O, Wolf J, Liverini V, Dupont E, Trinité V, Faist J and Wacker A 2015 *Opt. Express* **23** 5201–5212
[22] Winge D O, Lindskog M and Wacker A 2014 *Opt. Express* **22** 18389–18400
[23] Amanti M I, Scalari G, Terazzi R, Fischer M, Beck M, Faist J, Rudra A, Gallo P and Kapon E 2009 *New J. Phys.* **11** 125022
[24] Winge D O, Lindskog M and Wacker A 2012 *Appl. Phys. Lett.* **101** 211113

Journal of Physics: Conference Series **696** (2016) 012014 doi:10.1088/1742-6596/696/1/012014

Non-equilibrium slave bosons approach to quantum pumping in interacting quantum dots

Roberta Citro[1,2] and Francesco Romeo[1,2]

[1]Dipartimento di Fisica "E. R. Caianiello", University of Salerno, Via Giovanni Paolo II, Fisciano I-84084, Italy

[2]Spin-CNR UoS of Salerno, Fisciano I-84084, Italy

E-mail: `citro@sa.infn.it`

Abstract. We review a time-dependent slave bosons approach within the non-equilibrium Green's function technique to analyze the charge and spin pumping in a strongly interacting quantum dot. We study the pumped current as a function of the pumping phase and of the dot energy level and show that a parasitic current arises, beyond the pure pumping one, as an effect of the dynamical constraints. We finally illustrate an all-electrical mean for spin-pumping and discuss its relevance for spintronics applications.

1. Introduction

The idea of parametric quantum pumping regards a nanostructure (scattering region) connected to two external leads (reservoirs) in which a dc current is produced in absence of bias voltage by time periodic modulation of two system parameters. If the parameters change slowly as compared to all internal time scales of the system, the pumping is *adiabatic*, and the average charge per period does not depend on the details of the time dependence of the pumping parameters $X_i(t)$[1]. Using the concept of emissivity proposed by Büttiker et al.[3], Brouwer[4] related the charge pumped in a period to the derivatives of the instantaneous scattering matrix of the nanostructure with respect to the time-varying parameters. Since then, a general framework to compute the pumped charge through a conductor has been developed for noninteracting electrons[5]. The interest in the pumping phenomenon has shifted then to the experimental investigations of confined nanostructures, as quantum dots, where the realization of the periodic time-dependent potential can be achieved by modulating gate voltages applied to the structure[6]. In case of interacting electrons the computation of the pumped charge becomes rather involved and few works have addressed this issue for different systems and in specific regimes[7]. As for the case of interacting quantum dots, the pumped charge in a period was calculated by Aono[8] by exploiting the zero-temperature mapping of the Kondo problem within a non-equilibrium Green's function approach (NEGF). A very general formalism was developed in Ref.[9] where an adiabatic expansion of the self-energy based on the average-time approximation was used to calculate the dot Green's function, while a linear response scheme was employed in Ref.[10]. Alternatively to the NEGF approach, a powerful diagrammatic technique to treat the interacting quantum pumping based on a generalized-master equation method was instead proposed in Ref.[11]. In more recent years the interest on quantum pumping in interacting quantum dots has moved to the comprehension of the role of spin-orbit interaction[12]

Journal of Physics: Conference Series **696** (2016) 012014 doi:10.1088/1742-6596/696/1/012014

for achieving spin-current, and to non-adiabatic effects[13]. While the role of coupling of the quantum dot to elastic deformations was analyzed in various papers in the limit of interaction going to zero[14, 15].

Here we review a non-equilibrium slave boson approach for an interacting quantum dot within the NEGF method to study the quantum pumping effects and which was introduced in Ref.[16]. In particular we focus on the strongly interacting regime $U \to \infty$ and derive first, a time-dependent mean field equation for the slave boson and the constraint in the presence of a time-dependent tunnel barrier then, we derive the expression of the pumped current in the leads in terms of the Keldysh Green's function of the dot which depends on the dynamical constraint. In our treatment we are able to obtain an expression of the pumped current beyond the adiabatic (i.e. $\omega \to 0$) approximation, i.e. at finite frequency ω of the pump. Our main result is that the effect of the dynamical constraint generates a dynamical phase which couples to the phase of the external parameters φ giving rise to a parasitic current beyond the genuine pumping one. Finally, we discuss the proposal of a quantum pump able to produce spin current without polarized electrodes in the presence of magnetic tunnel barriers. Let us note that in the literature only few generalization of the slave-boson approach to a time-dependent case are present. Among them there is the generalization of slave-boson mean-field theory for t-J model to the time-dependent regime[19] to study the Landau quasiparticle transport and the study of nonlinear transport in QD connected to reservoirs with arbitrary strength of the Coulomb interaction under finite magnetic fields[20].

The organization of the paper is the following: In Sec.2, we introduce the model Hamiltonian and derive the effective Hamiltonian by means of the slave boson approach in the strongly interacting regime. In Sec.3 we give the dot Green's functions within the Keldysh approach and the expression of the non-equilibrium pumped current, focussing on the single photon approximation. Based on an approximate solution of the constraints equation, in Sec.4 we discuss the results obtained for the pumped current both in the case of a charge quantum pump and in the case of a spin quantum pump. The conclusions are given in Sec.5.

2. The interacting quantum dot model

We consider an interacting quantum dot (QD) coupled to noninteracting leads in which the current is generated by means of the temporal modulation of two out-of-phase gate voltages controlling the transparency of the tunneling barriers.
The Hamiltonian of the system is:

$$H = \sum_{k\sigma\alpha} \varepsilon_k c^\dagger_{k\sigma\alpha} c_{k\sigma\alpha} + \sum_\sigma \epsilon\, d^\dagger_\sigma d_\sigma + U n_\uparrow n_\downarrow + \sum_{k\sigma\alpha} (V_{k\alpha\sigma}(t) c^\dagger_{k\sigma\alpha} d_\sigma + h.c.), \tag{1}$$

where the operator $c^\dagger_{k\sigma\alpha}$ creates an electron of momentum k and spin σ in the lead α, while $d^\dagger$ is the creation operator of an electron state on the interacting quantum dot, ϵ being the energy of a single occupied electron state. The third term is the electron-electron (e-e) interaction energy U which comes into play when the dot screening length λ_s is bigger than the typical size of the quantum dot. The last term describes the tunneling between the quantum dot and the leads. We focus on the strongly interacting limit ($U \to \infty$) where the isolated QD can be only in the empty and in the singly occupied states $\{|0\rangle, |\sigma\rangle\}$ (where $\sigma = \uparrow, \downarrow$) while double occupancy is completely suppressed. By introducing the reference state $|\Omega\rangle$ and the operators b (slave-boson) and f_σ (pseudofermion) we may define: $|0\rangle = b^\dagger|\Omega\rangle$ and $|\sigma\rangle = f^\dagger_\sigma|\Omega\rangle$. Thus the creation/annihilation original operators can be written in terms of bosonic (i.e. b) and quasi-fermionic (i.e. f_σ) operators[17, 18] as: $d_\sigma \to b^\dagger f_\sigma$, $d^\dagger_\sigma \to f^\dagger_\sigma b$ with the extra constraint on the single occupation of the dot which can be introduced by a Lagrange multiplier λ.

Within the slave boson representation the Hamiltonian becomes:

$$H_{SB} = \sum_{k\sigma\alpha} \varepsilon_k c^\dagger_{k\sigma\alpha} c_{k\sigma\alpha} + \sum_\sigma \epsilon f^\dagger_\sigma f_\sigma + \sum_{k\sigma\alpha} (V_{k\alpha\sigma}(t) c^\dagger_{k\sigma\alpha} b^\dagger f_\sigma + h.c.) + \lambda(b^\dagger b + \sum_\sigma f^\dagger_\sigma f_\sigma - 1). \quad (2)$$

Here the Lagrange multiplier λ is fixed by the equation:

$$\partial_\lambda \langle H^{dot}_{SB} \rangle = 0 \rightarrow \sum_\sigma \langle f^\dagger_\sigma f_\sigma \rangle + \langle b^\dagger b \rangle - 1 = 0, \quad (3)$$

while the slave boson operator b evolves in time according to the equation of motion $i\hbar\dot{b} = [b, H_{SB}]$[21]:

$$i\hbar\partial_t b = \lambda b + \sum_{k\sigma\alpha} V_{k\alpha\sigma}(t) c^\dagger_{k\sigma\alpha} f_\sigma. \quad (4)$$

We treat the slave boson operator b within the mean field approximation ($\langle b \rangle = \mathcal{B}$, $\langle b^\dagger \rangle = \mathcal{B}^*$) and the original problem containing strong correlations is replaced by a constrained free-fermions-like theory whose dynamics is completely described by the following Hamiltonian:

$$\begin{aligned} H_{SBMF} &= H_{leads} + \sum_\sigma (\epsilon + \lambda(t)) f^\dagger_\sigma f_\sigma + \sum_{k\sigma\alpha} (V_{k\alpha\sigma}(t) \mathcal{B}^*(t) c^\dagger_{k\sigma\alpha} f_\sigma + h.c.) + \\ &+ \lambda(t)(|\mathcal{B}(t)|^2 - 1), \end{aligned} \quad (5)$$

with the dynamical equation for the constraints:

$$i\hbar\dot{\mathcal{B}}(t) = \lambda(t)\mathcal{B}(t) + \sum_{k\sigma\alpha} V_{k\alpha\sigma}(t) \langle c^\dagger_{k\sigma\alpha} f_\sigma \rangle \quad (6)$$

$$\sum_\sigma \langle f^\dagger_\sigma f_\sigma \rangle + |\mathcal{B}(t)|^2 - 1 = 0. \quad (7)$$

As shown in (5) the interaction renormalizes both the QD energy level ($\epsilon \rightarrow \epsilon + \lambda(t)$) and the tunneling amplitudes ($\bar{V}_{k\alpha\sigma}(t) \rightarrow V_{k\alpha\sigma}(t)\mathcal{B}^*(t)$). Let us note that due to the time evolution of the slave boson field, which is governed by the equation (6), a dynamical phase shift with respect to the phase of the external driving signals will appear.

3. Non-equilibrium Green's function method

3.1. The quantum dot Green's function

To derive the current flowing through the system via a quantum pumping mechanism we employ the non-equilibrium Green's functions (NEGF) formalism[22]. In our scheme of pumping we modulate the tunneling rates $\Gamma^\alpha_\sigma(t) = \Gamma^\alpha_{\sigma,0} + \Gamma^\alpha_{\sigma,\omega} \sin(\omega t + \varphi_\alpha)$ defined by $\Gamma^\alpha_\sigma(t) = 2\pi\rho_\alpha |V_{k_F\alpha\sigma}(t)|^2$. The retarded GF of the quantum dot (QD) uncoupled from the external leads is ($\hbar = 1$):

$$g^r_{sp}(t,t') = -i\delta_{sp}\theta(t-t') \exp\{-i \int_{t'}^t dt_1 (\epsilon + \lambda(t_1))\}, \quad (8)$$

being $s, p \in \{\uparrow, \downarrow\}$ the spin index. It has a form similar to that of the non-interacting system except for the appearance of the Lagrange multiplier $\lambda(t_1)$ in the exponential. When the QD is coupled to the leads, we must take into account the transition rate of an electron through the system and thus:

$$G^r_{sp}(t,t') = g^r_{sp}(t,t') \exp\{-\int_{t'}^t dt_1 \sum_{\alpha=l,r} \frac{\Gamma^\alpha_s(t_1)}{2} |\mathcal{B}(t_1)|^2\}, \quad (9)$$

where the exponential comes from the retarded self-energy:

$$\Sigma^r_{sp}(t_1,t_2) = -i\delta_{sp}\delta(t_1-t_2)\sum_{\alpha=l,r}\frac{\Gamma^\alpha_s(t_1)}{2}|\mathcal{B}(t_1)|^2, \tag{10}$$

which depends on the renormalization factor $|\mathcal{B}(t)|^2$. The corresponding advanced quantities, i.e. G^a and Σ^a, are computed using the symmetry properties of the two-times Green's function and are obtained directly from the relation $\chi^a_{sp}(t_1,t_2) = \chi^r_{ps}(t_2,t_1)^*$, χ being G or Σ. Finally, the Langreth rules[23] can be employed to compute the *lesser* self-energy as

$$\Sigma^<_{sp}(t_1,t_2) = i\delta_{sp}f(t_1-t_2)\sum_{\alpha=l,r}\Gamma^\alpha_s(t_1)|\mathcal{B}(t_1)|^2, \tag{11}$$

where $f(t_1-t_2)$ is the anti-Fourier transform of the Fermi function $f(E)$ (notice that the chemical potential is the same for both the leads).

In order to calculate the current flowing through the QD we need to calculate the *lesser* GF $G^<_{sp}(t_1,t_2)$ of the QD exploiting the Keldysh equation[24] and by using $G^{r/a}$ and $\Sigma^<$. Together with the single particle Green's function we need yet to solve the constraint equation. Explicitly we can rewrite Eq.(6) and (7) as:

$$\begin{aligned} i\partial_t\mathcal{B}(t) &= \Big[\lambda(t)+\sum_\sigma G^<_{\sigma\sigma}(t,t)\frac{\Gamma_\sigma(t)}{2}\Big]\mathcal{B}(t)+ \qquad (12)\\ &+ \sum_\sigma\int dt_1 G^r_{\sigma\sigma}(t,t_1)\mathcal{B}(t_1)\Gamma_\sigma(t_1)f(t_1-t)\\ |\mathcal{B}(t)|^2 &= 1+i\sum_\sigma G^<_{\sigma\sigma}(t,t). \qquad (13)\end{aligned}$$

Previous equations along with the H_{SBMF} completely describe the physics of the interacting quantum dot in the infinite-U limit.

3.2. Non-equilibrium currents

Within the non-equilibrium theory the spin resolved current pumped in the lead α can be written in terms of the QD Green's function in the following form:

$$\begin{aligned} I^\sigma_\alpha(t) &= \frac{2e}{\hbar}\sum_{\sigma'}Re\Big\{\int\frac{dE_1dE_2dE_3}{(2\pi)^3}e^{i(E_3-E_1)t}[G^r_{\sigma\sigma'}(E_1,E_2)\Sigma^<_{\alpha,\sigma'\sigma}(E_2,E_3)+ \qquad (14)\\ &+ G^<_{\sigma\sigma'}(E_1,E_2)\Sigma^a_{\alpha,\sigma'\sigma}(E_2,E_3)]\Big\},\end{aligned}$$

where the two-times Fourier transform of the GF are needed in a non-equilibrium situation since translational time invariance is broken. Here we are interested in deriving the pumping current in an almost adiabatic situation, i.e. at small ($\omega \ll U/\hbar$) but finite frequency of the pump, thus we expand all the relevant time dependent quantities by considering only the so called *single-photon* contributions as follows:

$$\begin{aligned} \lambda(t) &= \sum_{n=0,\pm1}\lambda_n\exp\{-in\omega t\} \qquad (15)\\ \Gamma^\alpha_\sigma(t) &= \sum_{n=0,\pm1}\Gamma^\alpha_{\sigma;n}\exp\{-in\omega t\}\\ \mathcal{B}(t) &= \sum_{n=0,\pm1}\mathcal{B}_n\exp\{-in\omega t\},\end{aligned}$$

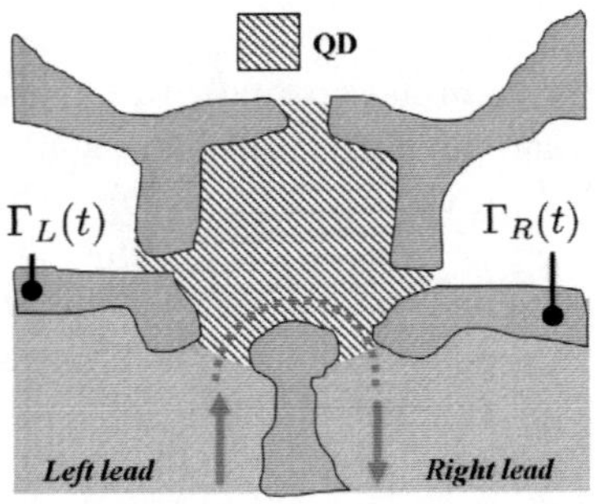

Figure 1. The system described in the main text: A Quantum dot (dashed region) coupled to external leads via tunnel barriers whose transparency can be modulated in time via top gates.

while the Green's function and the self-energy become of the form:

$$\chi^j_{sp}(E_1, E_2) = \delta_{sp} \sum_{\eta=0,\pm1} \chi^j_{s;\eta}(E_1)\delta(E_1 - E_2 + \eta\omega), \tag{16}$$

where χ is G or Σ, $j =<, r, a$. As will be carefully explained in Sec. 3.4, the single photon approximation is well justified under the assumptions of (a) adiabatic ($\omega \ll U/\hbar$) driving signals and (b) small modulation amplitudes (weak pumping). Assuming simultaneous validity of (a) and (b), the structure of the effective Hilbert space, lacking of the double occupied state of the quantum dot, is suitable to get a correct description of the time-dependent case. Within the single-photon approximation, the spin resolved dc current $\bar{I}^\sigma_\alpha$ pumped in the lead α takes the form:

$$\bar{I}^\sigma_\alpha = (2e/\hbar) \sum_{\eta=\pm1} Re[\int \frac{dE_1}{(2\pi)^3}[G^r_{\sigma;\eta}(E_1)\Sigma^<_{\alpha\sigma;-\eta}(E_1 + \eta\omega) + G^<_{\sigma;\eta}(E_1)\Sigma^a_{\alpha\sigma;-\eta}(E_1 + \eta\omega)], \tag{17}$$

which contains information about the pumping cycle and the absortion/emission processes of one photon. Within the single-photon approximation the behavior of the pumped current is mainly determined by integrals of products of the function:

$$\mathcal{D}^\sigma_\eta(E) = [E + \eta\omega - (\epsilon + \lambda_0) + i(\gamma_\sigma/2)]^{-1}, \tag{18}$$

where the $\eta = 0, \pm1$, λ_0 is the static part of the Lagrange multiplier and γ_σ is a renormalized linewidth depending on the slave boson field:

$$\gamma_\sigma = \sum_\alpha \Big[|\mathcal{B}_0|^2\Gamma^\alpha_{\sigma;0} + 2Re\{|\mathcal{B}_0|^2\Gamma^\alpha_{\sigma;1} + \Gamma^\alpha_{\sigma;0}(\mathcal{B}^*_0\mathcal{B}_1 + \mathcal{B}_0\mathcal{B}^*_{-1})\}\Big]. \tag{19}$$

As shown in (19) the finite frequency effects modifies the mean lifetime of the electron on the QD due to the interference effects originated by the slave boson field $\mathcal{B}_{\pm1}$. Moreover, let us note that the Fourier components of the slave boson field are a complex quantities, $\mathcal{B}_{\pm1,0} = |\mathcal{B}_{\pm1,0}| \exp\{i\phi^{\pm,0}_b\}$, i.e. they can be expressed in term of a phase, thus interference terms of the form $\cos(\phi^\pm_b - \phi^0_b)$ appear in $\mathcal{D}^\sigma_\eta(E)$. The existence of such dynamical phases (and the corresponding interference terms) provides a decoherence source for the electron transport. Furthermore, since $\Gamma^\alpha_{\sigma;\pm1}$ picks up a phase due to the external driving signals $\Gamma^\alpha_{\sigma;\pm1} = \pm i\frac{\Gamma^\alpha_{\sigma,\omega}}{2} \exp\{\mp i\varphi_\alpha\}$, the GF of the problem depends separately, on $\varphi_{L,R}$ and not on the phase difference $\varphi = \varphi_R - \varphi_L$. This will imply the presence of *parasitic* currents, as we will see below.

3.3. Constraint equations within the single photon approximation

As we have seen above, the current pumped through the quantum dot can be calculated from the Green's function. Since it depends parametrically on $\{\lambda_0, \lambda_{\pm 1}; \mathcal{B}_0, \mathcal{B}_{\pm 1}\}$ one has first to solve the constraint equations. By exploiting the Fourier expansions of Eqs.(12)-(13) within the one-photon approximation one gets a set of coupled equations that can be approximately solved. In particular, one considers the corrections beyond the adiabatic limit. From the analysis of the parametric expression of the current pumped one observes that the relevant quantities to be computed are: λ_0, λ_1, $|\mathcal{B}_0|^2$ and $\mathcal{B}_0^*\mathcal{B}_1 + \mathcal{B}_0\mathcal{B}^*_{-1}$. Since we are modulating the tunneling amplitudes via top gates, we can derive an approximate expression of $\lambda_{\pm 1}$ and $\mathcal{B}_0^*\mathcal{B}_1 + \mathcal{B}_0\mathcal{B}^*_{-1}$ to the first order in $\Gamma_{\sigma;1}$. The constraint equations to the lowest order are:

$$\lambda_0 = -\sum_\sigma \Gamma_{\sigma;0} \int \frac{dE}{2\pi} f(E) Re\{\mathcal{D}_0^\sigma(E)\} \tag{20}$$

$$|\mathcal{B}_0|^2 = 1 - \sum_\sigma |\mathcal{B}_0|^2 \Gamma_{\sigma;0} \int \frac{dE}{2\pi} f(E) |\mathcal{D}_0^\sigma(E)|^2.$$

These equations have the same structure as in the adiabatic case[9] but implicitly depend on $\mathcal{B}_{\pm 1}$ contained in $\mathcal{D}_0^\sigma(E)$, thus higher order constraints have to be considered in order to solve the problem. In particular, the equation which needs to be considered is the one for $\mathcal{B}_0^*\mathcal{B}_1 + \mathcal{B}_0\mathcal{B}^*_{-1}$ whose formal solution is[16]:

$$\mathcal{B}_0^*\mathcal{B}_1 + \mathcal{B}_0\mathcal{B}^*_{-1} = \frac{-\sum_\sigma |\mathcal{B}_0|^2 \Gamma_{\sigma;1} \int \frac{dE}{2\pi} \mathcal{D}_0^\sigma(E) \mathcal{D}_{-1}^{\sigma*}(E) f^{(-)}(E)}{1 + \sum_\sigma \Gamma_{\sigma;0} \int \frac{dE}{2\pi} \mathcal{D}_0^\sigma(E) \mathcal{D}_{-1}^{\sigma*}(E) f^{(-)}(E)}, \tag{21}$$

where $f^{(\pm)}(E) = f(E \pm \omega)$. Expanding the r.h.s. of Eq.(21) up to the first order in $\Gamma_{\sigma;1}$ and taking the limit $\omega \to 0$, the dependence on $\mathcal{B}_{\pm 1}$ disappears and thus Eqs.(20)-(21) can be solved independently from the other equations. Following a similar procedure, one can also solve the equation for λ_1 whose solution is:

$$\lambda_1 \approx \frac{\lambda_0}{2} \sum_\sigma |\mathcal{B}_0|^2 \Gamma_{\sigma;1} \int \frac{dE}{2\pi} |\mathcal{D}_0^\sigma(E)|^2 f(E) \Big|_{\mathcal{B}_{\pm 1} \to 0}. \tag{22}$$

The solution for the constraints are then plugged in the expression (17) to numerically obtain the current.

3.4. Validity of the non-equilibrium slave-boson approximation

Before presenting the results for the pumped current, let us comment on the validity of the time-dependent slave boson approximation. It is well established that the slave boson method works well in the strong U-limit for a static problem. As for the time-dependent case, we consider the weak-modulation (small amplitude of the pumping parameter) and almost adiabatic pumping limit which ensure that the double occupied quantum dot state remains empty, while the net electron flow is driven from one reservoir to another by absorbing or emitting an energy quantum $\bar{~}h\omega \ll U$ from the driving signals. In general, an arbitrary time-dependent driving signal allows multiple photon-assisted tunneling (mPAT) processes where the absorption/emission of n photons takes place, with an absorbed/emitted energy $n\hbar\omega$. When $n\hbar\omega$ is comparable to U, the quantum dot double occupation becomes allowed and the slave boson picture described in our work is not applicable. However, under the simultaneous assumptions of: (a) $\hbar\omega \ll U$ and (b) small amplitude modulation of the pumping signals the time-dependent slave boson procedure should provide a (non-trivial) perturbation of the static slave boson theory. Here, the assumption (b) ensures that multi-photon processes are less probable compared to the single-photon absorption/emission. It's also interesting to notice that for a small quantum dot (tens

Journal of Physics: Conference Series **696** (2016) 012014 doi:10.1088/1742-6596/696/1/012014

of nanometers in length) the charging energy U takes values in the meV range. This implies that $U/\hbar$ takes value 1.5 THz (assuming $U=$ 1 meV) and thus pumping frequencies ω up to the GHz range (i.e. the relevant experimental range) can be safely considered as adiabatic. Within this framework the single photon approximation is well justified and the structure of the Hilbert space, lacking of the double occupied quantum dot state, remains similar to that of the static case. In this respect, the time-dependent slave boson theory presented here provides small, but relevant, perturbations to the well established slave-boson static method. In general, the method is not applicable to the generic time-dependent case (strong amplitude pumping, non-adiabatic signals).

4. The pumped current

4.1. The charge current

Here we discuss the results of the current pumped in the right-lead by assuming $\Gamma_0 = 10\ \mu eV$ as unit of energy, while the current is normalized by $i_0 = -e\Gamma_0/\hbar$ ($i_0 \approx 2.56$ nA) and the temperature is fixed to zero.

In Fig.2 we show in the panel (d) the current pumped in the right lead I_R (continuous line) as a function of the position of the bare level of the dot ϵ and by setting the remaining parameters as follows: $\Gamma^{\alpha=L/R}_{\sigma;0} = 1.5$, $\Gamma^{\alpha=L/R}_{\sigma;\omega} = 0.05$, $\omega = 0.05$, $\varphi_L = 0$, $\varphi_R = \pi/4$. Moreover, the tunneling does not depend on the spin, and thus $I_R = I^{\sigma}_R$, i.e. the current is spin-independent. The lower curve in panel (d) of Fig.2 represents the contribution to the current due to the modulation of the constraint $\lambda(t)$, while the upper curve is the current pumped by the modulation of the tunnel barriers. As shown when $\epsilon \gg \mu$ the current induced by $\lambda(t)$ is vanishing and the total current is dominated by the contribution of the tunnel amplitudes; on the other hand, when the energy level of the QD ϵ is much lower then the Fermi energy, the current induced by the modulation of $\lambda(t)$ is the dominating one. The behavior of the current shown in Fig.2 can be explained when looking at the behavior of the slave boson parameters as a function of $(\epsilon-\mu)/\Gamma_0$. Panel (a) of Fig.2 shows $|\mathcal{B}_0|^2$ *vs* $(\epsilon - \mu)/\Gamma_0$ which is proportional to $(1-n)$ where n is the average occupation on the QD. The average occupation is going to zero when the energy level ϵ is well above the Fermi energy and the tunneling amplitudes is weakly renormalized, while a strong renormalization is obtained in presence of a non-vanishing electron density on the QD, i.e. when its occupation is close to one. A similar effect is evident in panel (b) of Fig.2 where λ_0 is plotted as a function of $(\epsilon - \mu)/\Gamma_0$. When $\epsilon \leq \mu$ the effect of the constraints becomes dominant. Another effect of the constraints becomes clear when we look at the absolute value of the oscillation amplitude of $\lambda(t)$, i.e. $|\lambda_1|$ (see panel (c)). As shown, below the Fermi energy, $|\lambda_1|$ becomes comparable to the oscillation amplitudes of the tunneling rates Γ^{α}_{σ} ($\approx 0.05\Gamma_0$) and thus the pump starts to be affected by the dynamics of the constraints. Indeed, comparing panel (c) and (d) one observes that the λ-induced current (red curve) is enhanced when $|\lambda_1|$ is close to its maximum value.

In Fig. 3 the pumped charge current in the left and right lead is shown as a function of the pumping phase φ_R. Differently from the adiabatic ($\omega \to 0$) and noninteracting case, the external phase difference $\varphi_R - \varphi_L$ is not the only relevant phase. Actually, in the strongly interacting limit terms of the form $\sin(\varphi_\alpha \pm (\phi^{\pm}_b - \phi^0_b))$ appear in the charge current. These interaction-induced terms are of the same order of magnitude of the perturbation induced by the external driving signals and thus strongly affect the usual $\sin(\varphi)$-behavior. In particular, the modification of the current-phase dependence due to the interaction produces rectification contributions ($I_{rect}(\varphi_L = 0, \varphi_R = 0) \neq 0$) similar to those reported in a different context in Ref. [25]. Here we do observe that at varying the bare QD energy level ϵ the system is guided through the resonance ($\epsilon+\lambda_0$) and the current changes its sign (see lower red dashed line). Thus in the strongly interacting case the presence of new phase differences in the pumped current is the main modification compared to the free-fermion case.

Journal of Physics: Conference Series **696** (2016) 012014 doi:10.1088/1742-6596/696/1/012014

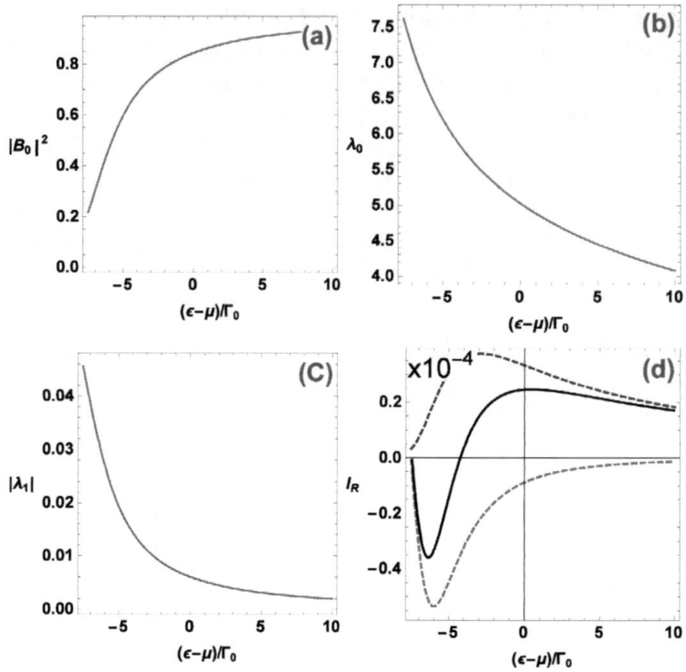

Figure 2. The top-panels (a)-(b) show the holes density $|\mathcal{B}_0|^2$, and λ_0 as a function of $(\epsilon-\mu)/\Gamma_0$, while the lower panels (c)-(d) show $|\lambda_1|$ and the total charge current I_R (full line) as a function of $(\epsilon-\mu)/\Gamma_0$ computed setting the remaining parameters as follows: $\Gamma_{\sigma;0}^{\alpha=L/R} = 1.5$, $\Gamma_{\sigma;\omega}^{\alpha=L/R} = 0.05$, $\omega = 0.05$, $\varphi_L = 0$, $\varphi_R = \pi/4$. The dashed lower curve (red) in panel (d) represents the current induced by the modulation of λ, while the upper curve (dashed-blue curve) depends only on the modulation of the tunneling amplitudes.

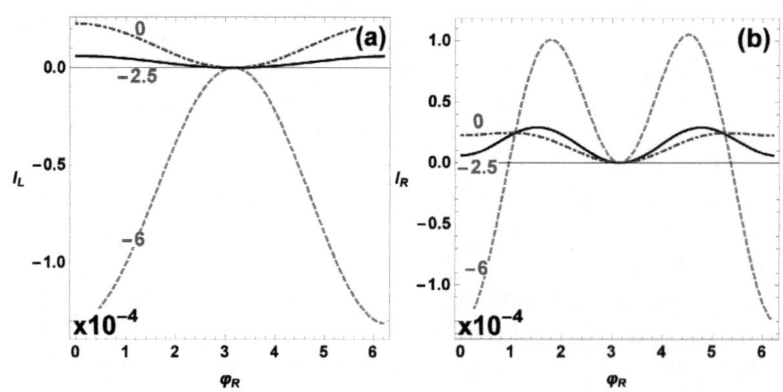

Figure 3. Charge current I_L (left panel) and I_R (right panel) as a function of the pumping phase φ_R computed fixing the remaining parameters as follows: $\Gamma_{\sigma;0}^{\alpha=L/R} = 1.5$, $\Gamma_{\sigma;\omega}^{\alpha=L/R} = 0.05$, $\omega = 0.05$, $\varphi_L = 0$. For both panels the curves are obtained for different values of the QD energy level $(\epsilon-\mu)/\Gamma_0$ as indicated in the figure labels. Data are rescaled by a factor 10^{-4}.

4.2. Spin pumping

A spin-pump can be obtained when the tunneling amplitudes depend explicitly on the spin e.g. as: $\Gamma^{\alpha}_{\sigma}(t) = \Gamma^{\alpha}(t)(1+p\sigma)$, where p is a polarization. This situation can be realized when the QD is connected to the external non-magnetic leads via thin magnetic tunnel barriers[26, 27]. The magnetic barriers, in fact, work as a spin-filter (SF) whose effect is controlled by the polarization $p \in [0, 1]$ of the barriers. We assume that the polarization p is independent on time, while the tunneling amplitude $\Gamma^{\alpha}(t)$ is taken of the form:

$$\Gamma^{\alpha}(t) = \Gamma^{\alpha}_{0} + \Big[\frac{\Gamma^{\alpha}_{\omega}\exp(i\varphi_{\alpha})}{2i}\exp(i\omega t) + c.c.\Big]. \tag{23}$$

In the presence of the ac modulations of the external gates, along with the spin-selective action of the barriers, a spin current $I_s = \sum_{\sigma} \sigma I_{\sigma}$ is generated beyond charge current $I_c = \sum_{\sigma} I_{\sigma}$. In Fig.4 we study the charge (full line) and spin (dashed-dotted line) currents as a function of the bare energy level ϵ of the QD setting the remaining parameters as in the figure caption. For non-vanishing polarizations (i.e. $p = 0.4$) of the barriers a spin current is observed. Furthermore when $\epsilon \approx \mu - 5\Gamma_0$ the charge current takes negligible values and the pump works as a pure spin current pump (Fig.4a). Let us note that the pure spin current obtained here is generated in absence of polarized leads thus avoiding the problem of the spin injection. Our analysis shows that an all-electrical control of the spin current is possible and is potentially useful in spintronics devices.

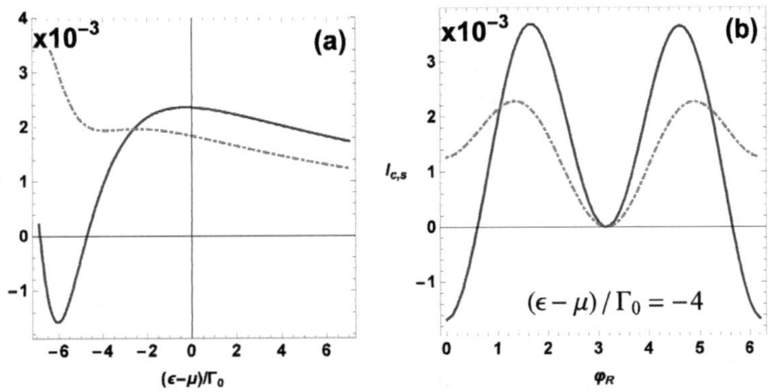

Figure 4. Charge current I_c (full line) and spin current I_s (dashed-dotted line) pumped in the right lead as a function of $(\epsilon - \mu)/\Gamma_0$ (left panel) and of the phase φ_R (right panel) by setting the remaining parameters as follows: $\Gamma^{\alpha=L}_{0} = 1.5$, $\Gamma^{\alpha=R}_{0} = 1.3$, $\Gamma^{\alpha=L/R}_{\omega} = 0.3$, $\omega = 0.05$, $p = 0.4$, $\varphi_L = 0$. In the left panel the phase bias is fixed as $\varphi_R = \pi/4$. Data are rescaled by a factor 10^{-3}.

5. Conclusions

We reviewed a time-dependent slave-bosons approach within a non-equilibrium Green's function formalism to study the quantum pumping of charge and spin in an interacting quantum dot in the limit of infinite Coulomb repulsion. The pump based mechanism is the time periodic modulation of two tunnel barriers whose transparencies are governed by the external top gates. In particular, we showed an equation of motion of the slave boson field $\mathcal{B}(t)$ which is equivalent to an infinite series of constraints, generalizing the adiabatic case. By using a finite set of constraints (single photon approximation) the expressions of the relevant Green's functions and

of the current have been derived. We showed that the pumped charge current contains, beyond a term related to the modulation of the out-of-phase external parameters, an additional term due to the internal dynamics of the Lagrange multiplier and the slave boson field (parasitic pumping current). While for a noninteracting quantum dot model the current depends only on the phase difference between the two external parameters, $\varphi_R - \varphi_L$, in the strongly interacting limit it depends separately on $\varphi_{L/R}$. The additional pumping contributions are originated by the phase difference of the external parameter phase (for instance φ_R) and the phase of an internal degrees of freedom (the slave boson field or the Lagrange multiplier). This mechanism is similar to that obtained in the presence of dynamical effects of a mechanical degree of freedom coupled to the QD[14].
Finally we showed how in the case of spin-dependent tunnel barriers with a finite polarization, the pump works as a spin current generator. Furthermore, by changing the energy level of the QD the pump can work as a pure spin current generator. This effect shows the possibility of all-electrical control of the spin current pumped through a QD and is relevant for spintronics.

ACKNOWLEDGMENTS

R. C. acknowledges financial support from the project FIRB-2012-HybridNanoDev (Grant No. RBFR1236VV).

6. References

[1] Thouless D (1983) *Phys. Rev. B* **27**, 6083
[2] Altshuler B and Glazman L (1999), *Science* **283**, 1864
[3] Büttiker M, Thomas H, and Prêtre A (1994), *Z. Phys. B* **94**, 133
[4] Brouwer P (1998), *Phys. Rev. B* **58**, 10135
[5] Zhou F, Spivak B, and Altshuler B (1999), *Phys. Rev. Lett.* **82**, 608; Makhlin Yu and Mirlin A D (2001), *Phys. Rev. Lett.* **87**, 276803; Entin-Wohlman O, Aharony A, and Levinson Y (2002), *Phys. Rev. B* **65**, 195411; Moskalets M and Büttiker M (2002), *Phys. Rev. B* **66**, 035306; *ibid.***66**, 205320
[6] Switkes M, Marcus C M, Campman K, and Gossard A C (1999), *Science* **283**, 1905; Watson S K, Potok R M, Marcus C M, and Umansky V (2003), *Phys. Rev. Lett.* **91**, 258301
[7] Pothier H, Lafarge P, Urbina C, Esteve D, and Devoret M H (1992), *Europhys. Lett.* **17**, 249; Aleiner I L and Andreev A V (1998), *Phys. Rev. Lett.* **81**, 1286; Citro R, Andrei N, and Niu Q (2003), *Phys. Rev. B***68**, 165312; Brouwer P W, Lamacraft A, and Flensberg K (2005), *Phys. Rev. B* **72**, 075316
[8] Aono T (2004), *Phys. Rev. Lett.* **93**, 116601 (2004)
[9] Splettstoesser J, Governale M, König J, and Fazio R (2005), *Phys. Rev. Lett.* **95**, 246803
[10] Sela E and Oreg Y (2006), *Phys. Rev. Lett.* **96**, 166802
[11] Splettstoesser J, Governale M, König J, Fazio R (2006), *Phys. Rev. B***74**, 085305; Cavaliere F, Governale M, König J (2009), *Phys. Rev. Lett.* **103**, 136801; Winkler N, Governale M, König J (2009), *Phys. Rev. B* **79**, 235309
[12] Rojek S, König J, Shnirman A (2013), *Phys. Rev. B* **87**, 075305
[13] Croy A, Saalmann U, Hernndez A R, and Lewenkopf C H (2012), *Phys. Rev. B* **85**, 035309
[14] Romeo F and Citro R (2009), *Phys. Rev. B* **80** 235328
[15] Perroni C A, Romeo F, Nocera A, Marigliano Ramaglia V , Citro R, Cataudella V (2014), *J. Phys.: Condens. Matter* **26** 365301
[16] Romeo F and Citro R, (2010) *Phys. Rev. B* **82**, 165321
[17] Coleman P (1984), *Phys. Rev. B* **29**, 3035
[18] Finnemore D K, Shelton R N, Clem J R, McCallum R W, Ku H C, McCarley R E, Chen S C, Klavins P, and Kogan V (1987), *Phys. Rev. B* **35**, 5319
[19] Ng T K (2004) *Phys. Rev. B* **69**, 125112
[20] Dong B and Lei X L (2001) *Phys. Rev. B* **63**, 235306
[21] Wu B H and Cao J C (2008), *Phys. Rev. B* **77**, 233307
[22] Jauho A P, Wingreen N S, and Meir Y (1994), *Phys. Rev. B* **50**, 5528
[23] Langreth D C, in: J. T. Devreese, E. van Doren (Eds.), *Linear and Nonlinear Electron Transport in Solids*, (Plenum Press, New York, 1976).
[24] Haug H and Jauho A P, *Quantum Kinetics in Transport and Optics of Semiconductors*, (Springer, Berlin, 1996)

[25] Braun M and Burkard G (2008), *Phys. Rev. Lett.* **101**, 036802
[26] Ruzdiński w and Barnaś J (2001), *Phys. Rev. B* **64**, 085318
[27] Souza F M, Egues J C, and Jauho A P (2007), *Phys. Rev. B* **75**, 165303
[28] Miao Guo-Xing, Müller M, and Moodera J S (2009), *Phys. Rev. Lett.* **102**, 076601

Green's function approach to the non-equilibrium superconductivity near the critical line

Pavel Lipavský

Faculty of Mathematics and Physics, Charles University, KB Karpov 3, 12116 Prague 2, Czech Republic

E-mail: lipavsky@karlov.mff.cuni.cz

Abstract. In spite of the absent friction of super-currents, normal currents affect the superconducting condensate. The BCS approach with Bogoliubov-Valutin quasiparticles is not suited for description of the normal current near the critical line. We review an alternative theory of the non-equilibrium superconductivity dealing exclusively with normal-state quasiparticles. It is based on the Thouless T-matrix criterion, which is extended to non-equilibrium. The problem of selfconsistency of the T-matrix in the superconducting state is examined and solved with the help of the multiple-scattering theory.

1. Introduction

The non-equilibrium superconductivity is traditionally described with the help of Green's functions. As Gor'kov has shown, the theory of Bardeen, Cooper and Schrieffer (BCS) is recovered if the set of usual single-particle functions is extended by anomalous functions correlating two creation (annihilation) operators of opposite spin. These anomalous functions vanish in the normal state while in the superconducting state they are finite and carry the essential part of physical properties being responsible for the non-dissipative current.

The presence of anomalous functions results in the gap Δ in the single-particle energy spectrum

$$E_{\mathbf{k}}^{(1,2)} = \sqrt{\epsilon_{\mathbf{k}}^2 + |\Delta|^2}, \tag{1}$$

where $\epsilon_{\mathbf{k}} = k^2/2m - k_{\mathrm{F}}^2/2m$. Superscripts remind that there are two branches of equal energy corresponding to two spin branches in the normal state. Single-particle excitations with energies (1) are called Bogoliubov-Valutin quasiparticles. As one can see, in the BCS approach the gap is a part of the unperturbed effective Hamiltonian, accordingly it is suited for description of well developed superconducting states sufficiently far from the critical temperature.

The phenomenological theory of Ginzburg and Landau (GL) approaches the superconductivity from the other side. Its derivation is restricted to a narrow vicinity of the critical temperature, where the BCS gap is very small and its effect can be treated in the power expansion. The free energy of Ginzburg and Landau goes to the power $|\Delta|^4$.

In spite of different regions on applicability, Gor'kov [1] has recovered the GL theory from anomalous Green's functions as the limit of small Δ. In equilibrium this limit is well defined and Gor'kov approach provides sound microscopic justification of the phenomenological theory.

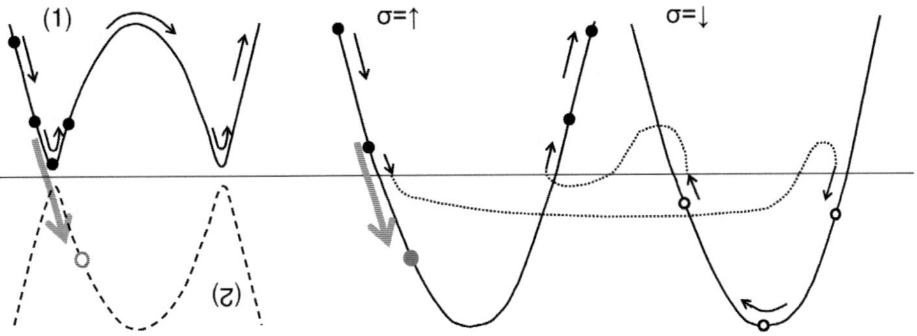

Figure 1. Motion of the Bogoliubov-Valutin quasiparticle in the energy band. The quasiparticle $\Psi_{\mathbf{k}}^{(1)\dagger}$ driven by a time-dependent vector potential, $\mathbf{k} = \mathbf{k}_0 + e\mathbf{A}$, adiabatically (thin black arrows) follows the quasiparticle energy $E^{(1)}$ in the left panel. The central and right panel show the same motion in the normal-state energy band. The Bogoliubov-Valutin quasiparticle starts as an electron of momentum $|\mathbf{k}| > k_{\mathrm{F}}$ and spin $\uparrow$. Crossing the Fermi level it changes into a hole of momentum $-\mathbf{k}$ with $|-\mathbf{k}| < k_{\mathrm{F}}$ and spin $\downarrow$. The reverse conversion happens as it leaves the Fermi sphere. By a non-adiabatic Zener-type tunneling across the gap (grey thick arrow in the left panel) the Bogoliubov-Valutin quasiparticle emerges as a quasi-hole with energy $|\mathbf{k}| < k_{\mathrm{F}}$ in the $E^{(2)}$-branch (dashed line rotated by 180°). This Zener-type tunneling corresponds to the continuous motion of the electron with spin $\uparrow$ (grey thick arrow in the central panel).

Out of equilibrium the situation becomes less transparent. Problems appear on both sides. The GL phenomenology does not apply since there are no well defined thermodynamical functions. Moreover, even if one constructs an approximative thermodynamical function, the state of the system is not in its minimum and the dynamics of the relaxation process towards the minimum is not covered by variational methods. The non-equilibrium version of the GL theory thus has to be derived microscopically [2, 3, 4].

A convenient microscopic picture of processes in the non-equilibrium superconductor provides the distribution of Bogoliubov-Valutin quasiparticles. Corresponding time evolution is covered by the kinetic equation of Boltzmann type in which the $E_{\mathbf{k}}^{(1,2)}$ enter as the quasiparticle energy. Bogoliubov-Valutin quasiparticles in superconductors are, however, confusing entities, see Fig. 1. Unlike in the Fermi liquid, their charge depends on momentum [5]

$$q_{\mathbf{k}} = -e\frac{\epsilon_{\mathbf{k}}}{E_{\mathbf{k}}}. \tag{2}$$

For $\epsilon_{\mathbf{k}} \gg |\Delta|$, the Bogoliubov-Valutin quasiparticle $\Psi_{\mathbf{k}}^{(1)\dagger}$ is like an electron with spin $\uparrow$ and momentum $\mathbf{k}$. Reducing its momentum to $\epsilon_{\mathbf{k}} \ll -|\Delta|$ it turns into a hole of spin $\downarrow$ and momentum $-\mathbf{k}$. Such quasiparticle picture is justified only for changes which are slow on the time scale of the inverse BCS gap. Under these conditions, the cross element between quasiparticles on branches $E^{(1)}$ and $E^{(2)}$ vanishes by dephasing, because it oscillates with frequency $E^{(1)}+E^{(2)}$, which is greater than $2|\Delta|$ and thus fast on the scale of changes. The adiabatic evolution of Bogoliubov-Valutin quasiparticles holds only for sufficiently large gap, which is not satisfied near the critical temperature.

For a small gap or fast perturbations the cross element has to be included. The coupling between branches results in the evolution of one quasiparticle into its counterpart, say quasiparticle goes from branch $E^{(1)}$ into branch $E^{(2)}$. In the dialect of the historical semiconductor model of superconductors, this jump corresponds to the Zener tunneling across the superconducting gap. The tunneling reveals peculiarities of the Bogoliubov-Valutin representation: the quasiparticle $\Psi_{\mathbf{k}}^{(1)\dagger}$ tunnels from its branch into the quasihole $\Psi_{-\mathbf{k}}^{(2)}$ in the second branch.

The above described complicated Zener-type tunneling corresponds to the usual evolution of Landau quasiparticle $\Psi_{\mathbf{k}}^{\uparrow\dagger}$ in the normal state. Indeed, $\Psi_{\mathbf{k}}^{(1)\dagger} = \Psi_{\mathbf{k}}^{\uparrow\dagger}$ for $\epsilon_{\mathbf{k}} \gg |\Delta|$ and $\Psi_{-\mathbf{k}}^{(2)} = \Psi_{\mathbf{k}}^{\uparrow\dagger}$ for $\epsilon_{\mathbf{k}} \ll -|\Delta|$. In the critical region, where the system approaches the normal state and the Δ becomes very small, perturbations are fast compared to $1/|\Delta|$ so that the tunneling prevails over adiabatical evolution in the vicinity of the energy gap. In such situation the gap function Δ should not be included in the unperturbed Hamiltonian. We have to leave the Bogoliubov-Valutin representation and formulate the superconductivity in terms of the 'normal' Landau quasiparticles treating the gap as a perturbation. Expansion in powers of Δ is closer in spirit to the original approach of Ginzburg and Landau.

In this paper we do not benefit from the anomalous functions which are ultimately linked with the Bogoliubov-Valutin representation. Instead we formulate the theory of superconductivity with the help of the T-matrix approximation of the selfenergy. The present approach is advantageous in the critical region, because the single-particle Green function in the limit of small Δ naturally matches the normal-state function and a corresponding evolution of single-particle excitations agrees with the customary Landau quasiparticle picture.

2. Thouless criterion

The theory dealing exclusively with 'normal' quasiparticles was proposed by Thouless [6] soon after discovery of the Cooper pairing. He described the pairing of electrons in the ladder approximation and analyzed properties of the T-matrix. When system reaches conditions for the onset of the superconductivity, the T-matrix becomes singular. The Thouless criterion

$$\frac{1}{T^R(0,\mathbf{0})} = 0 \tag{3}$$

establishes critical temperature of the superconducting state from the retarded T-matrix in the normal state for pair energy $\Omega = 0$ and pair momentum $\mathbf{Q} = \mathbf{0}$.

Unfortunately, the selfconsistent ladder approximation (known as the Galitskii approximation in solid state physics [7] and called standard approximation in the non-relativistic nuclear physics) fails in the superconducting state, because it does not yield the BCS gap. The Thouless criterion thus merely provides conditions in which the superconducting condensate nucleates, but it cannot be used to evaluate the magnitude of the gap even in the asymptotic region $\Delta \to 0$. In the rest of this section we generalize the Thouless criterion to non-equilibrium. In the next section we show how to overcome problems of the Galitskii approximation and evaluate the magnitude of the gap.

2.1. Galitskii approximation

Recently the Thouless criterion has been employed under non-equilibrium conditions [8]. To treat non-local contributions to the nucleating condensate, the criterion has to be expressed in the space-time variables. The ladder approximation of the T-matrix satisfies

$$T^R(12,34) = V(1234) + V(12\bar{5}\bar{6})G^R_{(2)}(\bar{5}\bar{6},\bar{7}\bar{8})T^R(\bar{7}\bar{8},34), \tag{4}$$

where V is the pairing potential, $G_{(2)}$ is the two-particle propagator, numbers are cumulative arguments $1 \equiv \mathbf{r}_1, t_1$, groups of arguments restricted to equal time are not separated by comma, and bar denotes space-time integrals. We tacitly assume that spin $\uparrow$ is associated with odd arguments and $\downarrow$ with even ones.

In the Galitskii approximation the selfenergy is the mean value of the T-matrix obtained by averaging over accessible interaction partners $\Sigma(1,3) = -iT(1\bar{2},3\bar{4})G(\bar{4},\bar{2})$. This time-contour relation yields the retarded selfenergy

$$\Sigma^R(1,3) = T^R(1\bar{2},3\bar{4})G^<(\bar{4},\bar{2}) - T^<(1\bar{2},3\bar{4})G^A(\bar{4},\bar{2}). \tag{5}$$

Analytic pieces of the T-matrix T^R and $T^<$ become singular under identical conditions. Their amplitudes of the singularity in the pole, however, essentially differ. While the retarded function has usual pole which appears with any bounded state, the pole of $T^<$ is enhanced by the Bose-Einstein condensation. Indeed, in the equilibrium $T^<(\Omega,\mathbf{Q}) = -2\text{Im}\ T^R(\Omega,\mathbf{Q})/(e^\Omega - 1)$. The distance of the pole from zero behaves as 1/volume so that the occupation of the condensation mode $\mathbf{Q} = \mathbf{0}$ is proportional to the volume of the sample, while occupations of other modes are independent of volume. To describe the effect of the condensate on the single-particle spectrum it is sufficient to keep the more divergent term

$$\Sigma^R(1,3) \approx -T^<(1\bar{2},3\bar{4})G^A(\bar{4},\bar{2}). \tag{6}$$

The correlation part of the T-matrix satisfies

$$T^<(12,34) = T^R(12,\bar{5}\bar{6})G^<_{(2)}(\bar{5}\bar{6},\bar{7}\bar{8})T^A(\bar{7}\bar{8},34). \tag{7}$$

When the T-matrix becomes singular, its value is much larger than the potential, $|T^R| \gg V$, therefore from (4) follows

$$T^R(12,34) \approx V(12\bar{5}\bar{6})G^R_{(2)}(\bar{5}\bar{6},\bar{7}\bar{8})T^R(\bar{7}\bar{8},34). \tag{8}$$

Using approximation (8) in Eq. (7) we obtain

$$T^<(12,34) = V(12\bar{5}\bar{6})G^R_{(2)}(\bar{5}\bar{6},\bar{7}\bar{8})T^<(\bar{7}\bar{8},34). \tag{9}$$

The set of equations for the Galitskii approximation is completed by the Dyson equation $G^R = G^R_0 + G^R_0\Sigma^R G^R$ and the generalized Kadanoff-Baym equation $G^< = G^R\Sigma^< G^A$, where $\Sigma^<(1,3) = T^<(1\bar{2},3\bar{4})G^>(\bar{4},\bar{2})$. We will focus on Eq. (9).

2.2. Linearized time-dependent Ginzburg-Landau theory

The condensate is described by the element $T^<$. One can equally well use $T^>$, because their Bose factor differ by unity which has to be compared with the macroscopic number of Cooper pairs in the system.

The BCS model of the interaction is point like, $V(1234) = V\delta(1-2)\delta(2-3)\delta(3-4)$, so that $T^<(12,34) = T^<(1,3)\delta(1-2)\delta(3-4)$ and the equation (9) simplifies as

$$T^<(1,3) = K(1,\bar{2})T^<(\bar{2},3), \tag{10}$$

where $K(1,2) = VG^R_{(2)}(11,22)$ is the integration kernel.

Following Gor'kov, we assume that the integration kernel extends over a small space and time region. This allows us to expand the gap in the lowest gradients

$$T^<(2,3) \approx T^<(1,3) + \frac{\partial T^<(1,3)}{\partial t_1}(t_2 - t_1) + \frac{\partial T^<(1,3)}{\partial \mathbf{r}_1}(\mathbf{r}_2 - \mathbf{r}_1) + \frac{1}{2}\frac{\partial^2 T^<(1,3)}{\partial \mathbf{r}_1^2}(\mathbf{r}_2 - \mathbf{r}_1)^2, \tag{11}$$

which leads to the equation

$$\alpha T^< + \Gamma\frac{\partial T^<}{\partial t_1} + \mathbf{S}\frac{\partial T^<}{\partial \mathbf{r}_1} - \frac{1}{2m^*}\frac{\partial^2 T^<}{\partial \mathbf{r}_1^2} = 0 \tag{12}$$

Journal of Physics: Conference Series **696** (2016) 012015 doi:10.1088/1742-6596/696/1/012015

with the Ginzburg-Landau (GL) parameter $\alpha(1) = K(1,\bar{2})/C(1)$, the relaxation rate $\Gamma(1) = -K(1,\bar{2})(t_{\bar{2}} - t_1)/C(1)$, the vector coefficient $\mathbf{S}(1) = K(1,\bar{2})(\mathbf{r}_{\bar{2}} - \mathbf{r}_1)/C(1)$, and the weight of the isotropic tensor $1/m^*(1) = -K(1,\bar{2})(\mathbf{r}_{\bar{2}} - \mathbf{r}_1)^2/C(1)$. The norm C is introduced for convenience of the final rearrangement.

Note that the differential equation acts only on the left argument of the correlation function. A conjugated equation can be derived for the right argument. Both equations are solved by the separable function

$$T^<(1,2) = \Delta^*(1)\Delta(2), \tag{13}$$

where the star denotes the complex conjugation. This result agrees with numerical observations that at the singularity the T-matrix becomes separable [9]. Due to the separability Eq. (12) simplifies to the linearized time-dependent Ginzburg-Landau (TDGL) equation

$$\alpha\Delta^* + \Gamma\frac{\partial\Delta^*}{\partial t} + \mathbf{S}\frac{\partial\Delta^*}{\partial\mathbf{r}} - \frac{1}{2m^*}\frac{\partial^2\Delta^*}{\partial\mathbf{r}^2} = 0. \tag{14}$$

Further progress requires to specify Green's functions. The time-contour two-particle propagator $G_{(2)}(12,34) = iG_\uparrow(1,3)G_\downarrow(2,4)$ has the retarded component

$$G^R_{(2)}(12,34) = -i\theta(t_1 - t_3)\left(G^>_\uparrow(1,3)G^>_\downarrow(2,4) - G^<_\uparrow(1,3)G^<_\downarrow(2,4)\right). \tag{15}$$

We assume all functions to be diagonal in the spin. Thouless evaluated the single-particle functions in the quasiparticle approximation of the spectral function, $A_\uparrow(\omega,\mathbf{k},\mathbf{r},t) \approx 2\pi\delta(\omega - \varepsilon_\uparrow(\mathbf{k},\mathbf{r},t))$, where $\varepsilon_\uparrow(\mathbf{k},\mathbf{r},t)$ is the energy of quasiparticle of spin $\uparrow$ and momentum $\mathbf{k}$ at point $\mathbf{r}$ and time t. Having only a single pole, the corresponding correlation functions can be expressed in the form of the Kadanoff-Baym ansatz $G^<_\uparrow(\omega,\mathbf{k},\mathbf{r},t) \approx f_\uparrow(\mathbf{k},\mathbf{r},t)\ A_\uparrow(\omega,\mathbf{k},\mathbf{r},t)$, which allows for a convenient extension to non-equilibrium.

The actual evaluation of coefficients α, Γ, $\mathbf{S}$ and m^* is rather lengthy and we refer the reader for details to [8]. Let us list four essential steps. First, the distribution f was obtained from the (normal state) Boltzmann equation in the relaxation time approximation allowing for the electric and magnetic fields, and the temperature gradient. Second, integrals giving coefficients were evaluated in the mixed Wigner representation. Except for explicit gradients one has to include internal gradients needed to shift the center of the kernel. For example $K(\mathbf{r}_1,\mathbf{r}_2) \equiv K(\mathbf{r},\xi)$ has the center coordinate $\mathbf{r} = \frac{1}{2}(\mathbf{r}_1 + \mathbf{r}_2)$ and the difference coordinate $\xi = \mathbf{r}_1 - \mathbf{r}_2$. In the TDGL equation we need material parameters at point $\mathbf{r}_1$. The coefficient thus includes the gradient correction $K(\mathbf{r}_1,\xi) \approx K(\mathbf{r},\xi) - (\partial K/\partial\mathbf{r})\xi/2$. Third, the gap function was scaled to the GL order parameter, $\psi = \Delta(\sqrt{n}/k_\mathrm{B}T)\sqrt{-\int_0^\infty dx(1/2x)(\partial^2(1/e^x+1)/\partial x^2)}$. The density n and temperature T are generally functions of time and space, giving additional gradients. In reality, these gradient cancel with internal gradient terms obtain in the second step. The equation for the order parameter thus has simpler parameters. Fourth, following Gor'kov, the space gradient were minimized by the choice of gauge so that the vector coefficient $\mathbf{S}$ vanishes. Going back to the original gauge one arrives at the linearized TDGL equation

$$\frac{1}{2m^*}\left(-i\frac{\partial}{\partial\mathbf{r}} - m^*\mathbf{v}_n - 2e\mathbf{A}\right)^2\psi + \alpha\psi = -\Gamma\left(\frac{\partial}{\partial t} - 2ie\phi\right)\psi. \tag{16}$$

In order to write the GL theory in its common form, equation (16) is the complex conjugate of Eq. (14). The electrostatic potential ϕ makes the theory gauge invariant. Units with $\hbar = 1$ are used.

Compared to studies based on the Gor'kov formalism, the Thouless approach gives a new term represented by the mean velocity of normal electrons $\mathbf{v}_n = \mathbf{j}_n/(en)$, where $\mathbf{j}_n$ is the

Journal of Physics: Conference Series **696** (2016) 012015 doi:10.1088/1742-6596/696/1/012015

electric current in the background of the nucleating condensate. Its contribution is rather small, nevertheless it can be observed via the fast far infra-red spectroscopy [10, 11] and according to numerical studies it modifies a motion of kinematic vortices [12]. In Ref. [12] it is shown that the new term is purely transient, therefore in stationary situations like in the conversion region near contacts, the term $m^*\mathbf{v}_n$ has no effect in spite of large normal currents.

3. T-matrix approaches to the superconductivity

It is desirable to extend the above treatment from the critical point into the superconducting region so that finite values of the GL order parameter or the BCS gap are covered. To this end we have to include the effect of the condensate on the single-particle spectrum and allow for the formation of the BCS gap. Let us first examine how the Galitskii approximation fails below the critical temperature.

3.1. Missing gap in the Galitskii approximation

The approximative retarded selfenergy obtains from Eqs. (6) and (13)

$$\Sigma^R(1,3) = -\Delta^*(1)\Delta(3)G^A(3,1). \tag{17}$$

Problems of the Galitskii approximation appear in equilibrium as well as out of equilibrium. In this section we restrict our discussion to the simple homogeneous equilibrium system. The gap function is then a constant which can be taken real and the selfenergy (17) is conveniently expressed in the energy-momentum representation

$$\Sigma^R(\omega, \mathbf{k}) = -\Delta^2 G^A(-\omega, -\mathbf{k}). \tag{18}$$

Similarly $\Sigma^A(-\omega,-\mathbf{k}) = -\Delta^2 G^R(\omega,\mathbf{k})$. From the Dyson equation for the retarded propagator

$$G^R(\omega,\mathbf{k}) = \frac{1}{\omega - \epsilon_{\mathbf{k}} - \Sigma^R(\omega,\mathbf{k})} = \frac{1}{\omega - \epsilon_{\mathbf{k}} + \Delta^2 G^A(-\omega,-\mathbf{k})} \tag{19}$$

and its advanced counterpart $G^A(-\omega,-\mathbf{k}) = 1/(-\omega - \epsilon_{-\mathbf{k}} + \Delta^2 G^R(\omega,\mathbf{k}))$ one finds a propagator

$$G^R(\omega,\mathbf{k}) = \frac{\omega + \epsilon_{-\mathbf{k}}}{2\Delta^2} - \sqrt{\left(\frac{\omega + \epsilon_{-\mathbf{k}}}{2\Delta^2}\right)^2 - \frac{1}{2\Delta^2}\frac{\omega + \epsilon_{-\mathbf{k}}}{\omega - \epsilon_{\mathbf{k}}}}. \tag{20}$$

This peculiar propagator is far from the simple BCS structure. The BCS spectral function has two δ-functions at $\omega = \pm E_{\mathbf{k}}$ given by energies (1). Taking for example $\epsilon_{\mathbf{k}} > 0$ and $\omega > 0$, the spectral function $-2\mathrm{Im}G^R(\omega,\mathbf{k})$ obtained from (20) has non-zero values in the wide interval of energies $\omega < \sqrt{\epsilon_{\mathbf{k}}^2 + 2\Delta^2}$. Instead of the energy gap we have obtained a continuous spectrum of states in the vicinity of the Fermi energy.

The missing BCS gap restricts applicability of Galitskii approximation to the normal state. In spite of the failure in the superconducting state, the conditions of the nucleation of the superconducting condensate result from the Galitskii approximation correctly.

3.2. Prang paradox

The gap problem was first noticed by Prang. Unfortunately, we were not able to trace down his original paper. His work is mentioned by Wild [13] who recovered Prang's result. Prang also noticed that the gap problem is cured making the loop of the selfenergy non-selfconsistent, i.e., replacing the 'even-argument' Green function by its bare value in the two-particle propagator

$$\hat{G}^R_{(2)}(12,34) = -i\theta(t_1 - t_3)\left(G^>(1,3)G^{0>}(2,4) - G^<(1,3)G^{0<}(2,4)\right) \tag{21}$$

and similarly the Green's function closing the loop

$$\hat{\Sigma}^R(1,3) = \hat{T}^R(1\bar{2},3\bar{4})G^{0<}(\bar{4},\bar{2}) - \hat{T}^<(1\bar{2},3\bar{4})G^{0A}(\bar{4},\bar{2}). \tag{22}$$

It is striking that the Galitskii approximation which is Φ-derivable and satisfies all demands on reliable theories fails while the weird theory with non-symmetrical treatment of interacting particles provides the well established result. Tolmachev calls this unexpected result Prang's paradox [14].

3.3. Kadanoff-Martin theory

The approximation tested by Prange is now known as the Kadanoff-Martin theory. This name might be misleading for newcomers, because Kadanoff and Martin [15] have derived another approximation with the fully selfconsistent two-particle T-matrix and the bare Green's function in the closing relation (22). Recently Kremp *et al* [16] have derived a renormalized version of the Kadanoff-Martin approximation. Their two-particle T-matrix is also fully selfconsistent, and the Green function in the closing relation includes binary correlations on the level of the ladder approximation with the gap excluded.

It is easy to see that non-selfconsistent closing relation (22) helps. The modified propagator (19) is

$$\hat{G}^R(\omega,\mathbf{k}) = \frac{1}{\omega - \epsilon_{\mathbf{k}} + \Delta^2 G^{0A}(-\omega,-\mathbf{k})} = \frac{1}{\omega - \epsilon_{\mathbf{k}} - \frac{\Delta^2}{\omega+\epsilon_{-\mathbf{k}}}} = \frac{\omega+\epsilon_{\mathbf{k}}}{\omega^2 - \epsilon_{\mathbf{k}}^2 - \Delta^2}, \tag{23}$$

where we have used $\epsilon_{-\mathbf{k}} = \epsilon_{\mathbf{k}}$. For a given gap Δ, this propagator agrees with the BCS propagator.

The gap is given by the kernel $K = VG_{(2)}$. The fully selfconsistent two-particle T-matrix of the original Kadanoff-Martin theory provides incorrect gap values. It can be seen near the critical temperature, where $G_{(2)}$ can be expanded in powers of Δ. Gor'kov version of the BCS theory has $G_{(2]} = G_0 G \approx G_0 G_0 + G_0 G_0 \bar{\Delta} G_0 \Delta G_0$. The second term yields the nonlinear 'potential' of the Ginzburg-Landau expansion $\hat{\beta}|\Delta|^2$, where the hat stands for normalization to Δ rather than the Ginzburg-Landau function ψ used below. In the original Kadanoff-Martin, $G_{(2]} = GG \approx G_0 G_0 + G_0 G_0 \bar{\Delta} G_0 \Delta G_0 + G_0 \bar{\Delta} G_0 \Delta G_0 G_0$, which yields twice larger 'potential' $\hat{\beta}_{\mathrm{KM}}|\Delta|^2 = 2\hat{\beta}|\Delta|^2$. All linear terms are identical, therefore the equilibrium gap obtained from the Kadanoff-Martin theory near critical temperature $|\Delta|^2 = -\hat{\alpha}/\hat{\beta}_{\mathrm{KM}} = -\hat{\alpha}/2\hat{\beta}$ is by factor $1/\sqrt{2}$ smaller than the BCS value.

Using Kadanoff-Martin out of equilibrium one can expect various artifacts, because the loop line of includes matching of energies in unequal approximation which harms the energy conservation. It was corrected ten years later by Patton [17]. In his unpublished thesis he proposed to use the bare functions also in the loop line of the T-matrix. This additional non-selfconsistency converts the Kadanoff-Martin theory into the approximation discussed by Prang which yields the BCS gap.

To obtain reasonable results from the Kadanoff-Martin theory even with Patton's correction, one has to neglect all other parts of the selfenergy except for the singular term $\hat{\Sigma}^R(1,3) = -\Delta^*(1)\Delta(3)G^{0A}(3,1)$. One can imagine non-physical properties which would result taking the theory literally. For example, the dressed propagator includes the full exchange-correlation selfenergy while the bare propagator depends on the mean-field electrostatic potential only. Energy bands in the mean-field approximation are far from the real ones being mostly shifted upwards. Due to the upward shift the bare Green's function yields a much lower electron density (for the fixed chemical potential). Since the bare function corresponds to the state of spin opposite to the dressed one, the approximation has a fake spin non-symmetry similar to the

Journal of Physics: Conference Series **696** (2016) 012015 doi:10.1088/1742-6596/696/1/012015

non-symmetry observed in ferromagnetic states. Like in magnetic systems, the different values of the Fermi momentum would destroy the Cooper pairing. These problems are corrected in the renormalized version of Kremp *et al* [16]. Our theory explained below can be simplified to the theory of Kremp *et al* with the Patton's restricted selfconsistency in the T-matrix.

3.4. Selfconsistency out of equilibrium

Out of equilibrium an implementation of the T-matrix approximation with the non-selfconsistent loop is hindered by its dependence on the non-equilibrium correlation function $G^{0<}$. It is easy to write down equation for $G^{0<}$ and eventually solve it, but such function has nothing to do with the real system. In the equilibrium one needs only propagators which represent a short-time behavior and all errors are forgotten on the quasiparticle life time. The non-equilibrium requires to solve the transport equation, where errors cumulate. To obtain a meaningful distribution of particles one has to work with fully selfconsistent correlation functions.

As long as one does not introduce additional non-systematic step like $G^{0<}$ constructed from the actual selfconsistent distribution of electrons and bare spectral function, the Kadanoff-Martin theory cannot be used out of equilibrium. It is possible to introduce the requested selfconsistency following the approach of Kremp *et al* based on expected macroscopic properties of the condensate. Here we start from microscopic considerations.

3.5. Pauli principle

Langer [18] argues that the problem of the Galitskii approximation in the superconducting state is hidden in unbalanced treatment of direct and exchange processes. Why the problem emerges only in the superconducting state?

Let us go back to the power expansion in the second quantization and assume a group of potentials from which we aim to evaluate the selfenergy. If this group includes a combination $\Psi^\dagger_{\uparrow\mathbf{k}}\Psi^\dagger_{\uparrow\mathbf{k}}\Psi_{\uparrow\mathbf{k}}\Psi_{\uparrow\mathbf{k}}$, it is zero in agreement with the Pauli principle. In the diagrammatic expansion, such contribution vanishes by cancelation of a direct and exchange process, as one can see from the decoupling of four operators

$$\langle\Psi^\dagger_{\uparrow\mathbf{k}'}\Psi^\dagger_{\uparrow\mathbf{k}}\Psi_{\uparrow\mathbf{k}}\Psi_{\uparrow\mathbf{k}'}\rangle = \langle\Psi^\dagger_{\uparrow\mathbf{k}'}\Psi_{\uparrow\mathbf{k}'}\rangle\langle\Psi^\dagger_{\uparrow\mathbf{k}}\Psi_{\uparrow\mathbf{k}}\rangle - \langle\Psi^\dagger_{\uparrow\mathbf{k}'}\Psi_{\uparrow\mathbf{k}}\rangle\langle\Psi^\dagger_{\uparrow\mathbf{k}}\Psi_{\uparrow\mathbf{k}'}\rangle, \tag{24}$$

which is non-zero for $\mathbf{k} \neq \mathbf{k}'$ but vanishes when $\mathbf{k} = \mathbf{k}'$. Neglect of the exchange process correspond to the neglect of the second term in decoupling (24).

In the normal state the neglect of exchange processes is well justified. The second term of decoupling (24) is zero except for $\mathbf{k} = \mathbf{k}'$, we can thus split dual sums over momenta as

$$\frac{1}{L^3}\sum_{\mathbf{k}}\frac{1}{L^3}\sum_{\mathbf{k}'}\dots = \frac{1}{L^3}\sum_{\mathbf{k}}\frac{1}{L^3}\sum_{\mathbf{k}\neq\mathbf{k}'}\dots + \frac{1}{L^6}\sum_{\mathbf{k}=\mathbf{k}'}\dots \tag{25}$$

For the $\Psi^\dagger_{\mathbf{k}}$ normalized to the plane wave $e^{i\mathbf{k}\cdot\mathbf{r}}$, sums over momenta are weighted with the inverse quantization volume $1/L^3$. The number of discrete $\mathbf{k}$-states in the Brillouin zone is proportional to L^3 so that the first term remains finite while the second one vanishes in the thermodynamical limit $L \to \infty$.

In the presence of the superconducting condensate the scaling with volume breaks. The Bose-Einstein occupation factor of the state with the condensate increases linearly with the volume so that the contribution of this single state remains finite under the thermodynamical limit. The neglect of exchange diagrams than might cause serious problems, in particular the missing superconducting gap.

Keeping non-zero $\Psi^\dagger_{\uparrow\mathbf{k}}\Psi^\dagger_{\uparrow\mathbf{k}}\Psi_{\uparrow\mathbf{k}}\Psi_{\uparrow\mathbf{k}}$ inside selfenergy results in various direct or indirect self-interactions. The self-interaction is in general a weak point of all diagrammatic approaches. It appears already in the mean-field, which includes the Coulomb action of a particle on itself. It eventually cancels with the self-exchange included in the diagrammatic Fock term.

It is often easy to avoid the self-interaction if we leave ties of diagrammatic rules. For example, the Coulomb self-interaction is absent in the Hartree potential. It should be stated that diagrammatic expansions are mostly the most convenient approach. Using the Hartree approximation as an example, we can see that it is formally a problem of a particle in the electrostatic potential, but unlike in the mean-field approximation each particle feels a different potential. For N particles one has to solve the Schrödinger equation with N different Hartree potentials. Moreover, the Hartree single-particle states do not form the orthogonal basis.

A systematic treatment of exchange diagrams on the level of the selfconsistent T-matrix is technically impossible. Langer [18] performed an alternative way of summation of fully anti-symmetrized power expansion and obtained the gap. His approach has no followers likely because of its complexity. Author [19] has eliminated indirect self-interactions due to missing exchange terms using the idea of the Watson-Lacelace-Fadeev multiple scattering theory. Morawetz [20] followed the Pauli principle and recovered the resulting theory from the decoupling of the three-particle correlation function.

4. Restricted selfconsistency

Among the above listed T-matrix theories of superconductivity, the multiple scattering approach has the structure best suited to describe the non-equilibrium superconductors. We will follow this line.

The multiple scattering theory aims to view binary encounters of particles as completed scattering events. Such viewpoint is clearly defined in two limiting cases. First, one can recollect orders of the interaction potential into sequence of individual binary events. These events are not affected by other particles, in fact, in strongly coupled system the binary collision is terminated by intervention of a third particle which takes over one of partners. Two selected particles collide again and again as they meet in the system. Corresponding T-matrices of binary events are constructed from bare propagators. This is the structure of the Fadeev equations, which are useful for very small systems (vast majority of application are just for three particles). It was not solved for the infinite system.

Second, one can use principles of the multiple scattering theory to eliminate double-counts of standard selfconsistent approximations. Such attitude was promoted by Soven [21] who showed that the Averaged T-matrix Approximation (ATA) of the scattering on impurities is overly selfconsistent, what limits its applicability to low concentrations. He proposed a simple correction scheme based on the multiple scattering theory. Like in the ATA, the mean scattering by randomly distributed impurities is described by the selfenergy. When one evaluates the T-matrix for the collision with some impurity, the selfenergy is reduced not to include the selected impurity (or the selected lattice site). In this way he obtained the Coherent Potential Approximation (CPA) which is as simple as the ATA having number of superior properties [22]. With respect to the superconductivity it is noteworthy, that the CPA describes a formation of a gap in the split-band limit, while the ATA fails.

In spite of the reduced selfconsistency, the CPA is based on the fully selfconsistent propagator. Accordingly, it can be extended to non-equilibrium systems [23], where the full selfconsistency is required. Following [19] we adopt Soven's approach to eliminate double-counts from the Galitskii approximation.

The Galitskii selfenergy (5) in the energy-momentum representation

$$\Sigma_{\uparrow}^{R}(\omega,\mathbf{k}) = \sum_{\mathbf{Q}} \sigma_{\uparrow\mathbf{Q}}^{R}(\omega,\mathbf{k}) \tag{26}$$

Journal of Physics: Conference Series **696** (2016) 012015 doi:10.1088/1742-6596/696/1/012015

is a sum over pair momentum $\mathbf{Q}$ of components

$$\sigma^R_{\uparrow\mathbf{Q}}(\omega,\mathbf{k}) = \int d\Omega\, T^R(\Omega,\mathbf{Q})G^<_\downarrow(\Omega-\omega,\mathbf{Q}-\mathbf{k}) - \int d\Omega\, T^<(\Omega,\mathbf{Q})G^A_\downarrow(\Omega-\omega,\mathbf{Q}-\mathbf{k}). \quad (27)$$

The Galitskii T-matrix is constructed from the single-particle Green's functions including $G^R_\uparrow$ which depends on $\sigma^R_{\uparrow\mathbf{Q}}$. The Galitskii approximation is thus overly selfconsistent. When one evaluates $\sigma^R_{\uparrow\mathbf{Q}}$, this $\mathbf{Q}$-component of the selfenergy ought to be eliminated from the effective medium not to act on itself.

4.1. Restricted T-matrix on the time contour

Let us allow for more general conditions in which perturbations make the energy-momentum representation of limited use. Nevertheless, we assume that we are able to identify channels of the T-matrix which we label by $\mathbf{Q}$. The selfenergy on the time contour is a sum over these channels

$$\Sigma_\uparrow(1,3) = \sum_{\mathbf{Q}} \sigma_{\uparrow\mathbf{Q}}(1,3). \quad (28)$$

The time integral run along the time contour. The subsidiary propagator

$$G_{\uparrow\mathbf{Q}}(1,3) = G_\uparrow(1,3) - G_\uparrow(1,\bar{5})\sigma_{\uparrow\mathbf{Q}}(\bar{5},\bar{7})G_{\uparrow\mathbf{Q}}(\bar{7},3) \quad (29)$$

is free of $\sigma_{\uparrow\mathbf{Q}}$. Indeed, from the Dyson equation

$$G_\uparrow(1,3) = G^0(1,3) + G^0(1,\bar{5})\Sigma_\uparrow(\bar{5},\bar{7})G_\uparrow(\bar{7},3) \quad (30)$$

one finds $G_{\mathbf{Q}} = G^0 + G^0(\sum_{\mathbf{Q}'\neq\mathbf{Q}}\sigma_{\mathbf{Q}'})G_{\mathbf{Q}}$. Using Eq. (29) we can construct the subsidiary propagator as a function of the dressed Green function G.

The T-matrix for each channel is in the ladder approximation

$$T_{\uparrow\mathbf{Q}}(1,2) = V + VG^{(2)}_{\uparrow\mathbf{Q}}(1,\bar{3})T_{\uparrow\mathbf{Q}}(\bar{3},2), \quad (31)$$

with internal $\uparrow$ lines free of the self-action

$$G^{(2)}_{\uparrow\mathbf{Q}}(1,2) = iG_{\uparrow\mathbf{Q}}(1,2)G_\downarrow(1,2). \quad (32)$$

Solving the ladder equation we obtain its $\mathbf{Q}$ channel. In the simple case of the homogeneous system, the $\mathbf{Q}$ component is identical to the Fourier component of the T-matrix. In the sum this component has a weight proportional to the inverse volume. Elimination of a single channel thus vanishes in the thermodynamical limit, volume $\to\infty$, unless the channel has macroscopic occupation due the Bose-Einstein condensation of Cooper pairs.

By averaging over collision partners we obtain the component of the averaged T-matrix

$$t_{\uparrow\mathbf{Q}}(1,3) = -iT_{\uparrow\mathbf{Q}}(1,3)G_\downarrow(3,1). \quad (33)$$

The T-matrix explores all orders of the interaction, therefore the same process cannot influence initial and final states of the scattering. The dressed Green's function thus reads

$$G_\uparrow(1,3) = G_{\uparrow\mathbf{Q}}(1,3) + G_{\uparrow\mathbf{Q}}(1,\bar{5})t_{\uparrow\mathbf{Q}}(\bar{5},\bar{7})G_{\uparrow\mathbf{Q}}(\bar{7},3). \quad (34)$$

The $\mathbf{Q}$ component of the selfenergy obtains from equations (29) and (34)

$$\sigma_{\uparrow\mathbf{Q}}(1,3) = t_{\uparrow\mathbf{Q}}(1,3) - \sigma_{\uparrow\mathbf{Q}}(1,\bar{5})G_{\uparrow\mathbf{Q}}(\bar{5},\bar{7})t_{\uparrow\mathbf{Q}}(\bar{7},3). \quad (35)$$

Journal of Physics: Conference Series **696** (2016) 012015 doi:10.1088/1742-6596/696/1/012015

The set of equations on the time contour is closed. Let us double-check its structure. Starting from some iteration of the selfenergy one evaluates the dressed Green's function via the Dyson equation (30) and the subsidiary Green's function via (29). These functions are used to prepare the two-particle propagator (32) from which one evaluates the **Q** channel of the two-particle T-matrix (31). Closing the upper loop of the two-particle T-matrix yields the averaged T-matrix (33) from which one obtains a higher iteration of the **Q** component of the selfenergy via (35). Evaluating all **Q** components the higher iteration of the selfenergy (28) is completed.

4.2. Non-equilibrium restricted T-matrix on the real time axis

Now we are ready to fulfill our basic aim to formulate a theory of non-equilibrium superconductivity which can be expanded in orders of the gap function. We convert the above set of equations into set of equations on the real axis using the generalized Kadanoff-Baym formalism in the notation of Ref. [24]. We will not write down advanced functions which are hermitian conjugates of their retarded counterparts, and the 'greater' correlation functions which are linked to corresponding 'smaller' function by the particle-hole symmetry, $>\longleftrightarrow<$. The spin $\downarrow$ functions obtain by the interchange $\downarrow\longleftrightarrow\uparrow$.

The selfenergy (28) is a sum

$$\Sigma_\uparrow^{<,R} = \sum_{\mathbf{Q}} \sigma_{\uparrow\mathbf{Q}}^{<,R} \tag{36}$$

with retarded components

$$\sigma_{\uparrow\mathbf{Q}}^R = t_{\uparrow\mathbf{Q}}^R - \sigma_{\uparrow\mathbf{Q}}^R G_{\uparrow\mathbf{Q}}^R t_{\uparrow\mathbf{Q}}^R \tag{37}$$

and correlation parts

$$\sigma_{\uparrow\mathbf{Q}}^< = \left(1 - t_{\uparrow\mathbf{Q}}^R G_{\uparrow\mathbf{Q}}^R\right) t_{\uparrow\mathbf{Q}}^< \left(1 - G_{\uparrow\mathbf{Q}}^A t_{\uparrow\mathbf{Q}}^A\right) - t_{\uparrow\mathbf{Q}}^R G_{\uparrow\mathbf{Q}}^< t_{\uparrow\mathbf{Q}}^A. \tag{38}$$

The T-matrix (33) has correlation parts

$$t_{\uparrow\mathbf{Q}}^<(1,3) = T_{\uparrow\mathbf{Q}}^<(1,3) G_\downarrow^>(3,1) \tag{39}$$

and retarded

$$t_{\uparrow\mathbf{Q}}^R(1,3) = T_{\uparrow\mathbf{Q}}^R(1,3) G_\downarrow^<(3,1) - T_{\uparrow\mathbf{Q}}^<(1,3) G_\downarrow^A(3,1). \tag{40}$$

The single-particle propagator obeys the Dyson equation

$$G_\uparrow^R = G^{0R} + G^{0R}\Sigma_\uparrow^R G_\uparrow^R \tag{41}$$

and the correlation part the generalized Kadanoff-Baym equation

$$G_\uparrow^< = G_\uparrow^R \Sigma_\uparrow^< G_\uparrow^A. \tag{42}$$

The retarded subsidiary function (29) has the retarded propagator

$$G_{\uparrow\mathbf{Q}}^R = G_\uparrow^R - G_\uparrow^R \sigma_{\uparrow\mathbf{Q}}^R G_{\uparrow\mathbf{Q}}^R \tag{43}$$

and the correlation part

$$G_{\uparrow\mathbf{Q}}^< = \left(1 - G_{\uparrow\mathbf{Q}}^R \sigma_{\uparrow\mathbf{Q}}^R\right) G_\uparrow^< \left(1 - \sigma_{\uparrow\mathbf{Q}}^A G_{\uparrow\mathbf{Q}}^A\right) - G_\uparrow^R \sigma_{\uparrow\mathbf{Q}}^< G_\uparrow^A. \tag{44}$$

The correlation part of the two-particle propagator (32) is

$$G_{\uparrow\mathbf{Q}}^{(2)<}(1,2) = G_{\uparrow\mathbf{Q}}^<(1,2) G_\downarrow^<(1,2) \tag{45}$$

giving the retarded propagator

$$G^{(2)R}_{\uparrow\mathbf{Q}}(1,2) = -i\theta(t_1-t_2)\left(G^{(2)>}_{\uparrow\mathbf{Q}}(1,2) - G^{(2)<}_{\uparrow\mathbf{Q}}(1,2)\right). \tag{46}$$

From the two-particle propagator one evaluates the retarded part of the two-particle ladder T-matrix (31)

$$T^R_{\uparrow\mathbf{Q}} = V + VG^{(2)R}_{\uparrow\mathbf{Q}}T^R_{\uparrow\mathbf{Q}}, \tag{47}$$

and its correlation part

$$T^<_{\uparrow\mathbf{Q}} = T^R_{\uparrow\mathbf{Q}}G^{(2)<}_{\uparrow\mathbf{Q}}T^A_{\uparrow\mathbf{Q}}. \tag{48}$$

The set of equations in this section is complete.

4.3. Non-equilibrium superconductivity

In the homogeneous system one can easily eliminate all self-actions. In the inhomogeneous case, the decomposition of the T-matrix becomes complicated. Treatment of all channels is important for small systems like quantum dots [25], but for bulk superconductors it is an unnecessary complication. The only channel, in which the self-action has to be eliminated is the condensation channel. The T-matrix approach in general allows for parallel condensation in many channels. The multi-gap structure might appear very far from equilibrium [26], but in weakly perturbed systems the situation is different. As shown in Ref. [27] the condensation in one channel blocks a condensation in all other channels. This is important for the stability of the condensate. Indeed, in the presence of persistent currents there are channels of smaller or none current, where the condensate would lead to a lower total energy of the system. Blocking of the parallel condensation makes the persistent current stable.

Let us simplify the theory neglecting the restricted selfconsistency in all but condensate channel. As shown above, the Bose-Einstein condensation leads to enhancement of a single element of the correlation part of the T-matrix. This element is separable, therefore

$$T^< = \Delta^*\Delta + \tilde{T}^<, \tag{49}$$

where the second term is a regular part. The retarded T-matrix has no singularity and satisfies the ladder equation of the Galitskii theory, $T^R = V + VG^R_{(2)}T^R$. The regular correlation is also identical to the Galitskii theory $\tilde{T}^< = T^RG^<_{(2)}T^A$.

Again, like in the Galitskii theory the two-particle function is constructed from dressed functions $G^<_{(2)}(1,2) = G^<_\uparrow(1,2)G^<_\downarrow(1,2)$ and $G^R_{(2)}(1,2) = -i\theta(t_1-t_2)\left(G^>_{(2)}(1,2) - G^<_{(2)}(1,2)\right)$. Since the full selfconsistency blocks condensation, the $\tilde{T}^<$ remains a regular function.

According to separation (49), the selfenergy splits into the 'superconducting' and regular part

$$\Sigma^R(1,3) = -\Delta^*(1)\tilde{G}^A(3,1)\Delta(3) + \tilde{\Sigma}^R(1,3) \tag{50}$$

and

$$\Sigma^<(1,3) = \Delta^*(1)\tilde{G}^>(3,1)\Delta(3) + \tilde{\Sigma}^<(1,3) \tag{51}$$

with the regular retarded Green's function $\tilde{G}^R = G^{0R} + G^{0R}\tilde{\Sigma}^R\tilde{G}^R$ and the correlation part $\tilde{G}^< = \tilde{G}^R\tilde{\Sigma}^<\tilde{G}^A$. The dressed Green's function includes the superconducting selfenergy

$$G^R(1,3) = \tilde{G}^R(1,3) - \tilde{G}^R(1,\bar{5})\Delta^*(\bar{5})\tilde{G}^A(\bar{7},\bar{5})\Delta(\bar{7})G^R(\bar{7},3) \tag{52}$$

and

$$G^<(1,3) = G^R(1,\bar{5})\left(\tilde{\Sigma}^<(\bar{5},\bar{7}) + \Delta^*(\bar{5})\tilde{G}^>(\bar{7},\bar{5})\Delta(\bar{7})\right)G^A(\bar{7},3). \tag{53}$$

Journal of Physics: Conference Series **696** (2016) 012015 doi:10.1088/1742-6596/696/1/012015

Equation for the singular part of $T^<$ implies the equation for the gap function

$$\Delta^*(1) = V\tilde{G}^R_{(2)}(1,\bar{2})\Delta^*(\bar{2}), \tag{54}$$

where the two-particle propagator $\tilde{G}^R_{(2)}(1,2) = -i\theta(t_1-t_2)\left(\tilde{G}^>_{(2)}(1,2) - \tilde{G}^<_{(2)}(1,2)\right)$ has restricted selfconsistency $\tilde{G}^<_{(2)}(1,2) = \tilde{G}^<_\uparrow(1,2)G^<_\downarrow(1,2)$.

The gap equation (54) corresponds to the Thouless criterion with the restricted selfconsistency on the gap in the two-particle propagator. This form extends applicability of the Thouless approach to non-equilibrium superconducting states. If we neglect the regular selfenergy or approximate it by energy-independent correction to the single-particle energy, $\epsilon_{\mathbf{k}}+\tilde{\Sigma}^R(\omega,\mathbf{k}) \approx \varepsilon_{\mathbf{k}}$ and $\tilde{\Sigma}^< = 0$, equation (54) becomes identical to the BCS gap equation.

In the above treatment we have not used the anomalous Green functions. We have avoided this useful tool to emphasis that the present approach is not based on them. For the single-valued condensate the singular separable part of the T-matrix leads to a separable of the two-particle propagator in which one can easily recognize the product of anomalous functions, see [28].

4.4. Time-dependent Ginzburg-Landau equation

In the limit of small gap $\Delta \to 0$, the present theory simplifies to the time-dependent Ginzburg-Landau theory. The gradient expansion of the integral kernel in the gap equation (54) yields linear terms which are identical to those obtained above from the Thouless criterion. Unlike in the Thouless theory, the two-particle propagator $\tilde{G}^R_{(2)}(1,2)$ in (54) depends on the gap Δ. Keeping terms to the second order of $\Delta \propto \psi$ one arrives at the TDGL equation

$$\frac{1}{2m^*}\left(-i\frac{\partial}{\partial \mathbf{r}} - m^*\mathbf{v}_n - 2e\mathbf{A}\right)^2\psi + \alpha\psi + \beta|\psi|^2\psi = -\Gamma\left(\frac{\partial}{\partial t} - 2i\phi\right)\psi. \tag{55}$$

Individual terms of this equation are conveniently derived from the Thouless criterion (linear terms) and the BCS theory (the cubic term). The T-matrix approach derive all these terms from a single microscopic ground.

5. Conclusions

We have discussed the effect of the normal current on the formation of the superconducting condensate. The BCS approach to this effect is troubled by the fact that the Bogoliubov-Valutin quasiparticle energies do not turn into single-electron energy bands in the zero-gap limit. In this representation the normal current is difficult to describe as it requires non-adiabatic processes.

The representation problem was circumvented by the Thouless criterion which deals with fully selfconsistent Green's function of the normal state. While the effect of the normal current on nucleation of the condensate is covered in the non-equilibrium Galitskii approximation, this approach fails to establish the amplitude of the condensate. We have shown that using the multiple-scattering corrections to the Galitskii approximation one obtains a theory applicable to the superconducting state and the gap amplitude results in agreement with the BCS theory.

5.1. Acknowledgments

The COST action MP1201 and associated MSTM project LD15062 are acknowledged.

References

[1] Gor'kov L P 1959 *Zh. Eksper. Teor. Fiz.* **36** 1918 [Sov. Phys. JETP **9**, 1364 (1959)]
[2] Abrahams E and Tsuneto T 1966 *Phys. Rev.* **152** 416
[3] Woo J W F and Abrahams E 1968 *Phys. Rev.* **169** 407
[4] Kopnin N B 2001 *Theory of Nonequilibrium Superconductivity* (Oxford: Claredon Press)

[5] Pals J A, Weiss K, van Attekum P M T M, Horstman R E and Wolter J 1982 *Phys. Rep.* **89** 324
[6] Thouless D J 1960 *Ann. Phys.* **10** 553
[7] Galitskii V M 1958 *Zh. Eksp. Teor. Fiz.* **34** 151 [Sov. Phys.–JETP **7**, 104 (1958)]
[8] Lin P J and Lipavský P 2008 *Phys. Rev. B* **77** 144505
[9] Bishop R F, Strayer M R and Irvine J M 1975 *J. of Low Temp. Phys.* **20** 573–584
[10] Lin P J and Lipavský P 2009 *Phys. Rev. B* **80** 212506
[11] Ikebe Y, Shimano R, Ikeda M, Fukumura T and Kawasaki M 2009 *Phys. Rev. B* **79** 174525
[12] Lipavský P, Elmurodov A, Lin P J, Matlock P and Berdiyorov G R 2012 *Phys. Rev. B* **86**(14) 144516
[13] Wild W 1960 *Z. Phys. A* **158** 322–346
[14] Tolmachev V V 1973 *Teorija Fermi-Gaza* (Moskva: Izdatelśtvo Moskovskogo universiteta)
[15] Kadanoff L P and Martin P C 1961 *Phys. Rev.* **124** 670–697
[16] Kremp D, Semkat D and Henneberger K 2010 *Journal of Physics: Conference Series* **220** 012004
[17] Patton B R 1971 *Ph.D. Thesis* Cornell University
[18] Langer J S 1964 *Phys. Rev.* **134** A553–A565
[19] Lipavský P 2008 *Phys. Rev. B* **78** 214506
[20] Morawetz K 2011 *Journal of Statistical Physics* **143** 482–500
[21] Soven P 1967 *Phys. Rev.* **156** 809–813
[22] Velický B, Kirkpatrick S and Ehrenreich H 1968 *Phys. Rev.* **175** 747–766
[23] Velický B 1969 *Phys. Rev.* **184**(3) 614–627
[24] Lipavský P, Morawetz K and Špička V 2001 *Kinetic equation for strongly interacting dense Fermi systems* (*Annales de Physique* vol 26,1) (Paris: EDP Sciences)
[25] Hopjan M and Lipavský P 2014 *Phys. Rev. B* **89**(9) 094507
[26] Bettelheim E 2010 *EPL (Europhysics Letters)* **90** 67002
[27] Lipavský P, Morawetz K, Šopík B and Männel M 2014 *The European Physical Journal B* **87** 8 ISSN 1434-6028
[28] Morawetz K 2010 *Phys. Rev. B* **82**(9) 092501

Time-dependent Landauer–Büttiker formalism for superconducting junctions at arbitrary temperatures

Riku Tuovinen[1], **Robert van Leeuwen**[1,4], **Enrico Perfetto**[2] and **Gianluca Stefanucci**[2,3,4]

[1] Department of Physics, Nanoscience Center, FIN 40014, University of Jyväskylä, Jyväskylä, Finland

[2] Dipartimento di Fisica, Università di Roma Tor Vergata, Via della Ricerca Scientifica 1, I-00133 Rome, Italy

[3] Laboratori Nazionali di Frascati, Istituto Nazionale di Fisica Nucleare, Via E. Fermi 40, 00044 Frascati, Italy

[4] European Theoretical Spectroscopy Facility (ETSF)

E-mail: `riku.m.tuovinen@jyu.fi`

Abstract. We discuss an extension of our earlier work on the time-dependent Landauer–Büttiker formalism for noninteracting electronic transport. The formalism can without complication be extended to superconducting central regions since the Green's functions in the Nambu representation satisfy the same equations of motion which, in turn, leads to the same closed expression for the equal-time lesser Green's function, i.e., for the time-dependent reduced one-particle density matrix. We further write the finite-temperature frequency integrals in terms of known special functions thereby considerably speeding up the computation. Simulations in simple normal metal – superconductor – normal metal junctions are also presented.

1. Introduction

The process of Andreev reflection[1] (AR) occurring at the interface between a normal metal (N) and a superconductor (S) is of great importance with applications in spintronics and quantum computing. An incoming electron from N to S produces a Cooper pair in S and a reflected hole in N[2–4]. In an NSN junction normal metal electrodes are spatially separated by a superconducting central region and an entangled electron–hole pair can be transported. This can be seen when the junction separation is of the order of the superconducting coherence length for the studied material, and when the incident electron energies are less than the superconducting gap for the AR process to occur[5].

The quantum transport problems are typically time dependent; there is no guarantee that the system would in an instant relax to a steady-state configuration once the junction is "switched on" (as in connecting different devices or driving them out of equilibrium by an external perturbation). In contrast, there are transient effects depending on, e.g., the system's geometry[6–9], its predisposition to external perturbations[10–13], the physical properties of the transported quanta and their mutual interactions[14–19]. Even if the transport mechanisms were discussed in an idealized noninteracting setting, it is therefore important to consider a fully time-dependent description of the studied processes.

The Landauer–Büttiker formalism is simple to understand as it relates to an intuitive physical picture of charge transport in a multiterminal junction[20, 21]. Including the transient description to the formalism by studying the nonequilibrium Green's function approach does not complicate the final result[22–27]; the physical picture is still clear and intuitive as different features of the transport setup can be directly linked to the time-dependence[28–30]. In this paper, we present an extension to earlier results for both superconducting junctions and arbitrary temperatures (Sec. 2). We present a formula for the time-dependent one-particle reduced density matrix (TD1RDM) as such since it is a closed, analytic expression which can readily be implemented for numerical model systems. Further details of the derivation are to be found in another work[31]. In Sec. 4 we illustrate the features of the formula by studying transients in simple NSN junctions.

2. Background and Nambu representation

We consider a quantum transport setup similar to one studied in the previous volume of this conference series[29]. In this setup, a noninteracting central region is connected between metallic leads, and the Hamiltonian takes the form

$$\begin{aligned}\hat{H} &= \hat{H}_{\text{leads}} + \hat{H}_{\text{central}} + \hat{H}_{\text{coupling}} \\ &= \sum_{k\alpha\sigma} [\epsilon_{k\alpha} + \theta(t)V_\alpha]\, \hat{c}^\dagger_{k\alpha\sigma}\hat{c}_{k\alpha\sigma} + \sum_{mn\sigma} T_{mn}\hat{c}^\dagger_{m\sigma}\hat{c}_{n\sigma} + \sum_{mk\alpha\sigma}\left(T_{mk\alpha}\hat{c}^\dagger_{m\sigma}\hat{c}_{k\alpha\sigma} + \text{h.c.}\right). \end{aligned} \quad (1)$$

The operators $\hat{c}^{(\dagger)}$ annihilate (create) electrons from (to) a region specified by the subscript indices: $k\alpha$ is the k-th basis element of the α-th lead, m, n label the basis elements of the central region, and $\sigma \in \{\uparrow, \downarrow\}$ is a spin-$\frac{1}{2}$ index. These operators moreover obey the fermionic anticommutation relations $\{\hat{c}_{x\sigma}, \hat{c}^\dagger_{y\sigma'}\} = \delta_{xy}\delta_{\sigma\sigma'}$. The Hamiltonian structure is determined by the single-particle levels in the leads $\epsilon_{k\alpha}$ and the tunneling matrices T between the states of the central region (T_{mn}) and between the states of the central region and the leads ($T_{mk\alpha}$). The system is driven out of equilibrium for times $t > 0$ by a sudden shift of the lead energy levels by V_α. In addition to a sudden bias, it is also possible to include time-dependent bias profiles without complicating the following derivations[32]. In order to describe a superconducting island, we will now add a pairing field operator $\hat{\Delta}$ to the Hamiltonian of the central region by[33, 34]

$$\hat{H}_{\text{central}} \to \sum_{mn\sigma} T_{mn}\hat{c}^\dagger_{m\sigma}\hat{c}_{n\sigma} + \sum_m \Delta_m \hat{c}^\dagger_{m\uparrow}\hat{c}^\dagger_{m\downarrow} + \sum_m \Delta^*_m \hat{c}_{m\downarrow}\hat{c}_{m\uparrow}. \quad (2)$$

The one-electron Green's function in the above setup can be defined via the *Nambu spinor* $\underline{\hat{\Phi}}_m = (\hat{\Phi}^1_m, \hat{\Phi}^2_m)^T = (\hat{c}_{m\uparrow}, \hat{c}^\dagger_{m\downarrow})^T$, which obeys the anticommutation relation in a tensor product sense $\{\hat{\Phi}^\mu_m, \hat{\Phi}^\nu_n\} = \delta_{mn}\delta^{\mu\nu}$, as a contour-ordered product[34, 35]

$$\underline{G}_{rs}(z, z') = -\mathrm{i}\langle \mathcal{T}_\gamma[\underline{\hat{\Phi}}_r(z) \otimes \underline{\hat{\Phi}}^\dagger_s(z')]\rangle \quad (3)$$

where the contour-ordering operator $\mathcal{T}_\gamma$ is taken for the variables z, z' on the Keldysh contour γ[28]. When the product in Eq. (3) is expanded, the elements in the resulting 2×2 *Nambu matrix* are the normal and the anomalous components of the Green's function[36, 37]. As already denoted above, we put an underline for quantities in the Nambu space. The matrix elements in the Green's function label the transport setup in the following block form [notice that the indices r, s in Eq. (3) may belong to any block and that the number of leads is arbitrary]

$$\underline{\boldsymbol{h}} = \begin{pmatrix} \underline{h}_{11} & 0 & \cdots & \underline{h}_{1C} \\ 0 & \underline{h}_{22} & \cdots & \underline{h}_{2C} \\ \vdots & \vdots & \ddots & \vdots \\ \underline{h}_{C1} & \underline{h}_{C2} & \cdots & \underline{h}_{CC} \end{pmatrix} ; \qquad \underline{\boldsymbol{G}} = \begin{pmatrix} \underline{G}_{11} & \underline{G}_{12} & \cdots & \underline{G}_{1C} \\ \underline{G}_{21} & \underline{G}_{22} & \cdots & \underline{G}_{2C} \\ \vdots & \vdots & \ddots & \vdots \\ \underline{G}_{C1} & \underline{G}_{C2} & \cdots & \underline{G}_{CC} \end{pmatrix} \quad (4)$$

with

$$(\underline{h}_{\alpha\alpha'})_{kk'}(t) = \begin{pmatrix} [\epsilon_{k\alpha}+\theta(t)V_\alpha]\delta_{\alpha\alpha'}\delta_{kk'} & 0 \\ 0 & -[\epsilon_{k\alpha}+\theta(t)V_\alpha]\delta_{\alpha\alpha'}\delta_{kk'} \end{pmatrix}, \tag{5}$$

$$(\underline{h}_{CC})_{mn} = \begin{pmatrix} T_{mn} & \Delta_m\delta_{mn} \\ \Delta_m^*\delta_{mn} & -T_{mn} \end{pmatrix}, \tag{6}$$

$$(\underline{h}_{C\alpha})_{mk\alpha} = \begin{pmatrix} T_{mk\alpha} & 0 \\ 0 & -T_{mk\alpha} \end{pmatrix}, \quad (\underline{h}_{\alpha C})_{k\alpha m} = \begin{pmatrix} T_{k\alpha m} & 0 \\ 0 & -T_{k\alpha m} \end{pmatrix} \tag{7}$$

for the leads, central region and couplings, respectively. It is important to notice that even though only the central region is superconducting, all the blocks in the Hamiltonian are written in the form of Bogoliubov–de Gennes[38, 39]. Including the pairing field in the Hamiltonian of the central region adds no extra complication to the evolution of the Green's function[31]; the only difference, compared to the earlier work in Refs. [28–30], is in the interpretation of the matrices in Nambu space. It is also possible to include non-local pairing field Δ_{mn} with arbitrary spin-coupling leading to 4-component Nambu spinors which includes the possibility to study also Majorana fermions[31, 40]. The Hamiltonian and the Green's function are connected via the equation of motion (with the boundary condition that the Green's function is antiperiodic along the contour, i.e., the Kubo–Martin–Schwinger boundary conditions[28, 41, 42])

$$\left[\mathrm{i}\frac{\mathrm{d}}{\mathrm{d}z}\underline{\mathbf{1}} - \underline{\boldsymbol{h}}(z)\right]\underline{\boldsymbol{G}}(z,z') = \delta(z,z')\underline{\mathbf{1}} \tag{8}$$

and the corresponding adjoint one. We describe the leads within the wide-band approximation (WBA), where the electronic levels of the central region are in a narrow range compared to the lead bandwidth which gives for the retarded embedding self-energy

$$\underline{\Sigma}^{\mathrm{R}}_{\alpha,mn}(\omega) = \sum_k (\underline{h}_{C\alpha})_{mk\alpha}\frac{1}{\omega - \underline{\epsilon}_{k\alpha} - \underline{V}_\alpha + \mathrm{i}\eta}(\underline{h}_{\alpha C})_{k\alpha n} \approx -\mathrm{i}\underline{\Gamma}_{\alpha,mn}/2, \tag{9}$$

where the bandwidth matrices satisfy $\underline{\Gamma} = \sum_\alpha \underline{\Gamma}_\alpha$. For the lead Green's function between the coupling Hamiltonians in Eq. (9) the structure is similar to that of Eq. (5). Approximating the embedding self-energy this way, as a purely imaginary constant, closes the equation of motion (8), and it can then be solved analytically[28–30]. As the solution we get the lesser Green's function (in region CC) in the equal-time limit and the TD1RDM by

$$\begin{aligned}(\underline{\rho}_{CC})_{mn}(t) = -\mathrm{i}(\underline{G}^<_{CC})_{mn}(t,t) &= -\mathrm{i}\begin{pmatrix} (G^<_{CC,\uparrow})_{mn}(t,t) & (-F^>_{CC})_{nm}(t,t) \\ (\bar{F}^<_{CC})_{mn}(t,t) & (-G^>_{CC,\downarrow})_{nm}(t,t) \end{pmatrix} \\ &= \begin{pmatrix} \langle \hat{c}^\dagger_{n\uparrow}(t)\hat{c}_{m\uparrow}(t)\rangle & \langle \hat{c}_{n\downarrow}(t)\hat{c}_{m\uparrow}(t)\rangle \\ \langle \hat{c}^\dagger_{n\uparrow}(t)\hat{c}^\dagger_{m\downarrow}(t)\rangle & \langle \hat{c}_{n\downarrow}(t)\hat{c}^\dagger_{m\downarrow}(t)\rangle \end{pmatrix}\end{aligned} \tag{10}$$

where the normal ($G^{\lessgtr}_\sigma$) and anomalous ($F^{\lessgtr}$) components[34] follow by expanding the product in Eq. (3). It is also possible to solve the equations of motion analytically for the two-time Keldysh components of the Green's function; this offers the possibility to extract not only densities and currents but other physical quantities such as noise from the solution[32].

Expressed in the left eigenbasis, $\langle\Psi^{\mathrm{L}}|\underline{h}_{\mathrm{eff}} = \epsilon\langle\Psi^{\mathrm{L}}|$ (the eigenvalues ϵ are in general complex), of the nonhermitian effective Hamiltonian $\underline{h}_{\mathrm{eff}} = \underline{h}_{CC} - \mathrm{i}\underline{\Gamma}/2$ the matrix elements of the TD1RDM take the explicit form[30, 31]

$$\begin{aligned}&\langle\Psi^{\mathrm{L}}_j|\underline{\rho}_{CC}(t)|\Psi^{\mathrm{L}}_k\rangle \\ &= \sum_\alpha\left\{\Gamma_{\alpha,jk}\Lambda_{\alpha,jk} + V_\alpha\Gamma_{\alpha,jk}\left[\Pi_{\alpha,jk}(t) + \Pi^*_{\alpha,kj}(t)\right] + V_\alpha^2\Gamma_{\alpha,jk}\mathrm{e}^{-\mathrm{i}(\epsilon_j-\epsilon_k^*)t}\Omega_{\alpha,jk}\right\},\end{aligned} \tag{11}$$

with

$$\Gamma_{\alpha,jk} = \langle \Psi_j^{\mathrm{L}} | \underline{\Gamma}_\alpha | \Psi_k^{\mathrm{L}} \rangle, \tag{12}$$

$$\Lambda_{\alpha,jk} = \frac{\mathrm{i}}{\epsilon_k^* - \epsilon_j} \left\{ \frac{1}{\mathrm{e}^{\beta(\epsilon_k^* - \mu_\alpha)} + 1} + \frac{1}{2\pi \mathrm{i}} \left[\psi\left(\frac{1}{2} - \frac{\beta(\epsilon_k^* - \mu_\alpha)}{2\pi\mathrm{i}}\right) - \psi\left(\frac{1}{2} - \frac{\beta(\epsilon_j - \mu_\alpha)}{2\pi\mathrm{i}}\right) \right] \right\}, \tag{13}$$

$$\begin{aligned} \Pi_{\alpha,jk}(t) = {} & \frac{\mathrm{i}}{(\epsilon_k^* - \epsilon_j)(\epsilon_k^* - \epsilon_j - V_\alpha)} \left\{ \frac{\mathrm{e}^{-\mathrm{i}(\epsilon_j - \epsilon_k^*)t}}{\mathrm{e}^{\beta(\epsilon_k^* - \mu_\alpha)} + 1} + \mathrm{i}\mathrm{e}^{-\pi t/\beta} \mathrm{e}^{-\mathrm{i}(\epsilon_j - \mu_\alpha)t} \times \right. \\ & \left. \left[\mathfrak{F}(\epsilon_k^* - \mu_\alpha, t, \beta) + \frac{\epsilon_k^* - \epsilon_j - V_\alpha}{V_\alpha} \mathfrak{F}(\epsilon_j - \mu_\alpha, t, \beta) - \frac{\epsilon_k^* - \epsilon_j}{V_\alpha} \mathfrak{F}(\epsilon_j - \mu, t, \beta) \right] \right\}, \end{aligned} \tag{14}$$

$$\begin{aligned} \Omega_{\alpha,jk} = {} & \frac{\frac{\mathrm{i}}{\mathrm{e}^{\beta(\epsilon_k^* - \mu)} + 1} - \frac{1}{2\pi} \left[\psi\left(\frac{1}{2} - \frac{\beta(\epsilon_j - \mu_\alpha)}{2\pi\mathrm{i}}\right) - \psi\left(\frac{1}{2} - \frac{\beta(\epsilon_k^* - \mu)}{2\pi\mathrm{i}}\right) \right]}{(\epsilon_k^* - \epsilon_j)(\epsilon_k^* - \epsilon_j + V_\alpha) V_\alpha} \\ & - \frac{\frac{\mathrm{i}}{\mathrm{e}^{\beta(\epsilon_k^* - \mu_\alpha)} + 1} - \frac{1}{2\pi} \left[\psi\left(\frac{1}{2} - \frac{\beta(\epsilon_j - \mu)}{2\pi\mathrm{i}}\right) - \psi\left(\frac{1}{2} - \frac{\beta(\epsilon_k^* - \mu_\alpha)}{2\pi\mathrm{i}}\right) \right]}{(\epsilon_k^* - \epsilon_j)(\epsilon_k^* - \epsilon_j - V_\alpha) V_\alpha}, \end{aligned} \tag{15}$$

where β is the inverse temperature, μ is the chemical potential and we also denoted the electrochemical potential as $\mu_\alpha = \mu + V_\alpha$. In the above expressions ψ is the digamma function[43] and we defined another special function by $\mathfrak{F}(z, t, \beta) \equiv \frac{1}{\mathrm{i}\beta z + \pi} \, {}_2F_1\left(1, \frac{1}{2} + \frac{\mathrm{i}\beta z}{2\pi}, \frac{3}{2} + \frac{\mathrm{i}\beta z}{2\pi}, \mathrm{e}^{-2\pi t/\beta}\right)$ with ${}_2F_1$ being the hypergeometric function[44]. Studying the asymptotic behaviour of the digamma and hypergeometric functions the results in Eqs. (13), (14) and (15) can be shown to reduce to those in Ref. [30] in the zero-temperature limit ($\beta \to \infty$)[31].

3. An introductory example

Let us motivate the discussion for the NSN setup by means of a simple example. Consider a single dot connected to two leads for which the Hamiltonian can be separated in parts for leads, tunneling and dot, respectively as

$$\begin{aligned} \hat{H} = {} & \sum_{k\alpha\sigma} \epsilon_{k\alpha} \hat{c}_{k\alpha\sigma}^\dagger \hat{c}_{k\alpha\sigma} + \sum_{k\alpha\sigma} t_{k\alpha 0} \hat{c}_{k\alpha\sigma}^\dagger \hat{c}_{0\sigma} + \sum_{k\alpha\sigma} t_{k\alpha 0}^* \hat{c}_{0\sigma}^\dagger \hat{c}_{k\alpha\sigma} + \epsilon_0 \sum_\sigma \hat{c}_{0\sigma}^\dagger \hat{c}_{0\sigma} \\ & + \Delta_0 \hat{c}_{0\uparrow}^\dagger \hat{c}_{0\downarrow}^\dagger + \Delta_0^* \hat{c}_{0\downarrow} \hat{c}_{0\uparrow} \end{aligned} \tag{16}$$

with $\epsilon_{k\alpha}$ giving the level structure of the leads $\alpha \in \{L, R\}$, $t_{k\alpha 0}$ corresponding to the tunneling strength between the leads and the dot, and ϵ_0, Δ_0 being the energy and the pairing strength in the dot, respectively. Let us introduce a new set of operators $\hat{\hat{c}}_{x\sigma} = \hat{c}_{x\sigma}^\dagger$ obeying the fermionic anticommutation relation. The Hamiltonian can now be rewritten in terms of the new and old operators as

$$\begin{aligned} \hat{H} = {} & \sum_{k\alpha} \epsilon_{k\alpha} \hat{c}_{k\alpha\uparrow}^\dagger \hat{c}_{k\alpha\uparrow} + \sum_{k\alpha} (-\epsilon_{k\alpha}) \hat{\hat{c}}_{k\alpha\downarrow}^\dagger \hat{\hat{c}}_{k\alpha\downarrow} + \sum_{k\alpha} \epsilon_{k\alpha} \\ & + \sum_{k\alpha} \left(t_{k\alpha 0} \hat{c}_{k\alpha\uparrow}^\dagger \hat{c}_{0\uparrow} + t_{k\alpha 0}^* \hat{c}_{0\uparrow}^\dagger \hat{c}_{k\alpha\uparrow} \right) + \sum_{k\alpha} \left[(-t_{k\alpha 0}) \hat{\hat{c}}_{0\downarrow}^\dagger \hat{\hat{c}}_{k\alpha\downarrow} + (-t_{k\alpha 0}^*) \hat{\hat{c}}_{k\alpha\downarrow}^\dagger \hat{\hat{c}}_{0\downarrow} \right] \\ & + \epsilon_0 \hat{c}_{0\uparrow}^\dagger \hat{c}_{0\uparrow} + (-\epsilon_0) \hat{\hat{c}}_{0\downarrow}^\dagger \hat{\hat{c}}_{0\downarrow} + \epsilon_0 + \Delta_0 \hat{c}_{0\uparrow}^\dagger \hat{\hat{c}}_{0\downarrow} + \Delta_0^* \hat{\hat{c}}_{0\downarrow}^\dagger \hat{c}_{0\uparrow} \end{aligned} \tag{17}$$

where two constant shifts $\sum_{k\alpha} \epsilon_{k\alpha}$ and ϵ_0 occur due to the anticommutation relations. Each term in Eq. (17) has a similar structure, $\hat{c}^\dagger \hat{c}$, and we may model the dot part as in Fig. 1 where

Journal of Physics: Conference Series **696** (2016) 012016 doi:10.1088/1742-6596/696/1/012016

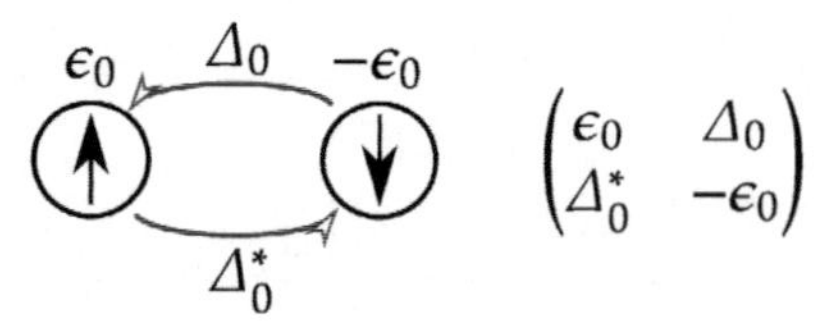

Figure 1: The dot viewed as a two-level system for different spins.

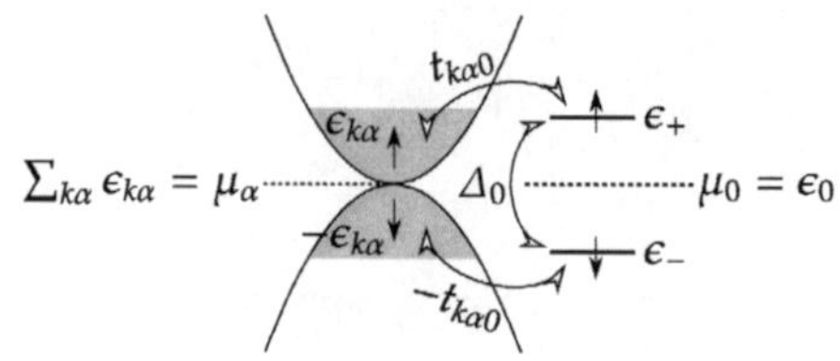

Figure 2: Couplings between the lead and the dot in a transport setup.

the matrix is of the form of Eq. (6) and the corresponding eigenvalues are $\epsilon_\pm = \pm\sqrt{\epsilon_0^2 + |\Delta_0|^2}$. The transport setup corresponding to Eq. (17) can then be viewed through the energy diagram in Fig. 2 where we notice the nature of the constant shifts in Eq. (17); they could be regarded as a chemical potential. The energy levels for the lead sector are raised so that the energy level continuum for the spin-up particles goes up from $\sum_{k\alpha} \epsilon_{k\alpha}$ and the energy level continuum for the spin-down particles goes down from $\sum_{k\alpha} \epsilon_{k\alpha}$. Similarly, for the dot sector we have the energy levels raised by ϵ_0. The coupling terms $t_{k\alpha 0}$ connect separately the spin-up and spin-down particles between the leads and the dot, and the pairing strength term Δ_0 acts as a hopping term flipping the spins within the dot.

According to this picture for the NSN setup and the Nambu structure in Eq. (10) we will evaluate the local bond currents for the more general structure in Eq. (6) by

$$J_{mn}(t) = -\sum_\sigma \left[T_{mn}(G^<_{CC,\sigma})_{nm}(t,t) + \text{h.c.}\right] \tag{18}$$

and the Cooper pair density by

$$P_m(t) = \mathrm{i}(F^>_{CC})_{mm}(t,t)\mathrm{e}^{2\mathrm{i}T_{mm}t} \tag{19}$$

satisfying the continuity equation[34]

$$\frac{\mathrm{d}}{\mathrm{d}t} n_m(t) = \sum_n J_{mn}(t) - 4\,\mathrm{Im}[\Delta_m^* P_m(t)\mathrm{e}^{-2\mathrm{i}T_{mm}t}] \tag{20}$$

where the site density is the expectation value $n_m = \langle \hat{n}_m \rangle$ of $\hat{n}_m = \sum_\sigma \hat{c}^\dagger_{m\sigma}\hat{c}_{m\sigma}$. In the Nambu representation of the lesser Green's function the diagonal blocks therefore give rise to the bond current whereas the off-diagonal blocks correspond to the Cooper pair density. In the continuity equation (20) the two different terms on the right-hand side can also be identified as the normal current and the super current.

4. TD response of a superconducting junction

As a first study we show a numerical confirmation of the presented formula; in the limit when the superconducting gap Δ and the temperature $1/\beta$ vanish we should recover equal results with the formula in Ref. [30]. For the sake of simple interpretation of the transients let us take, as an example, a 2-site tight-binding dimer coupled to two semi-infinite one-dimensional leads. Let the hopping parameter between the sites be equal everywhere: $t_\alpha = t_{\alpha C} = t_C =: \epsilon_0$ (for $\alpha = L, R$) leading to the tunneling rate $\Gamma_\alpha = 2t^2_{\alpha C}/|t_\alpha| = 2\epsilon_0$. This parameter is the strength of the bandwidth matrix elements in the form of Eq. (9). In fact, WBA is not a very good approximation in this case, as the resonances are comparatively rather wide, but we are only

Journal of Physics: Conference Series **696** (2016) 012016 doi:10.1088/1742-6596/696/1/012016

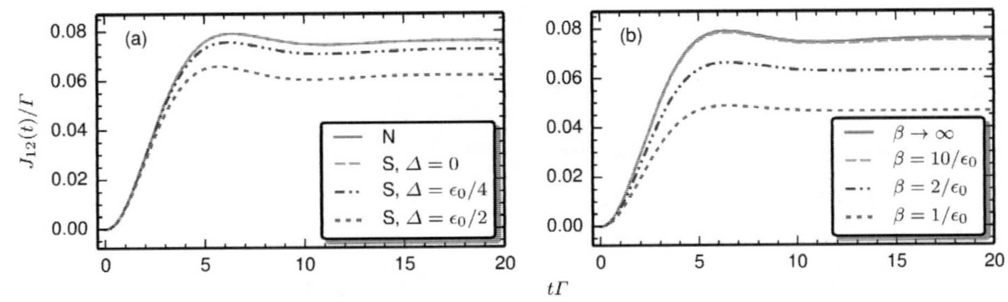

Figure 3: (Color online) Transient current through a 2-site dimer; comparison between different formulae and parameters. (a) Normal vs. superconducting central region at zero temperature; (b) normal central region at varying temperatures.

comparing different formulae within the WBA, so the results should only be taken as comparative. Let us also bias the leads symmetrically to $V_L = -V_R = \epsilon_0$ with respect to the chemical potential $\mu = 0$. We calculate, from the TD1RDM, the local bond current between the two sites in the dimer using Eq. (18). (Due to equal hopping parameters through the setup this is equal to the current through the lead interfaces modulo a minor time delay.)

In Fig. 3(a) we compare normal central region to a superconducting one by evaluating the TD1RDM from a normal Hamiltonian without the pairing field (non-Nambu), and from a Nambu Hamiltonian with varying pairing field strength at zero temperature. We see how the N and S ($\Delta = 0$) cases are on top of each other, and increasing the value for Δ decreases the absolute value of the current through the central region as the energy levels of the central region get raised by $\sqrt{\Delta^2 + \epsilon_0^2}$. In Fig. 3(b) we compare normal central regions at varying temperatures. In this comparative benchmark of varying temperature, we do not consider superconducting central regions as the temperature effects for the pairing field $\Delta(T)$ should also be taken into account according to the self-consistent gap equation[4]. (In further simulations, also these parameters are considered in more detail.) The zero-temperature limit, $\beta \to \infty$, is evaluated from the results in Ref. [30] which roughly agrees with an evaluation with $\beta = 10/\epsilon_0$. (Increasing β even more would naturally bring the curves exactly on top of each other.) Because the level structure of the studied system is symmetric around the chemical potential, increasing the temperature $1/\beta$ decreases the current due to broadening of the distribution function close to the Fermi level. In general, however, there is a possibility of enhancing the current by increasing the temperature if, for instance, the electronic levels were all above the Fermi level. In that case, the lead-states with energy higher than the energy of the levels get occupied, leading to an enhanced current.

After the comparisons presented above we can confidently conclude that the formulation of the NSN transport setup and the implementation of the formula for the TD1RDM is working properly. Now, we turn to a more concrete and physically relevant example, and we analyze the transient features in more detail. We consider a superconducting island made of a benzene-like molecule belonging to the class of quasi-one-dimensional polyacene chains[45]. Transport in simple island setups has been studied, e.g., in Refs. [46–50], in a single-electron-tunneling level, where Coulomb blockade region is explored, and it is shown how the superconducting gap Δ strongly and non-trivially affects the tunneling process. In polyacene samples (and in other carbon based materials, such as graphene) the superconductivity could be induced, e.g., by charge injection, chemical doping or using the proximity effect leading to critical temperatures ranging from 1 to 10 K[51–54]. Our setup is shown schematically in Fig. 4. We model the benzene molecule in a single π-orbital tight-binding framework with the hopping parameter $t_C = -2.7$ eV[55], and relate other energies to this scale. We also saturate the molecule's edges

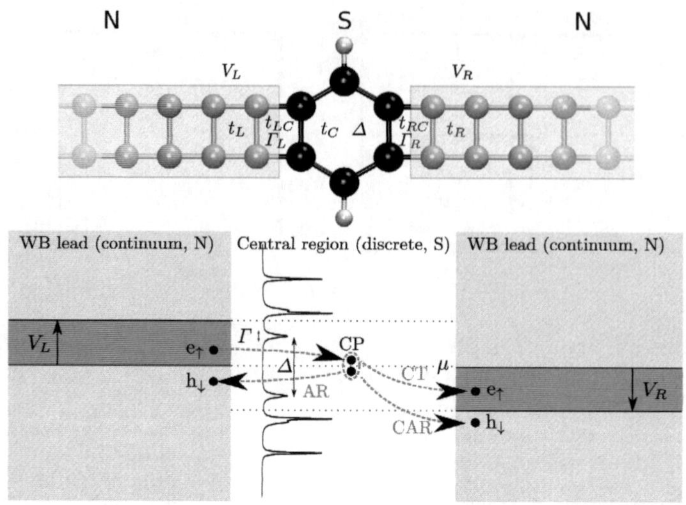

Figure 4: (Color online) Transport setup in an NSN junction. Normal metal leads of continuum states undergo a level shift due to the bias voltage $V_{L/R}$ with respect to the chemical potential $\mu = E_F$ (Fermi level). The discrete level structure of the central region is determined by the tight-binding and gap Δ parameters, and it is also broadened due to the coupling (Γ). Possible transition mechanisms are shown as CT, AR and CAR; see text for description.

(longitudinally, in the transport direction) by hydrogen with modified tight-binding parameters for hydrogen on-site energies and hydrogen–carbon hopping[56], respectively, so that there is no band gap in equilibrium. This condition is set because we want to isolate the effects from the superconducting gap Δ without complicating the spectrum with the semiconducting gap. The coupling strength between the molecule and the leads and the lead hopping are chosen so that we are in weak coupling regime $\Gamma = 0.2$ eV.

Looking at the setup in Fig. 4 more closely suggests different transition mechanisms depending on the parameters. The simplest case is when the bias voltage V_α is larger than the superconducting gap Δ. In this case, all the levels inside the bias window act as transport channels, and transitions through the superconducting states are disrupted since the energy for the incoming electrons is high enough to break possible Cooper pairs (CP); this is referred to as normal tunneling (NT). If the bias voltage is smaller than the superconducting gap, this opens a possibility for the formation of a CP in the central region. In this case, it is possible to observe Andreev reflection (AR) between an electron and a hole in the source (or drain) lead forming the CP in the center, or to observe a crossed Andreev reflection (CAR) where an electron from the source (drain) lead is coupled to a hole in the drain (source) lead through the CP in the center. Also, direct tunneling of an electron via the CP, referred to as cotunneling (CT), is a possible transmission channel. Next, we simulate these different processes by a suitable parameter choice.

We start with a familiar example by simulating NT: The condition is such that the bias window is larger than the gap. We will, in addition, fix the temperature, $\beta = 100/|t_C|$, well below the critical temperature so that the gap can be approximated as the (constant) value at zero temperature $\Delta(T=0)$. The sample is a benzene molecule consisting of 8 atomic sites (6 carbon and 2 hydrogen), coupled to the leads from four sites overall, see Fig. 4. The bias voltage is symmetrically set to $V_L = -V_R = 3|t_C|/2$ and the gap Δ is varied but kept smaller than or equal to this value. The transient currents through the sample [calculated by summing the bond currents from Eq. (18) transversally in the middle of the molecule], the corresponding Fourier

Journal of Physics: Conference Series **696** (2016) 012016 doi:10.1088/1742-6596/696/1/012016

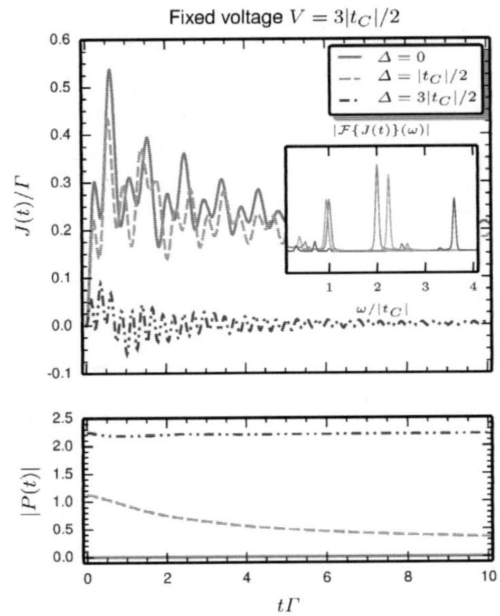

Figure 5: (Color online) Transient currents (top panel) and pair densities (bottom panel) in the molecule when varying Δ. The inset shows the absolute value of the Fourier-transformed current.

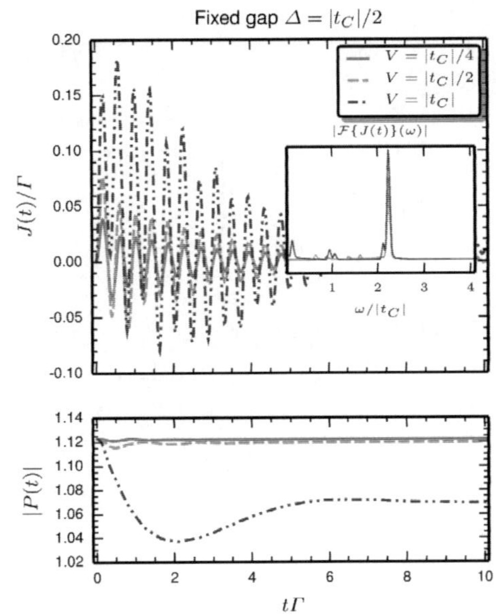

Figure 6: (Color online) Transient currents (top panel) and pair densities (bottom panel) in the molecule when varying V. The inset shows the absolute value of the Fourier-transformed current.

transforms and the pair densities [calculated by summing the pair densities within the molecule: $P(t) = \sum_m P_m(t)$ from Eq. (19)] can be seen in Fig. 5. Increasing the gap Δ decreases the overall current as the conducting states are being pushed away from the bias window. This also leads to shifts in the transient frequencies seen in the Fourier spectrum. By looking also at the spectral function $A(\omega) = -\frac{1}{\pi}\,\mathrm{Im}\,\mathrm{Tr}[G^{\mathrm{R}}(\omega)]$, where G^{R} is the normal Nambu component of the retarded Green's function, plotted in Fig. 8 we can further identify the transitions.

In Fig. 5, when $\Delta = 0$, we see two intramolecular transitions at frequencies $\omega = |t_C|$ and $\omega = 2|t_C|$ which move a little when the gap is increased to $\Delta = |t_C|/2$ corresponding to the shifted energy levels in the spectral function. We also observe two lead–molecule transitions at $\omega = |t_C|/2$ and $\omega = 5|t_C|/2$ when $\Delta = 0$; these frequencies shift with the peaks in the spectral function corresponding to the fixed bias window at $V = 3|t_C|/2$. The reason why we do not see a lead–molecule transition at $\omega = 3|t_C|/2$ when $\Delta = 0$, even though there is a zero-energy state in the molecule, is due to the fact that this state corresponds to the wavefunction's nodal planes being located exactly at the lead interface, and therefore it is an *inert state* not taking part to the transient dynamics[29, 30]. Also, as the conditions are for NT, we observe the pair density within the molecule going to zero from its equilibrium value when $\Delta < V$; this means that there are no out-of-equilibrium CPs forming in the central region, and we see no AR or CAR processes. When we set the gap equal to the bias window, we notice, first of all, that the steady-state current goes to zero since there are no transport channels within the bias window. Some transient oscillations are still present due to the states in the vicinity of the resonant window, which is seen as an intramolecular transition at $\omega \sim 7|t_C|/2$.

Next, we will adjust the transport conditions to visualize the AR and CAR processes. We have the same benzene molecule as the central region at the same temperature, $\beta = 100/|t_C|$.

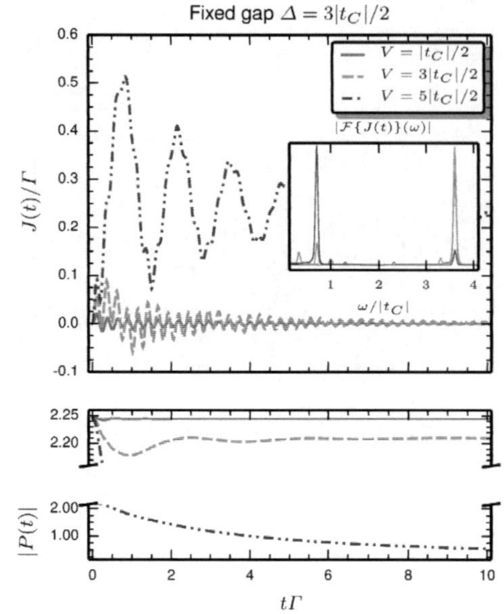

Figure 7: (Color online) Transient currents (top panel) and pair densities (bottom panel) in the molecule when varying V. The inset shows the absolute value of the Fourier-transformed current.

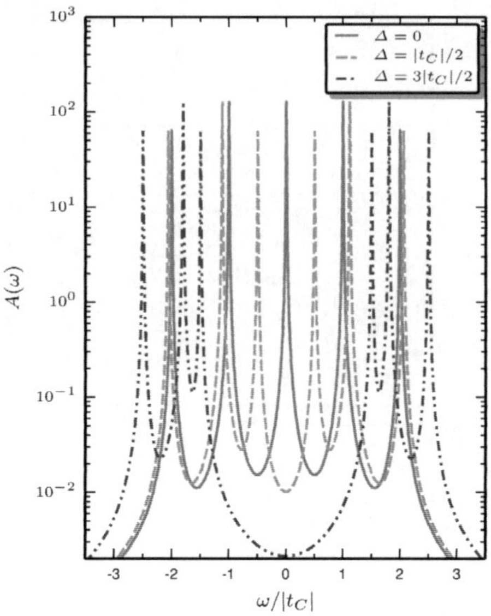

Figure 8: (Color online) Spectral functions of the coupled benzene molecule when varying Δ.

In Fig. 6 we choose the gap as $\Delta = |t_C|/2$, and in Fig. 7 we, on the other hand, choose the gap as $\Delta = 3|t_C|/2$. When $V \leq \Delta$ we observe, in both cases, the transient current oscillating towards a zero steady-state current. The oscillation frequencies seen in the Fourier spectrum can be interpreted from the spectral function in Fig. 8 mainly as intramolecular transitions (around $\omega = |t_C|$ and $\omega = 2|t_C|$). Interestingly, in Fig. 6 also, when we increase the bias voltage above the superconducting gap $V = |t_C|$ there still are no other states within the bias window except the inert state split by Δ. This resonant window does not add anything to the transient dynamics (due to the inert state), but the static part of the density matrix is modified leading to nonzero steady-state current. This is interpreted as CT where, in addition to AR and CAR, also an electron is transferred from one lead to another via the CP. This is also confirmed by looking at the pair density as it remains nonzero also for $V = |t_C|$. With the larger gap in Fig. 7 and for $V \leq \Delta$ we observe CP formation within the molecule but for $V > \Delta$ the pair density goes to zero. For the smaller voltages we mainly find the first intramolecular transition at around $\omega = 7|t_C|/2$. For larger voltages we also see the lead–molecule transitions at lower frequencies, and we recover again the NT regime as the bias voltage is high enough for breaking the CPs within the molecule.

In this transport setup, it is not, in general, easy to distinguish between AR and CAR processes as they involve multiple steps. One possible case is when an electron hops from the lead to the central region (molecule–lead transition), then "transfers" from the spin-up sector in the central region to the spin-down sector for which the probability is given by the pairing strength ("intramolecular" transition between the two branches of states split by Δ), and then finally hops back to the lead as a hole (molecule–lead transition). As all these transitions are visible in the transient oscillations, we may only conclude whether AR and CAR processes are present or not.

Journal of Physics: Conference Series **696** (2016) 012016 doi:10.1088/1742-6596/696/1/012016

5. Conclusions and outlook

We presented an extension to the time-dependent Landauer–Büttiker formalism, discussed in Refs. [28–30], to include superconducting central region in the transport setup, and to evaluate the TD1RDM at arbitrary temperatures. The derived formulae are analytic and closed expressions involving known special functions, and they can readily be implemented to study various quantum transport problems very efficiently and also at large temporal and spatial scales.

As an application of the presented formalism we simulated transport in a superconducting benzene-like molecule attached to two-dimensional normal metal leads. Assigning a proper parameter set for the transport window and the superconducting gap, we observed formation of Cooper pairs within the central molecule leading to Andreev reflection processes.

Acknowledgments

R.T. thanks the Väisälä Foundation of The Finnish Academy of Science and Letters for financial support. R.v.L. thanks the Academy of Finland for support. E.P. and G.S. acknowledge funding by MIUR FIRB Grant No. RBFR12SW0J.

References

[1] Andreev A 1964 *Sov. Phys. JETP* **19** 1228
[2] Cooper L N 1956 *Phys. Rev.* **104** 1189
[3] Bardeen J, Cooper L N and Schrieffer J R 1957 *Phys. Rev.* **106** 162
[4] Bardeen J, Cooper L N and Schrieffer J R 1957 *Phys. Rev.* **108** 1175
[5] Eiles T M, Martinis J M and Devoret M H 1993 *Phys. Rev. Lett.* **70** 1862
[6] Khosravi E, Stefanucci G, Kurth S and Gross E 2009 *Phys. Chem. Chem. Phys.* **11** 4535
[7] Perfetto E, Stefanucci G and Cini M 2010 *Phys. Rev. B* **82** 035446
[8] Vieira D, Capelle K and Ullrich C A 2009 *Phys. Chem. Chem. Phys.* **11** 4647
[9] Wang B, Li J, Xu F, Wei Y, Wang J and Guo H 2015 *Nanoscale* **7** 10030
[10] Kurth S, Stefanucci G, Khosravi E, Verdozzi C and Gross E K U 2010 *Phys. Rev. Lett.* **104** 236801
[11] Foieri F and Arrachea L 2010 *Phys. Rev. B* **82** 125434
[12] Ness H and Dash L K 2011 *Phys. Rev. B* **84** 235428
[13] Arrachea L, Mucciolo E R, Chamon C and Capaz R B 2012 *Phys. Rev. B* **86** 125424
[14] Wijewardane H O and Ullrich C A 2005 *Phys. Rev. Lett.* **95** 086401
[15] Verdozzi C, Stefanucci G and Almbladh C O 2006 *Phys. Rev. Lett.* **97** 046603
[16] Myöhänen P, Stan A, Stefanucci G and van Leeuwen R 2008 *EPL* **84** 67001
[17] Uimonen A M, Khosravi E, Stan A, Stefanucci G, Kurth S, van Leeuwen R and Gross E K U 2011 *Phys. Rev. B* **84** 115103
[18] Myöhänen P, Tuovinen R, Korhonen T, Stefanucci G and van Leeuwen R 2012 *Phys. Rev. B* **85** 075105
[19] Latini S, Perfetto E, Uimonen A M, van Leeuwen R and Stefanucci G 2014 *Phys. Rev. B* **89** 075306
[20] Landauer R 1957 *IBM J. Res. Dev.* **1** 233
[21] Büttiker M 1986 *Phys. Rev. Lett.* **57** 1761
[22] Caroli C, Combescot R, Nozières P and Saint-James D 1971 *J. Phys. C* **4** 916
[23] Caroli C, Combescot R, Lederer D, Nozières P and Saint-James D 1971 *J. Phys. C* **4** 2598
[24] Cini M 1980 *Phys. Rev. B* **22** 5887

[25] Meir Y and Wingreen N S 1992 *Phys. Rev. Lett.* **68** 2512

[26] Jauho A P, Wingreen N S and Meir Y 1994 *Phys. Rev. B.* **50** 5528

[27] Perfetto E, Stefanucci G and Cini M 2008 *Phys. Rev. B* **78** 155301

[28] Stefanucci G and van Leeuwen R, *Nonequilibrium Many-Body Theory of Quantum systems: A Modern Introduction* (Cambridge Univerisity Press, 2013).

[29] Tuovinen R, Perfetto E, van Leeuwen R and Stefanucci G 2013 *J. Phys.: Conf. Ser.* **427** 012014

[30] Tuovinen R, Perfetto E, Stefanucci G and van Leeuwen R 2014 *Phys. Rev. B* **89** 085131

[31] Tuovinen R, Perfetto E, Stefanucci G and van Leeuwen R In preparation

[32] Ridley M, MacKinnon A and Kantorovich L 2015 *Phys. Rev. B* **91** 125433

[33] Jauho A P 2000 Time-dependent transport in interacting mesoscopic systems *Progress in Nonequilibrium Green's Functions* ed Bonitz M, Nareyka R and Semkat D (World Scientific)

[34] Stefanucci G, Perfetto E and Cini M 2010 *Phys. Rev. B.* **81** 115446

[35] Zeng Z Y, Li B and Claro F 2003 *Phys. Rev. B.* **68** 115319

[36] Nambu Y 1960 *Phys. Rev.* **117** 648

[37] Stefanucci G, Perfetto E and Cini M 2010 *J. Phys.: Conf. Ser.* **220** 012012

[38] Bogoliubov N N 1958 *Sov. Phys. JETP* **34** 58

[39] de Gennes P G 1966 *Superconductivity of Metals and Alloys* (New York: Benjamin)

[40] Leijnse M and Flensberg K 2012 *Semicond. Sci. Technol.* **27** 124003

[41] Kubo R 1957 *J. Phys. Soc. Jpn.* **12** 570

[42] Martin P C and Schwinger J 1959 *Phys. Rev.* **115** 1342

[43] Weisstein E W Digamma function. From MathWorld—A Wolfram Web Resource URL `http://mathworld.wolfram.com/DigammaFunction.html`

[44] Weisstein E W Hypergeometric function. From MathWorld—A Wolfram Web Resource URL `http://mathworld.wolfram.com/HypergeometricFunction.html`

[45] Kivelson S and Chapman O L 1983 *Phys. Rev. B* **28** 7236

[46] Averin D V and Nazarov Y V 1992 *Phys. Rev. Lett.* **69** 1993

[47] Maisi V F, Lotkhov S V, Kemppinen A, Heimes A, Muhonen J T and Pekola J P 2013 *Phys. Rev. Lett.* **111** 147001

[48] Heimes A, Maisi V F, Golubev D S, Marthaler M, Schön G and Pekola J 2014 *Phys. Rev. B* **89** 014508

[49] Hekking F W J, Glazman L I, Matveev K A and Shekhter R I 1993 *Phys. Rev. Lett.* **70** 4138

[50] Hanke U, Galperin Y, Chao K A, Gisselfält M, Jonson M and Shekhter R I 1995 *Phys. Rev. B* **51** 9084

[51] Ojeda-Aristizabal C, Ferrier M, Guéron S and Bouchiat H 2009 *Phys. Rev. B* **79** 165436

[52] Komatsu K, Li C, Autier-Laurent S, Bouchiat H and Guéron S 2012 *Phys. Rev. B* **86** 115412

[53] Haberer D, Petaccia L, Fedorov A V, Praveen C S, Fabris S, Piccinin S, Vilkov O, Vyalikh D V, Preobrajenski A, Verbitskiy N I, Shiozawa H, Fink J, Knupfer M, Büchner B and Grüneis A 2013 *Phys. Rev. B* **88** 081401

[54] Yang S L, Sobota J A, Howard C A, Pickard C J, Hashimoto M, Lu D H, Mo S K, Kirchmann P S and Shen Z X 2014 *Nat Commun* **5** 3493

[55] Hancock Y, Uppstu A, Saloriutta K, Harju A and Puska M J 2010 *Phys. Rev. B* **81** 245402

[56] Robinson J P, Schomerus H, Oroszlány L and Fal'ko V I 2008 *Phys. Rev. Lett.* **101** 196803

Journal of Physics: Conference Series **696** (2016) 012017 doi:10.1088/1742-6596/696/1/012017

Calculation of the current response in a nanojunction for an arbitrary time-dependent bias: application to the molecular wire

Michael Ridley,[1,2] Angus MacKinnon,[2] and Lev Kantorovich[1]

[1]Department of Physics, King's College London, Strand, London, WC2R 2LS, United Kingdom

[2]Department of Physics, The Blackett Laboratory, Imperial College London, South Kensington Campus, London SW7 2AZ, United Kingdom

E-mail: michael.ridley12@imperial.ac.uk

Abstract. Recently [Phys. Rev. B **91**, 125433 (2015)] we derived a general formula for the time-dependent quantum electron current through a molecular junction subject to an arbitrary time-dependent bias within the Wide Band Limit Approximation (WBLA) and assuming a single particle Hamiltonian. Here we present an efficient numerical scheme for calculating the current and particle number. Using the Padé expansion of the Fermi function, it is shown that all frequency integrals occurring in the general formula for the current can be removed analytically. When the bias in the reservoirs is assumed to be sinusoidal it is possible to manipulate the general formula into a form containing only summations over special functions. To illustrate the method, we consider electron transport through a one-dimensional molecular wire coupled to two leads subject to out-of-phase biases. We also investigate finite size effects in the current response and particle number that result from the switch-on of this bias.

1. Introduction

The Landauer-Büttiker (LB) formalism has for some time provided a reliable method for modelling coherent transport through molecular junctions [1, 2]. In the original formulation, this method enabled computation of the current in the leads coupled to a molecular region C, when a bias in at least one of the leads causes a constant shift in the Fermi energies of the leads. Work done with the LB formalism has mainly focused on the calculation of steady-state properties, corresponding to a time regime that consigns the bias switch-on time to the distant past [3]. The LB expressions for the current and quantum noise may be derived from the Nonequilibrium Green's Function (NEGF) formalism, which is a more general alternative to the S-Matrix approach [4, 5]. This requires the solution of the Kadanoff-Baym equations for the lesser Green's function of the molecular region [4], from which the electric current in the leads and electron number in the molecule may be extracted. When the bias is time-independent, all GFs are functions of the time difference, and hence it is possible to work purely in the frequency domain when solving for the components that enter the expression for the current.

However, recent studies have opened up the possibility of calculating the time-dependent (TD) current in a multilevel nanojunction using a closed integral formula; expressions that include the transient effects of the switch-on (and hence the system preparation) have been derived for the cases of a constant [4, 6, 7] and time-dependent [8] bias. Common to both approaches is the assumption of the Wide-Band Limit Approximation (WBLA), which neglects the detailed electronic structure of the leads and renders the Kadanoff-Baym equations for the lesser GF analytically tractable. Computationally, the main cost of the method we presented in [8] lies with the evaluation of several integrals in the (ω, t) plane. The purpose of this paper is to present an efficient computational scheme for evaluating these integrals with no loss of accuracy making consideration of complex multi-level and multi-terminal systems within the WBLA numerically possible.

The structure of this paper is as follows: in Section 2, we introduce the model of the generic switch-on problem and the basic assumptions involved in the WBLA. In addition we prove a continuity equation for the current formula derived in [8]. In Section 3, the formula for the current is re-expressed using the Padé expansion of the Fermi function to remove all frequency integrals, forming the basis for our numerical implementation. Section 4 presents numerical results for the current through a molecular wire, modelled with a simple tight-binding Hamiltonian. In this implementation, expressions are worked out for a sinusoidal bias whereby Bessel function expansions are used to remove all $t-$integrals in the TD current. This enables simulations of the transport resulting from any combination of sinusoidal biases in the leads, including those which contain a symmetry-breaking phase. Furthermore it enables us to identify some novel finite size effects on the dynamics and to study the conservation of charge as the transient current dies out.

2. The Model

2.1. Hamiltonian and GF Components

The NEGF formalism is a method for calculating the evolution in time of ensemble averages, following an event at a particular time t_0 which breaks the physical symmetry of the system between times $t = \pm\infty$. In quantum transport one is primarily concerned with calculations of the time-dependent current following the switch-on of a bias in one of the leads. Here we model this switch-on by adding a spatially constant time-dependent shift to the lead energy levels.

We will in general be concerned with the following Hamiltonian:

$$\hat{H}(t) = \sum_{k\alpha} \varepsilon_{k\alpha}(t)\, \hat{d}^{\dagger}_{k\alpha}\hat{d}_{k\alpha} + \sum_{mn} V_{mn}\hat{d}^{\dagger}_{m}\hat{d}_{n} + \sum_{m,k\alpha}\left[V_{m,k\alpha}\hat{d}^{\dagger}_{m}\hat{d}_{k\alpha} + V_{k\alpha,m}\hat{d}^{\dagger}_{k\alpha}\hat{d}_{m}\right] \tag{1}$$

Here, $\hat{d}_{k\alpha}$, $\hat{d}_m$ and $\hat{d}^{\dagger}_{k\alpha}$, $\hat{d}^{\dagger}_m$ are annihilation and creation operators of leads and central system electronic states, where for simplicity spin degrees of freedom are neglected. We collect elements of this Hamiltonian into a matrix consisting of 'blocks' corresponding to each of the physical subsystems it describes. The first term is a Hamiltonian of the lead states k belonging to each lead α. The second term is the Hamiltonian of the molecule coupled to the lead; it contains inter-molecular hopping matrix elements V_{mn} between the m and n molecular orbitals, defining a molecular Hamiltonian matrix 'block' $\mathbf{h}_{CC}$ with elements V_{mn}. The third term describes the coupling of the molecule to the leads, and defines the $\alpha - C$ 'block' of the Hamiltonian matrix, $\mathbf{h}_{\alpha C}$ with elements $V_{k\alpha,m}$. The switch-on problem is specified by assigning values to the matrix elements of $\hat{H}(t)$ at times before and after t_0.

We assume that, prior to t_0, the system has equilibrated with lead-molecule couplings present in the Hamiltonian $\hat{H}(t < t_0) \equiv \hat{H}_0$, so that $\hat{H}_0$ is as in Eq. (1) with time-independent lead state energies $\varepsilon_{k\alpha}(t < t_0) = \varepsilon_{k\alpha}$. The system is therefore described by an equilibrium density operator $\hat{\rho}_0 = Z^{-1}e^{-\beta(\hat{H}_0 - \mu\hat{N})}$, where Z is the partition function, $\hat{N}$ the number operator, $\beta = 1/k_B T$ the inverse temperature and μ the chemical potential. At times following the switch-on, the lead energies acquire a time-dependent shift, $\varepsilon_{k\alpha}(t > t_0) = \varepsilon_{k\alpha} + V_\alpha(t)$, where the function $V_\alpha(t)$ is arbitrary. This is an example of a *partition-free* approach [9, 6, 8] to the switch-on problem,

Journal of Physics: Conference Series **696** (2016) 012017 doi:10.1088/1742-6596/696/1/012017

as the system Hamiltonian is not divided into decoupled subregions prior to t_0. *Partitioned* approaches [5] add the lead-molecule coupling at the switch-on time and therefore cannot be expected to give a physically meaningful transient.

The problem of calculating the current in response to this kind of switch-on is mapped to the solution of a set of coupled integro-differential equations for Green's Functions (GFs) defined on a complex time contour γ, which we refer to as the Konstantinov-Perel' contour [10]. One then solves for various components of the two-time GF of the central region CC 'block', whose rigorous definition may be found elsewhere [4, 8]. Here we simply note that the retarded GF satisfies the following equation of motion:

$$\left(i\frac{d}{dt_1} - \mathbf{h}_{CC}\right)\mathbf{G}^r_{CC}(t_1,t_2) = \mathbf{1}_{CC}\delta(t_1-t_2) + \int_{t_0}^{t} d\bar{t}\,\boldsymbol{\Sigma}^r_{CC}(t_1,\bar{t})\,\mathbf{G}^r_{CC}(\bar{t},t_2) \tag{2}$$

On the right-hand side of this expression, one encounters the retarded component of the embedding self-energy,

$$\left[\boldsymbol{\Sigma}^r_{CC}(t_1,t_2)\right]_{mn} = \sum_\alpha e^{-i\psi_\alpha(t_1,t_2)}\int\frac{d\omega}{2\pi}e^{i\omega(t_1-t_2)}\left[\Lambda_{\alpha,mn}(\omega) - \frac{i}{2}\Gamma_{\alpha,mn}(\omega)\right] \tag{3}$$

We have in this expression introduced the mn component of the level width matrix in lead α

$$\Gamma_{\alpha,mn}(\omega) = 2\pi\sum_k T_{m,k\alpha}T_{k\alpha,n}\delta(\omega-\epsilon_{k\alpha}) \tag{4}$$

which forms a Hilbert transform pair with $\Lambda_{\alpha,mn}(\omega)$. In addition we have defined the phase factor:

$$\psi_\alpha(t_1,t_2) = \int_{t_2}^{t_1} d\bar{t}\,V_\alpha(\bar{t}) \tag{5}$$

In the WBLA, one assumes that $\Gamma_{\alpha,mn}(\omega)$ is replaced with a frequency-independent value $\Gamma_{\alpha,mn} = \Gamma_{\alpha,mn}(\varepsilon^F_\alpha)$, where ε^F_α is the equilibrium Fermi energy of lead α [5]. This has the effect of mapping Eq. (2) onto an expression which is analytically tractable, resulting in the following exact expression:

$$\mathbf{G}^r_{CC}(t_1,t_2) = \int\frac{d\omega}{2\pi}e^{i\omega(t_1-t_2)}\left(\omega\mathbf{1}_{CC} - \mathbf{h}^{eff}_{CC}\right)^{-1} \equiv \int\frac{d\omega}{2\pi}e^{i\omega(t_1-t_2)}\mathbf{G}^r_{CC}(\omega) \tag{6}$$

In (6), $\mathbf{h}^{eff}_{CC} \equiv \mathbf{h}_{CC} - \frac{i}{2}\boldsymbol{\Gamma} \equiv \mathbf{h}_{CC} - \frac{i}{2}\sum_\alpha\boldsymbol{\Gamma}_\alpha$ is an effective Hamiltonian of the central region, whose eigenvalues correspond to unstable eigenmodes of the molecular structure, which have acquired a finite lifetime due to the presence of the leads. All other components of the GF can also be explicitly calculated in the time domain [8]. In particular, the greater and lesser GFs can be expressed as

$$\mathbf{G}^{\gtrless}_{CC}(t_1,t_2) = \mp i\int\frac{d\omega}{2\pi}f(\mp(\omega-\mu))\sum_\alpha\mathbf{S}_\alpha(t_1,t_0;\omega)\,\boldsymbol{\Gamma}_\alpha\mathbf{S}^\dagger_\alpha(t_2,t_0;\omega) \tag{7}$$

where we make the definition

$$\mathbf{S}_\alpha(t,t_0;\omega) \equiv e^{-i\mathbf{h}^{eff}_{CC}(t-t_0)}\left[\mathbf{G}^r_{CC}(\omega) - i\int_{t_0}^{t}d\bar{t}e^{-i\left(\omega\mathbf{1}-\mathbf{h}^{eff}_{CC}\right)(\bar{t}-t_0)}e^{-i\psi_\alpha(\bar{t},t_0)}\right] \tag{8}$$

and note that all information concerning the time-dependent bias $V_\alpha(t)$ of lead α is to be found in the phase factors ψ_α.

Journal of Physics: Conference Series **696** (2016) 012017 doi:10.1088/1742-6596/696/1/012017

2.2. Current and Continuity Equation

In [8], results were presented for the transport through a quantum dot using direct numerical integration to evaluate the current at each time $t > t_0$. The current in lead α is defined as the time derivative of the average charge in that lead, $I_\alpha(t) \equiv q\left\langle \frac{d\hat{N}_\alpha(t)}{dt} \right\rangle$, and may be expressed in the following rather compact form (taking electron charge $q = -1$ and a spin degeneracy of 2, i.e. $\hat{N}_\alpha = 2 \times \sum_k \hat{d}^\dagger_{k\alpha}\hat{d}_{k\alpha}$):

$$I_\alpha(t) = \frac{1}{\pi}\int d\omega f(\omega-\mu)\,\mathrm{Tr}_C\left[2\mathrm{Re}\left[ie^{i\omega(t-t_0)}e^{i\psi_\alpha(t,t_0)}\mathbf{S}_\alpha\mathbf{\Gamma}_\alpha\right] - \mathbf{\Gamma}_\alpha\sum_\beta \mathbf{S}_\beta\mathbf{\Gamma}_\beta\mathbf{S}^\dagger_\beta\right] \tag{9}$$

Here, the $\mathbf{S}_\alpha$ are given by Eq. (8). The molecule which has been sandwiched between macroscopic leads is fundamentally an open quantum system, so the total charge in the molecular region is not necessarily conserved even though the total amount of charge in the molecular region plus that of all the leads is fixed for all times. This is a condition automatically satisfied by the NEGF formalism [4]. The condition for local charge conservation is that the sum of all currents measured in the leads of a multiterminal device is equal to zero:

$$\sum_\alpha I_\alpha(t) = 0 \tag{10}$$

This condition states that the rate of charge entering the molecular region is balanced by the rate of charge leaving it. One should now recall that there is no direct coupling between leads, so that all charge transfer between leads is mediated by the central region, and also that the definition of the current in lead α was given as $I_\alpha(t) \equiv q\left\langle \frac{d\hat{N}_\alpha(t)}{dt} \right\rangle$, i.e. it was given as the rate of increase of charge in this lead. Eq. (10) is simply a statement of the condition that if charge increases in one lead connected to the molecule, then this increase is compensated exactly by a net decrease of charge from the other leads making up the junction. The total amount of charge in the molecular region at any given time is directly related to the lesser GF:

$$N_C(t) = -2i\mathrm{Tr}_C\left[\mathbf{G}^<_{CC}(t,t)\right] = \frac{1}{\pi}\sum_\beta\int d\omega f(\omega-\mu)\,\mathrm{Tr}_C\left[\mathbf{S}_\beta(t,t_0;\omega)\mathbf{\Gamma}_\beta\mathbf{S}^\dagger_\beta(t,t_0;\omega)\right] \tag{11}$$

The time derivative of the charge number $\frac{dN_C(t)}{dt}$ in the central region has been referred to as the *displacement current* in the literature [4]. To make precise the relation between this quantity and the currents measured in the leads, we take the time-derivatives of the matrix $\mathbf{S}_\beta \equiv \mathbf{S}_\beta(t,t_0;\omega)$ and its complex conjugate, e.g.

$$\frac{d\mathbf{S}_\beta}{dt} = -i\mathbf{h}^{eff}_{CC}\mathbf{S}_\beta - ie^{-i\omega(t-t_0)}e^{-i\psi_\beta(t,t_0)} \tag{12}$$

This expression, along with its complex conjugate, is substituted into the derivative of N_C in Eq. (11) to give

$$\frac{d\mathbf{N}_C(t)}{dt} = \frac{1}{\pi}\sum_\alpha\int d\omega f(\omega-\mu)\,\mathrm{Tr}_C\left[2\mathrm{Re}\left[ie^{i\omega(t-t_0)}e^{i\psi_\alpha(t,t_0)}\mathbf{S}_\beta\mathbf{\Gamma}_\beta\right] - \sum_\alpha\mathbf{\Gamma}_\alpha\mathbf{S}_\beta\mathbf{\Gamma}_\beta\mathbf{S}^\dagger_\beta\right] \tag{13}$$

where we use the fact that $\mathbf{h}^{eff}_{CC} - \mathbf{h}^{eff\dagger}_{CC} = -i\sum_\alpha\mathbf{\Gamma}_\alpha \equiv -i\mathbf{\Gamma}$. If we then sum over lead indices in the expression (9), the following theorem is established:

$$\frac{d\mathbf{N}_C(t)}{dt} = \sum_\alpha I_\alpha(t) \tag{14}$$

This is a statement of global charge conservation in the nanojunction. When the bias in every lead is constant, (11) reduces to the formula for the particle number given in [6]. In the long time limit the number of charges on the molecule is constant in time, i.e. an ideal stationary state is reached; hence the condition for local charge conservation (10) is trivially satisfied in this case. Although the condition of global charge conservation (14) is never violated, physical situations exist in which the condition of local charge conservation in the molecular (central) region of a nanojunction (10) is then violated, i.e. $dN_C(t)/dt \neq 0$.

3. Pade Expansion of the Time-Dependent Current

3.1. Exact Formula for the Current: Hurwitz-Lerch Functions

In previous work [8], numerical integration was used to compute the current formula (9) at different times for a single level system. We now seek to extend this work to a numerical method that facilitates time-dependent transport calculations in systems of real experimental interest, for instance carbon nanotube transistors [11]. To proceed with a numerical implementation of Eqs. (9) and (11), one can introduce the right and left eigenproblems for the renormalized Hamiltonian matrix $\mathbf{h}_{CC}^{eff}$ [12]:

$$\mathbf{h}_{CC}^{eff}\left|\varphi_j^R\right\rangle = \bar{\varepsilon}_j\left|\varphi_j^R\right\rangle \text{ and } \left\langle\varphi_j^L\right|\mathbf{h}_{CC}^{eff} = \bar{\varepsilon}_j\left\langle\varphi_j^L\right| \tag{15}$$

The eigenenergies $\bar{\varepsilon}_j$ contain an imaginary part that is strictly negative (as $\mathbf{\Gamma}$ is positive-definite), and the same value of $\bar{\varepsilon}_j$ corresponds to each of the left and right eigenvectors. Using the idempotency property

$$\sum_j \frac{\left|\varphi_j^R\right\rangle\left\langle\varphi_j^L\right|}{\left\langle\varphi_j^L \mid \varphi_j^R\right\rangle} = \mathbf{I} = \sum_j \frac{\left|\varphi_j^L\right\rangle\left\langle\varphi_j^R\right|}{\left\langle\varphi_j^R \mid \varphi_j^L\right\rangle}, \tag{16}$$

the formula (9) can be put into the form:

$$\begin{aligned} I_\alpha(t) = {} & \frac{1}{\pi}\sum_j \int dx\, f(x) \Bigg\{ 2\mathrm{Re}\Bigg[\frac{\left\langle\varphi_j^L\right|\mathbf{\Gamma}_\alpha\left|\varphi_j^R\right\rangle}{\left\langle\varphi_j^L\middle|\varphi_j^R\right\rangle}\Bigg(i\frac{e^{i\psi_\alpha(t,t_0)}e^{i(x+\mu-\bar{\varepsilon}_j)(t-t_0)}}{x+\mu-\bar{\varepsilon}_j} \\ & + \int_{t_0}^{t} d\bar{t}\, e^{i\psi_\alpha(t,\bar{t})}e^{i(x+\mu-\bar{\varepsilon}_j)(t-\bar{t})}\Bigg)\Bigg] - \sum_{\beta,k}\frac{\left\langle\varphi_k^R\right|\mathbf{\Gamma}_\alpha\left|\varphi_j^R\right\rangle\left\langle\varphi_j^L\right|\mathbf{\Gamma}_\beta\left|\varphi_k^L\right\rangle}{\left\langle\varphi_j^L\middle|\varphi_j^R\right\rangle\left\langle\varphi_k^R\middle|\varphi_k^L\right\rangle} e^{-i\left(\bar{\varepsilon}_j-\bar{\varepsilon}_k^*\right)(t-t_0)} \\ & \times\Bigg[\frac{1}{\left(x+\mu-\bar{\varepsilon}_j\right)\left(x+\mu-\bar{\varepsilon}_k^*\right)} + \int_{t_0}^{t} d\bar{t}\Bigg[i\frac{e^{i\psi_\beta(\bar{t},t_0)}e^{i\left(x+\mu-\bar{\varepsilon}_k^*\right)(\bar{t}-t_0)}}{x+\mu-\bar{\varepsilon}_j} + c.c_{j\leftrightarrow k}\Bigg] \\ & + \int_{t_0}^{t} d\bar{t}\int_{t_0}^{t} d\bar{t}'\, e^{-i\psi_\beta(\bar{t},\bar{t}')}e^{-i(x+\mu-\bar{\varepsilon}_j)(\bar{t}-t_0)}e^{i\left(x+\mu-\bar{\varepsilon}_k^*\right)(\bar{t}'-t_0)}\Bigg]\Bigg\} \end{aligned} \tag{17}$$

where $f(x)$ is the Fermi function and $c.c_{j\leftrightarrow k}$ denotes the complex conjugate of the preceding term with indices j and k exchanged. We therefore reduce $I_\alpha(t)$ to a sum of integrals over scalar functions in the (x, t) plane. One integral whose structure is repeated throughout (17) can be performed analytically, using a contour integral over a semi-circular contour in the upper-half of the complex plane:

$$\int dx \frac{f(x)\, e^{i(x-(\bar{\varepsilon}_j-\mu))(t-t_0)}}{x-(\bar{\varepsilon}_j-\mu)} = -e^{-\frac{\pi}{\beta}(t-t_0)} e^{-i(\bar{\varepsilon}_j-\mu)(t-t_0)} \Phi\left(e^{-\frac{2\pi}{\beta}(t-t_0)}, 1, \frac{1}{2} - \frac{\beta}{2i\pi}(\bar{\varepsilon}_j - \mu)\right) \tag{18}$$

Here we have introduced the so-called *Hurwitz-Lerch Transcendent* Φ [13]:

$$\Phi(z, s, a) \equiv \sum_{n=0}^{\infty} \frac{z^n}{(n+a)^s} \tag{19}$$

The formula (18) enables to replace several frequency integrals with a fast-converging series expansion. We note in passing that, if one follows the steps in the formal integration of the lesser GF $\mathbf{G}^{<}_{CC}(t_1, t_2)$ which enabled the derivation of Eq. (9) in [8], one uses a well-known [4] transformation of Matsubara summations into frequency integrals when evaluating a set of convolution integrals taken along the vertical branch of the Konstantinov-Perel' contour. However, these steps can be omitted altogether, and a direct route can be taken to an expression for $I_\alpha(t)$ in terms of Φ which retains the Matsubara summation at all stages.

3.2. Removal of Frequency Integrals: Padé Expansion

The Padé approximation to the Fermi function is a series expansion whose terms possess a simple pole structure [14, 15, 16]:

$$f(x) = \frac{1}{e^{\beta x}+1} = \frac{1}{2} - \lim_{N_p\to\infty} \sum_{l=1}^{N_p} \eta_l \left(\frac{1}{\beta x + i\zeta_l} + \frac{1}{\beta x - i\zeta_l}\right) \tag{20}$$

This is identical in structure to the Matsubara expansion, which has parameter values $\eta_l = 1$ and $\zeta_l = \pi(2l-1)$, but it can be shown to converge much more quickly than the Matsubara expansion as N_p increases [15, 16]. Unlike the Matsubara expansion, in which the poles are spaced at constant intervals of $\frac{2\pi}{\beta}$ along the imaginary axis, the poles $i\zeta_l/\beta$ in the Padé expansion are spaced unevenly along the imaginary axis, and the prefactors η_l are all real and positive-valued. In practice, the expansion $f_N(x)$ truncated at some finite value of N_p is used, with N_p chosen such that for values of $L \gg \varepsilon$, where ε denotes the typical energy scale of the problem, the deviation $\delta f_N(L) \equiv |f(x) - f_N(x)|_{\beta x = L} < 10^{-p}$, where p can be chosen to give arbitrary accuracy [15, 16, 17]. The rate of convergence of the Padé expansion to the Fermi function at all energies increases as the temperature is increased.

The expansion (20) is substituted into (17), resulting in x-integrals that can be evaluated analytically with the residue theorem, for example:

$$\int dx f(x)\, e^{ix(t-\bar{t})} = \pi\left[\delta(t-\bar{t}) - \frac{2i}{\beta} \lim_{N_p\to\infty} \sum_{l=1}^{N_p} \eta_l e^{-\frac{\zeta_l}{\beta}(t-\bar{t})}\right] \tag{21}$$

Using this method and the identity (18) one finally obtains an expression for the current in lead α that is asymptotically exact (as $N \to \infty$), and which has all frequency integrals eliminated:

$$
\begin{aligned}
I_\alpha(t) \simeq \frac{1}{\pi}\sum_j \Bigg[2\mathrm{Re}\Bigg[\frac{\langle \varphi_j^L | \mathbf{\Gamma}_\alpha | \varphi_j^R \rangle}{\langle \varphi_j^L | \varphi_j^R \rangle} \Big(-ie^{i\psi_\alpha(t,t_0)} e^{-i(\bar{\varepsilon}_j-\mu)(t-t_0)} \bar{\Phi}\left(\beta, -(\bar{\varepsilon}_j-\mu)\right) \\
- \sum_{l=1}^{N_p} \frac{2\eta_l \pi i}{\beta} \int_{t_0}^{t} d\bar{t} e^{-i\left(\bar{\varepsilon}_j-\mu-i\frac{\varsigma_l}{\beta}\right)(t-\bar{t})} e^{i\psi_\alpha(t,\bar{t})} \Big) \Bigg] \\
- \sum_{\beta,k} \frac{\langle \varphi_k^R | \mathbf{\Gamma}_\alpha | \varphi_j^R \rangle \langle \varphi_j^L | \mathbf{\Gamma}_\beta | \varphi_k^L \rangle}{\langle \varphi_j^L | \varphi_j^R \rangle \langle \varphi_k^R | \varphi_k^L \rangle} e^{-i\left(\bar{\varepsilon}_j-\bar{\varepsilon}_k^*\right)(t-t_0)} \Bigg[\frac{\Psi\left(\frac{1}{2}-\frac{\beta}{2i\pi}(\bar{\varepsilon}_j-\mu)\right) - \Psi\left(\frac{1}{2}+\frac{\beta}{2i\pi}(\bar{\varepsilon}_k^*-\mu)\right)}{\bar{\varepsilon}_j-\bar{\varepsilon}_k^*} \\
\int_{t_0}^{t} d\bar{t} \left[-ie^{-i\left(\bar{\varepsilon}_k^*-\mu\right)(\bar{t}-t_0)} e^{i\psi_\beta(\bar{t},t_0)} \bar{\Phi}\left(\beta, -(\bar{\varepsilon}_j-\mu)\right) + c.c_{j\leftrightarrow k} \right] \\
+ \sum_{l=1}^{N_p} \frac{2\eta_l \pi i}{\beta} \int_{t_0}^{t} d\bar{t} d\bar{t}' e^{i(\bar{\varepsilon}_j-\mu)(\bar{t}-t_0)} e^{-i(\bar{\varepsilon}_k^*-\mu)(\bar{t}'-t_0)} e^{-i\psi_\beta(\bar{t},\bar{t}')} \left[\theta\left(\bar{t}-\bar{t}'\right) e^{-\frac{\varsigma_l}{\beta}(\bar{t}-\bar{t}')} - \theta\left(\bar{t}'-\bar{t}\right) e^{-\frac{\varsigma_l}{\beta}(\bar{t}'-\bar{t})} \right] \Bigg] \Bigg]
\end{aligned} \tag{22}
$$

Here we have introduced the *digamma function* Ψ, defined as the logarithmic derivative of the complex gamma function [5], and the simplified notation $\bar{\Phi}(\beta, z) \equiv \exp\left(-\frac{\pi}{\beta}(t-t_0)\right) \Phi\left(e^{-\frac{2\pi}{\beta}(t-t_0)}, 1, \frac{1}{2}+\frac{\beta z}{2i\pi}\right)$. Importantly, the difference of two digamma functions appearing in the expression for $I_\alpha(t)$ can be converted into a well converging numerical series thanks to the property:

$$
\frac{\Psi(z_1) - \Psi(z_2)}{z_1 - z_2} = \sum_{n=0}^{\infty} \frac{1}{(n+z_1)(n+z_2)} \tag{23}
$$

The general result (22) forms the basis for all subsequent numerical work involving the calculation of the current response to an arbitrary time-dependent bias. In practice, the time integrals may be performed numerically or removed analytically given the assumption of a particular functional form for $V_\alpha(t)$. Several of the decaying modes in the current can be directly identified by inspection of (22), although the exact nature of its long-time behaviour depends on the bias chosen. In particular, if we separate out the real and imaginary parts of the complex eigenvalues, $\bar{\varepsilon}_j = \lambda_j - i\gamma_j$, we see that modes appearing in the single summation decay as $e^{-\gamma_j(t-t_0)}$, whereas the prefactor governing the decay of modes in the double sum is $e^{-(\gamma_j+\gamma_k)(t-t_0)}$. This defines a timescale of $\tau \equiv Max\{1/\gamma_j\}$ as the time taken for transient behaviour to vanish following the switch-on.

4. Results

We shall now apply the method which was outlined in Section 3 to a molecular wire coupled to two leads, illustrated schematically in Fig. 1(a). We describe this system using a tight-binding model for the wire with nearest-neighbor hopping, as described in [18, 19]. Specifically, we assume that the leads are connected by a wire of 5 sites, with one state per site, and that the interaction of the chain orbitals with those in the leads is only via the end sites. Within the WBLA, this translates into the condition that only the Γ_{11} and Γ_{55} elements in the level-width matrix are non-zero and are given by $\Gamma_{11} = 2\pi \sum_{i\in L} |V_{i1}|^2 \delta\left(\varepsilon_L^F - \varepsilon_{iL}\right)$ and $\Gamma_{55} = 2\pi \sum_{j\in R} |V_{Nj}|^2 \delta\left(\varepsilon_R^F - \varepsilon_{jR}\right)$, respectively. For simplicity, we assume that for each site in the chain there is a single energy level, so that effectively we are describing a chain of quantum dots coupled to each other with a nearest-neighbor hopping, similarly to the system studied in [20], resulting in a tridiagonal molecular Hamiltonian matrix $\mathbf{h}_{CC}$, with elements $[\mathbf{h}_{CC}]_{k,k} \equiv E$ and $[\mathbf{h}_{CC}]_{k,k+1} = [\mathbf{h}_{CC}]_{k+1,k} \equiv \tau$. Additionally, we choose the chemical potential μ to be zero.

We now focus on the case of a sinusoidal AC bias turned on in each lead, with a frequency, amplitude, constant part and phase shift that depends upon the lead:

$$V_\alpha(t) = V_\alpha + A_\alpha \cos(\Omega_\alpha (t - t_0) + \phi_\alpha) \tag{24}$$

This enables us to remove analytically all the remaining integrals in (22), by use of the identity:

$$e^{i\psi_\alpha(t_1,t_2)} = e^{iV_\alpha(t_1-t_2)} \sum_{r,s=-\infty}^{\infty} J_r\left(\frac{A_\alpha}{\Omega_\alpha}\right) J_s\left(\frac{A_\alpha}{\Omega_\alpha}\right) e^{i(r-s)\phi_\alpha} e^{ir\Omega_\alpha(t_1-t_0)} e^{-is\Omega_\alpha(t_2-t_0)} \tag{25}$$

where J_r denotes a Bessel function of the first kind of order r. When Eq. (25) is substituted into (22), the single and double time integrals in (22) can be removed analytically and the problem of calculating the time-dependent current response is reduced to a simple embedded summation over left/right eigenvectors, the Padé parameters and Bessel functions. For the special case of the sinusoidal bias (24), this summation can be expressed entirely in terms of the special functions (19) and (23). We confirm that in the case of a single quantum dot our method reproduces exactly the $I - t$ characteristic curves published in [8], which were obtained by direct numerical integration of (9).

In this work, we apply the bias (24) to the leads L and R in the system shown in Fig. 1(a). The amplitudes A_α and constant bias shifts V_α are identical for each lead, as is the driving frequency of the bias, Ω_α. The physical symmetry of this transport problem is then broken by a relative phase-shift of $\phi_R = -\pi/2$ of $V_R(t)$ with respect to $V_L(t)$, whose phase is fixed at $\phi_L = 0$. The tridiagonal central region Hamiltonian has all diagonal terms equal to 1, and all off-diagonal terms $\tau = 0.1$. Other parameters in the model are $V_L = 5 = V_R$, $A_L = 4 = A_R$, $\Gamma_{11} = 0.5$, $\Gamma_{55} = 0.5$, and the inverse temperature $\beta = 10$. We evaluate Eq. (22) using the Padé parameters, and for this temperature we find good convergence with a Padé expansion order of $N_p \sim 20$. The resulting current in each lead is plotted in Fig. 1(b) (thick lines), along with the suitably rescaled bias applied to the leads (dotted lines). In addition, we plot the steady-state value of the current at each time, using the LB formula. This serves as a comparison between the best predictions of the steady-state formalism and our fully time-dependent method, from which it is instantly apparent that the TD current amplitude is an order of magnitude greater than that of the LB current. There is a much sharper peak in the transient current in lead L than in R, because the bias applied in L has the value $V_L + A_L$ at the switch-on time $t = 0$, whereas the corresponding bias in lead R has the value V_R. In the long-time regime, the current signal in each lead has the same form but the signal in R is phase-shifted by $-\pi/2$ with respect to the signal in L. Additionally, 'ringing' oscillations in the current occurring at higher frequencies than the driving frequency Ω_α can be seen, as also observed in [5] and [8]. We identify the physical source of these frequencies in the energy gap $|V_\alpha + \mu - E|$ between onsite energies and the constant bias shift in the leads.

One can also see from Fig. 1(b) that, whereas the LB formalism preserves the condition of charge conservation, $I_L(t) + I_R(t) = 0$, at all times, this condition is not satisfied by the two TD currents. We understand this as follows: electrons propagate in the lead at the finite Fermi velocity v_α^F, so that the effects of a bias applied locally will take a finite time to propagate to other parts of the junction [4]. If a constant bias is switched on in each lead, then following a time delay equal to the transmission time of this signal across the junction, the rate of flow of electrons from C to L (I_L) must equal the rate of electron transport from R to C ($-I_R$), as in

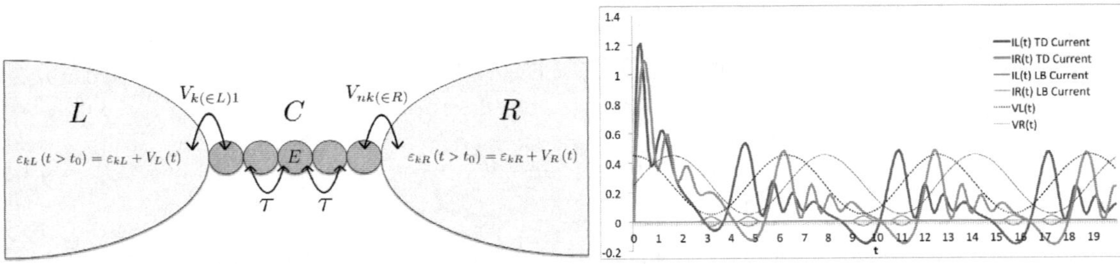

(a) (b)

Figure 1: (a) Schematic of the molecular chain coupled to two reservoirs, modelled as a sequence of quantum dots with inter-dot hopping and lead-dot hopping terms in the Hamiltonian. (b) Current through each lead of a molecular wire junction in which the symmetry is broken due to biases phase-shifted by $\pi/2$, i.e. $\phi_L = 0$, $\phi_R = -\pi/2$. The number of molecular sites is 5, and other parameters chosen are $k_BT = 0.1$, $V_L = 5 = V_R$, $A_L = 4 = A_R$, $\Omega_L = 1 = \Omega_R$, $\Gamma_{11} = 0.5$, $\Gamma_{55} = 0.5$, $E = h_{kk} = 1$, $\tau = h_{kk+1} = h_{k+1k} = 0.1$.

this case the number of electrons in the molecular region $N_C(t) = -2i\text{Tr}\left[\mathbf{G}^<_{CC}(t,t)\right]$ is constant in time. If the bias is time-dependent, then the Hamiltonian is no longer symmetric under time translations and, in general, $N_C(t)$ also varies with time. Therefore local charge conservation does not apply in this case, a fact that is reflected in the current characteristics of Fig. 1(b).

It is a point of concern with Fig. 1(b) that it appears to indicate an unbroken net flow of charge into the molecule, because after the usual transient 'ringing' regime is over, the average value of $I_L(t)+I_R(t)$ *per cycle* appears to be positive. This is a property which, if sustained indefinitely, means that there will always be a charge increase in the central region of the junction - clearly an unphysical situation. To investigate what is occurring here, we compute in Fig. 2 the transport properties over a much longer time range of the same 5-site model as was considered in Fig. 1(b). The long time plots of both the sum of the currents (blue curve) and the particle number N_C (black) are displayed in Fig. 2, for this system, where to compute N_C we have expanded (11) into the left/right eigenbasis in a similar fashion to (22). Certain qualitative features of the plots stand out, in particular a periodic 'sloshing' of the charge in the molecular region as the number of particles increases there, before levelling off at a value below the maximum charge of 10. Simultaneously with this dynamic filling of the molecular levels, the average value of $I_L(t)+I_R(t)$ drops until the integral under this curve evaluates to zero over a full cycle. Thus in addition to the fast decay of the initial transient, there is a secondary underlying decay which occurs over a much longer time scale. Note that this decay over two time scales is not possible with a single-level model, so that it is fundamentally a finite-size effect.

In Fig. 2 we compare the results obtained for the sparse tight-binding level-width matrix with the same plots for level-widths that are relatively non-sparse. Specifically, we take $\Gamma_{L,mn} = \delta_{mn}0.5 = \Gamma_{R,mn}$, giving a level-width with a constant term along the main diagonal. The timescale for decay of all transient modes in formula (22) is $\tau \equiv Max\{1/\gamma_j\}$, where the γ_j are defined as imaginary parts of the complex eigenvalues, $\bar{\varepsilon}_j = \lambda_j - i\gamma_j$. For the sparse tight-binding level-width $\tau = 61.9$, whereas in the non-sparse case we have a decay time of $\tau = 2$. This order-of-magnitude difference is well illustrated by the extremely fast relaxation of the non-sparse number density (red curve) to a steady oscillation below $N_C = 10$. This implies that the integral of $I_L(t)+I_R(t)$ over one cycle is zero, implying no long-term change in the sum of currents, as shown by the orange curve in Fig. (2). Additionally, we note that despite the qualitatively similar 'sloshing' between two values of the electron number in the molecule after $t \sim \tau$ has elapsed, there is a difference in the amplitude of the oscillating signal for the

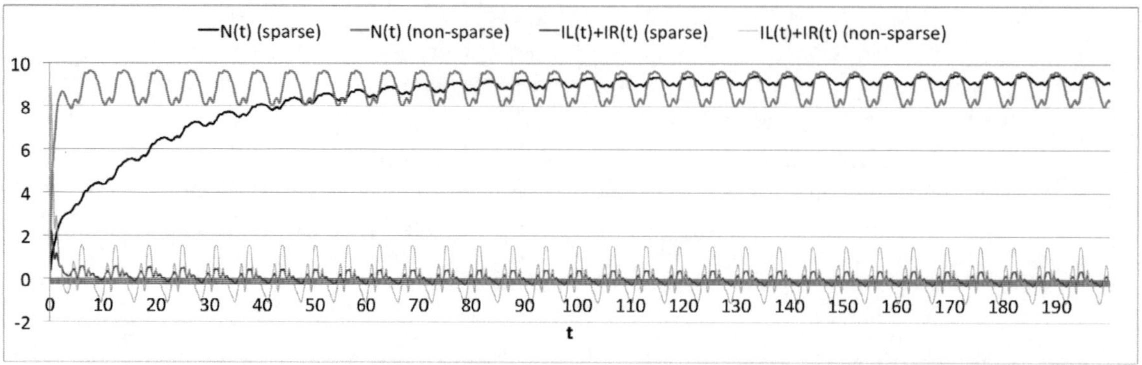

Figure 2: The electron number (black line) and sum of currents (blue line) for the sparse tight-binding hamiltonian is compared with the electron number (red line) and sum of currents (orange line) for a hamiltonian in which each lead is coupled with the same strength to every site in the chain.

sparse and non-sparse cases. This is simply due to the fact that the overlap integral prefactors (for example $\left\langle \varphi_j^L \middle| \mathbf{\Gamma}_\alpha \middle| \varphi_j^R \right\rangle$) are smaller on average when $\mathbf{\Gamma}_\alpha$ is sparse.

5. Conclusions

In this work we have shown that the general formula for the current in a multi-terminal junction derived in [8] can be manipulated into another form which is very convenient for numerical calculations. Using the Padé approximation for the Fermi function it is possible to convert infinite frequency integrals into fast-converging summations. In the case of a sinusoidal bias applied to the leads all the time integrals can be removed analytically and expressed in terms of special functions. To illustrate this formalism, we considered a molecular chain in a two-terminal junction with a $-\pi/2$ relative phase shift in the sinusoidal biases applied to both leads. The calculated current demonstrates similar transient and ringing effects to those observed previously [5, 8] and reveals novel effects due to this dynamical symmetry-breaking. Specifically, we demonstrate the violation of local charge conservation in the molecular region. We anticipate that in future we will be able to apply this formalism to the study of quantum 'pumping', where symmetry-breaking parameters in the periodic driving signal lead to a net transfer of charge through the system with zero net bias [21]. We also find that for extended systems there can be two distinct decay processes in the current characteristics: an initial 'ringing' transient following the bias switch-on, and also an underlying decay of the entire signal implied by global charge conservation. The first type of decay is a well-known effect, and has been studied before for static [7] and TD [8] biases. This latter type of decay is characterized by the fact that the signal retains its shape throughout, and also by a decay time which is related to sparseness of the level-width matrix. We anticipate that this double decay effect will be observable in more realistic extended systems than the one considered here, and that the formalism developed here provides a fast way to access transient properties in those systems.

Acknowledgements Michael Ridley was supported through a studentship in the Centre for Doctoral Training on Theory and Simulation of Materials at Imperial College funded by the Engineering and Physical Sciences Research Council under grant number EP/G036888/1.

References

[1] Landauer R 1970 *Phil. Mag.* **21** (172) 863

[2] Büttiker M 1992 *Phys. Rev. B* **46** (19) 12485
[3] Di Ventra M 2008 *Electrical transport in nanoscale systems* CUP
[4] Stefanucci G and van Leeuwen R 2013 *Nonequilibrium Many-Body Theory of Quantum Systems* CUP
[5] Jauho A -P, Wingreen N S and Meir Y 1994 *Phys. Rev. B* **50** (8) 5528
[6] Tuovinen R, van Leeuwen R, Perfetto E, and Stefanucci G 2013 *J. Phys.: Conf. Ser.* **427** (1) 012014
[7] Tuovinen R, Perfetto E, Stefanucci G, and van Leeuwen R 2014 *Phys. Rev. B* **89** (8) 085131
[8] Ridley M, MacKinnon A and Kantorovich L 2015 *Phys. Rev. B* **91** (12) 125433
[9] Stefanucci G and Almbladh C O 2004 *Phys. Rev. B* **69** (19) 195318
[10] Konstantinov O V and Perel' V I 1961 *Sov. Phys. JETP* **12** (1) 142
[11] Li S D, Yu Z, Yen S F, Tang W C, and Burke P J 2004 *Nano Lett.* **4** (4) 753
[12] Latini S, Perfetto E, Uimonen A M, van Leeuwen R and Stefanucci G 2014 *Phys. Rev. B* **89** (7) 075306
[13] Lerch M 1887 *Acta Math.* **11** (1-4) 19–24
[14] Ozaki T 2007 *Phys. Rev. B* **75** (3) 035123
[15] Hu J, Xu R, and Yan Y 2010 *J. Chem. Phys.* **133** (10) 101106
[16] Hu J, Luo M, Jiang F, Xu R and Yan Y 2011 *J. Chem. Phys.* **134** (24) 244106
[17] Zhang Y, Chen S and Chen G 2013 *Phys. Rev. B* **87** (8) 085110
[18] Mujica V, Kemp M and Ratner M A 1994 *J. Chem. Phys.* **101** (8) 6849
[19] Mujica V, Kemp M and Ratner M A 1994 *J. Chem. Phys.* **101** (8) 6856
[20] Chen S, Xie H, Zhang Y, Cui X and Chen G 2013 *Nanoscale* **5** (1) 169
[21] Strass M, Hänggi P and Kohler S 2005 *Phys. Rev. Lett.* **95** (13) 130601

Journal of Physics: Conference Series **696** (2016) 012018 doi:10.1088/1742-6596/696/1/012018

Effective bias and potentials in steady-state quantum transport: A NEGF reverse-engineering study

Daniel Karlsson

Department of Physics, Nanoscience Center P.O.Box 35 FI-40014 University of Jyväskylä, Finland

E-mail: daniel.l.e.karlsson@jyu.fi

Claudio Verdozzi

Department of Physics, Division of Mathematical Physics, Lund University, 22100 Lund, Sweden; and European Theoretical Spectroscopy Facility (ETSF)

E-mail: claudio.verdozzi@teorfys.lu.se

Abstract. Using non-equilibrium Green's functions combined with many-body perturbation theory, we have calculated steady-state densities and currents through short interacting chains subject to a finite electric bias. By using a steady-state reverse-engineering procedure, the effective potential and bias which reproduce such densities and currents in a non-interacting system have been determined. The role of the effective bias is characterised with the aid of the so-called exchange-correlation bias, recently introduced in a steady-state density-functional-theory formulation for partitioned systems. We find that the effective bias (or, equivalently, the exchange-correlation bias) depends strongly on the interaction strength and the length of the central (chain) region. Moreover, it is rather sensitive to the level of many-body approximation used. Our study shows the importance of the effective/exchange-correlation bias out of equilibrium, thereby offering hints on how to improve the description of density-functional-theory based approaches to quantum transport.

1. Introduction

The method of Non-Equilibrium Green's functions (NEGF) is a general and powerful tool to describe out-of-equilibrium quantum phenomena [1, 2, 3]. Nowadays, NEGF are extensively used in many areas of physics, and new fields of application emerge continually [4]. In this work, we use NEGF to gain insight into another theoretical approach. We address conceptual aspects of Density-Functional Theory (DFT) [5, 6] and Time-Dependent DFT (TDDFT) [7] in the context of quantum transport phenomena.

In equilibrium DFT, the particle density n is the fundamental variable. There exists an invertible map between n and the external potential v, which means that all observables are functionals of n. In time-dependent DFT, it is the time-dependent density $n(t)$ that plays this role. Several more extensions of DFT exists, like current DFT [8], quantum electrodynamics DFT [9], and so on. All these functional theories make use of the concept of reduced quantities. Furthermore, in DFT-type theories, the interacting problem is mapped into an independent-particle one, the so-called Kohn-Sham (KS) system [6], where the non-interacting particles experience an effective one-particle potential v_{KS} (commonly referred to as the KS potential).

A geometrical arrangement often considered in quantum-transport studies is a small (nanoscale) central region, where inter-particle interactions are explicitly taken into account, connected to macroscopic metallic (free-electron-like) contacts. This setup prompts the notion of partitioned systems in quantum transport [10, 11, 12, 13, 14, 15] for which a TDDFT description is a natural choice [12, 13].

However, even if the interactions are confined to the central region, the exact TDDFT KS potential v_{KS} is nonzero in the leads, due to the non-local nature of exchange-correlation contributions. In equilibrium v_{KS} tends to zero deep in the leads, while for steady-state regimes, as reached at the end of a TDDFT time evolution (henceforth referred to as steady-state TDDFT), v_{KS} tends to a constant value [12]. In the latter case, the limit of v_{KS} can be interpreted as an effective bias, b_{KS}. With this perspective, one can exactly describe the system (within a given numerical accuracy) as an enlarged (with respect to the physical device) central region, including part of the leads where v_{KS} is not constant in space, plus effective leads with a renormalized, but constant, bias and a current I flowing through the whole system. For a schematic, see Figure 1.

The above considerations can be translated into a numerical reverse-engineering scheme to find v_{KS} and b_{KS} in the entire system. For the aforementioned extended region, of size L, we gauge away the potential in, say, the right right lead, and thus we remain with $L+1$ quantities to determine (v_{KS} and b_{KS} in the left lead), using the L NEGF densities in the central region and the current I as constraints [16]. For this scheme, it is essential that the KS bias in the leads is approximately constant. Hence the necessity of achieving convergence by progressively increasing the size of an enlarged central region, such that all spatial variations of v_{KS} are inside the region). When converged, b_{KS} from this scheme will be v_{KS} deep in the leads, according to the discussion above.

What just described can be rigorously justified in terms of a recently introduced DFT-like description of steady-state quantum transport [17]. In this formalism, the reduced quantities are the steady-state current I (hence i-DFT) and the densities in the central region only. The quantities one needs to consider are the external potentials in the central region v_c and the bias b. Close to equilibrium, i.e. for sufficiently small biases, it has been shown [17] that the map $(v_c, b) \rightarrow (n, I)$ is locally invertible. The map also allows for the construction of a KS system in this scheme. In i-DFT, $v_{KS,c}$ and b_{KS} depend on the specific partitioning, even though I and n in the central region do not. In the central region I and n from steady-state TDDFT and i-DFT will agree, but the potential and bias will not. Accordingly, outside the central region the density from i-DFT will differ from the exact density. Results from our reverse engineering procedure and from i-DFT agree when the latter is applied to an extended region large enough so that $v_{KS,c}$ deep in the leads and b_{KS} agree.

In this work, we perform a reverse-engineering study of the map $(v_{KS}, b_{KS}) \rightarrow (n, I)$ for finite biases [16]. Historically, determining KS potentials via numerical reverse engineering has provided valuable conceptual insight into static and time-dependent DFT approaches for continuum [18, 19], lattice [20, 21, 22, 23, 24, 25] and continuum-lattice systems [26]. For the lattice case (the kind of systems studied here), the procedure, mostly applied to exact solutions from finite systems but also from NEGF treatments [27], can face so-called v-representability issues [20, 21, 22, 23, 28] (i.e., a map inversion may not be possible).

Additional hurdles may arise when performing reverse engineering at finite bias. For example, if the system can exhibit negative differential conductance [1] the inversion procedure may yield multiple solutions, in contrast to the case of the low-bias limit [17].

The reverse-engineering study performed here provides us with two main results: a) Out of equilibrium, the effective bias is strongly reduced compared to the applied bias and b) the size of

[1] Perhaps this is most easily seen when we apply a bias to leads with finite bands, where the current becomes small if the band overlap is small.

Journal of Physics: Conference Series **696** (2016) 012018 doi:10.1088/1742-6596/696/1/012018

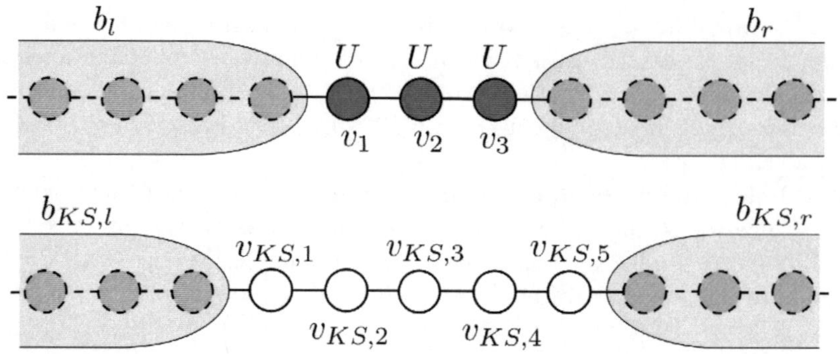

Figure 1. Schematic of the lead-chain-lead system. The upper system is the interacting system, taken for sake of illustration with three interacting sites. The lower system is the non-interacting KS system, with v_{KS} and b_{KS} chosen such as to obtain the same (I, n) as in the interacting system. To illustrate the construction used for steady-state TDDFT, an extended central region containing also one site from each lead is depicted. In an i-DFT description, the minimal allowed size of the central region should be the same as in the upper system.

this correction is heavily dependent on the system parameters. Additionally, our study provides some insight into aspects related to i-DFT.

We begin by introducing the model, and then we describe briefly steady-state NEGF for interacting and non-interacting systems. We then describe the reverse engineering algorithm, present our results, and summarize our conclusions.

2. Model

We consider a finite central region contacted to macroscopic leads. All interactions are confined to the central region. The Hamiltonian of the entire system is

$$\hat{H} = \hat{H}_c + \hat{H}_{leads} + \hat{H}_{cl}, \tag{1}$$

where the three terms correspond to the central region, to the leads, and to the couplings between the central region and the leads, respectively.

The central region consists of a small interacting 1D chain with L lattice sites, described by

$$\hat{H}_c = \sum_{ij\sigma} T_{ij}\hat{c}^\dagger_{i\sigma}\hat{c}_{j\sigma} + \sum_{i\sigma} v_i\hat{n}_{i\sigma} + U\sum_i \hat{n}_{i\uparrow}\hat{n}_{i\downarrow}, \tag{2}$$

where $\hat{c}_{i\sigma}$ ($\hat{c}^\dagger_{i\sigma}$) is the annihilation (creation) operator on site i with spin σ, and $\hat{n}_{i\sigma} = \hat{c}^\dagger_{i\sigma}\hat{c}_{i\sigma}$ is the density operator. The first term describes tunneling between the sites in the chain. We choose the hopping matrix $(\mathbf{T})_{ij} = T_{ij} = -T$ if i is adjacent to j, and 0 otherwise. The second term is the onsite potential with elements $(\mathbf{v})_{ij} = \delta_{ij}v_i$ and the third is the interaction with strength U.

The two 1D leads are semi-infinite and have the chemical potential μ. Their Hamiltonian is $\hat{H}_{leads} = \sum_{\alpha=r,l}\hat{H}_\alpha$, where $\alpha = r(l)$ refers to the right (left) lead, and

$$\hat{H}_\alpha = \sum_{ij\in\alpha,\sigma} T^\alpha_{ij}\hat{c}^\dagger_{i\sigma}\hat{c}_{j\sigma} + b_\alpha(t)\hat{N}_\alpha. \tag{3}$$

$b_\alpha(t)$ is the (site-independent) bias in lead α, and $\hat{N}_\alpha = \sum_{i\in\alpha,\sigma}\hat{n}_{i\sigma}$ is the number operator in lead α. The matrix elements are $T^\alpha_{ij} = -T^\alpha$ if i is adjacent to j, and 0 otherwise.

The lead-chain coupling is

$$\hat{H}_{cl} = -T_{cl} \sum_{\sigma} \left(\hat{c}^{\dagger}_{1_l \sigma} \hat{c}_{1_c \sigma} + \hat{c}^{\dagger}_{1_r \sigma} \hat{c}_{L_c \sigma} \right) + h.c., \tag{4}$$

that is, the rightmost site in the left lead is connected via T_{cl} to the leftmost site in the chain, and similarly for the right lead.

For simplicity, we choose all tunneling parameters to be equal (transparent contacts), meaning $T = T^{\alpha} = T_{cl}$. We furthermore set $T = 1$, which defines the unit of energy. For our steady-state treatment, we define the applied bias $b_{\alpha} = b_{\alpha}(t \to \infty)$, and $b = b_L - b_R$. Furthermore, we consider the non-magnetic case, meaning that $n_{\uparrow} = n_{\downarrow} = n$ is enforced, and we choose to work at zero temperature.

3. NEGF in steady-state

In this section we briefly describe NEGF in the steady state for both the interacting case and for the non-interacting KS scheme. More details will be given for the second scheme, where we also describe the reverse engineering $NEGF \to KS$ which is the novel part of this work.

3.1. NEGF in steady-state: The interacting system

Using a NEGF description of quantum transport for an interacting system, the retarded Green's function for our central region can be written as an $L \times L$ matrix in the site basis [2, 15],

$$\mathbf{G}^R(\omega) = \frac{1}{\omega \mathbf{1} - \mathbf{T} - \boldsymbol{v} - \boldsymbol{\Sigma}^R_{MB}(\omega) - \boldsymbol{\Sigma}^R_{emb,l}(\omega) - \boldsymbol{\Sigma}^R_{emb,r}(\omega)}. \tag{5}$$

The so-called embedding self-energies $\boldsymbol{\Sigma}^R_{emb,\alpha}$ account in an exact way for the effects of the leads, allowing for a finite matrix description of an infinite system. All interaction effects from the central region are contained in the many-body self-energy $\boldsymbol{\Sigma}^R_{MB}$.

In equilibrium, knowledge of $\mathbf{G}^R(\omega)$ is enough to determine properties of the system. In steady-state, we need one more quantity and here we choose the lesser Green's function $\mathbf{G}^<(\omega)$. It is directly related to the density, and is defined by the Keldysh equation

$$\mathbf{G}^<(\omega) = \mathbf{G}^R(\omega) \left(\boldsymbol{\Sigma}^<_{MB}(\omega) + \boldsymbol{\Sigma}^<_{emb,l}(\omega) + \boldsymbol{\Sigma}^<_{emb,r}(\omega), \right) \mathbf{G}^A(\omega). \tag{6}$$

where $\mathbf{G}^A = (\mathbf{G}^R)^{\dagger}$.

As approximations to $\boldsymbol{\Sigma}_{MB}$, we will make use of two standard approximations: 2nd Born (see e.g. [29, 30]) that takes into account all diagrams of the self-energy up to second order, and the particle-particle T-Matrix Approximation (TMA) [31, 32, 33, 34]. In these approximations, $\boldsymbol{\Sigma}^{R/<}_{MB} = \boldsymbol{\Sigma}^{R/<}_{MB}[\mathbf{G}^R, \mathbf{G}^<]$, and thus the equations have to be solved self-consistently. Both approximations are conserving in the Kadanoff-Baym sense [35, 36], i.e. certain conservation laws are fulfilled. For us, the most important one is the continuity equation, which in the context of steady-state transport implies that the current through each lead is the same, $I_l = -I_r = I$ (we define I_{α} to be the current out from lead α).

The embedding self-energies are given by

$$\boldsymbol{\Sigma}^<_{emb,\alpha} = if(\omega - \mu - b_{\alpha}) \boldsymbol{\Gamma}^{\alpha}(\omega), \quad \boldsymbol{\Gamma}^{\alpha}(\omega) = -2\Im m \boldsymbol{\Sigma}^R_{emb,\alpha}(\omega). \tag{7}$$

The matrix structure of $\boldsymbol{\Sigma}^R_{emb,\alpha}$ is given by

$$(\boldsymbol{\Sigma}^R_{emb,\alpha})_{ij}(\omega) = \Sigma^R_{emb}(\omega - b_{\alpha}) \delta_{ij} \left(\delta_{1i} \delta_{\alpha l} + \delta_{Li} \delta_{\alpha r} \right).$$

Since the leads are semi-infinite and tight-binding, $\Sigma^R_{emb}(\omega)$ has an analytic expression [2].

This defines the Green's functions of Eq. (5) and Eq. (6), which are solved self-consistently. The convergence was improved by using the Pulay mixing scheme [30, 37]. The site density is obtained directly from the lesser Green's function, while the current is obtained from the Meir-Wingreen formula [38]:

$$n_k = \int_{-\infty}^{\infty} \frac{d\omega}{2\pi i} (\mathbf{G}^<)_{kk}(\omega) \tag{8}$$

$$I = \int_{-\infty}^{\infty} \frac{d\omega}{2\pi i} Tr\left[\mathbf{\Gamma}^L(\omega)\left(\mathbf{G}^<(\omega) - 2\pi i f(\omega - \mu - b_L)\mathbf{A}(\omega)\right)\right], \tag{9}$$

where the spectral function $2\pi\mathbf{A} = i(\mathbf{G}^R - \mathbf{G}^A)$ and $f(\omega) = \theta(-\omega)$ is the Fermi function at zero temperature. n and I are then used as an input into the KS reverse engineering scheme.

3.2. NEGF in steady-state: Kohn-Sham Green's functions

We define the KS Green's functions as the Green's functions $\mathbf{G}^R_{KS}, \mathbf{G}^<_{KS}$ that yield the same density and current as in the original system, but which pertain to a non-interacting system with an effective (diagonal) potential $\mathbf{v}_{KS}$. The defining equations are

$$\mathbf{G}^R_{KS}(\omega) = \frac{1}{\omega\mathbf{1} - \mathbf{T} - \mathbf{v}_{KS} - \mathbf{\Sigma}^{R,KS}_{emb,l}(\omega) - \mathbf{\Sigma}^{R,KS}_{emb,r}(\omega)} \tag{10}$$

$$\mathbf{G}^<_{KS}(\omega) = \mathbf{G}^R_{KS}(\omega)\left(\mathbf{\Sigma}^{<,KS}_{emb,l}(\omega) + \mathbf{\Sigma}^{<,KS}_{emb,r}(\omega)\right)\mathbf{G}^A_{KS}(\omega) \tag{11}$$

$$\mathbf{\Sigma}^{<,KS}_{emb,\alpha} = if(\omega - \mu - b_{KS,\alpha})\mathbf{\Gamma}^\alpha_{KS}(\omega) \tag{12}$$

$$\mathbf{\Gamma}^\alpha_{KS}(\omega) = -2\Im m\mathbf{\Sigma}^{R,KS}_{emb,\alpha}(\omega) \tag{13}$$

$$(\mathbf{\Sigma}^{R,KS}_{emb,\alpha})_{ij}(\omega) = \Sigma^R_{emb}(\omega - b_{KS,\alpha})\delta_{ij}\left(\delta_{1i}\delta_{\alpha l} + \delta_{Li}\delta_{\alpha r}\right). \tag{14}$$

The KS system are described by the same hopping matrices as the original system [22, 28], resulting in identical shapes for the embedding self-energies. The difference is that it is $b_{KS,\alpha}$, instead of b, that enters them. [2]

Since we can add a constant potential to the KS system without changing the physical properties, we restrict, without loss of generality, $b_{KS,R} = 0$, and we define $b_{KS} = b_{KS,L}$. This fixes the gauge, which is crucial for the reverse engineering to converge.

The densities and current from the KS system, equal to the ones from the original system, are given by the same equations as for the interacting case,

$$n_k = \int_{-\infty}^{\infty} \frac{d\omega}{2\pi i} (\mathbf{G}^<_{KS})_{kk}(\omega) \tag{15}$$

$$I = \int_{-\infty}^{\infty} \frac{d\omega}{2\pi i} Tr\left[\mathbf{\Gamma}^L_{KS}(\omega)\left(\mathbf{G}^<_{KS}(\omega) - 2\pi i f(\omega - \mu - b_{KS})\mathbf{A}_{KS}(\omega)\right)\right]. \tag{16}$$

However, since the KS system is a non-interacting one, the Meir-Wingreen formula can be written as an integral over a transmission function,

$$I = \int_{\mu}^{\mu + b_{KS}} \frac{d\omega}{2\pi} \mathcal{T}_{KS}(\omega), \tag{17}$$

[2] It is, however, possible to consider a KS lattice with different hopping matrices, but one can then run into severe $v-$representability issues [23], and thus we do not consider this option any further.

Journal of Physics: Conference Series **696** (2016) 012018 doi:10.1088/1742-6596/696/1/012018

where $\mathcal{T}_{KS}(\omega) = Tr\left[\mathbf{\Gamma}^L_{KS}(\omega)\mathbf{G}^R_{KS}(\omega)\mathbf{\Gamma}^R_{KS}(\omega)\mathbf{G}^A_{KS}(\omega)\right]$ is the KS transmission function. At this point, we wish to emphasise that i) Eq. (17) is a non-equilibrium formula, and all Green's functions need to be evaluated at the bias b_{KS} ii) Eq. (17) gives the correct current, provided that b_{KS} and v_{KS} exists.

In short, the current from a NEGF treatment of the original many-body system is now recast in the form of an integral over a non-interacting transmission function $\mathcal{T}_{KS}$. The physical interpretation of $\mathcal{T}_{KS}$ is not clear though, since it is the transmission function of a fictitious image (KS) system: Physical meaning should not be attributed to $\mathcal{T}_{KS}$ itself, but rather to its integral. On the other hand, for low bias $\mathcal{T}_{KS}$ gives the KS conductance which, as shown next, can be related to the physical conductance.

As usual in DFT-like schemes, v_{KS} can be decomposed into the external potential v, the Hartree potential Un_i, and the exchange-correlation potential v_{xc}. Likewise, we define the effective bias to be the sum of the external bias and the exchange-correlation bias b_{xc}, (this latter quantity was recently introduced in i-DFT [17] as one of the basic variables):

$$v_{KS}(i) = v_i + Un_i + v_{xc}(i) \tag{18}$$
$$b_{KS} = b + b_{xc}. \tag{19}$$

If $U \to 0$, then $v_{xc}, b_{xc} \to 0$. Since b_{KS} and b_{xc} differ by the external bias, in the following we will use them interchangeably. The introduction of b_{xc} allows for a rewriting of the current expression, Eq. (17), for low bias. We assume that $\mathcal{T}_{KS}$ tends smoothly, for small b_{KS}, to the equilibrium one. This yields

$$\left(\frac{dI}{db_{KS}}\right)_{b_{KS}\to 0} = \frac{1}{2\pi}\mathcal{T}_{KS}(\omega = \mu, b_{KS} \to 0) = \sigma_{KS} \tag{20}$$

where σ_{KS} is usually referred to as the KS conductance. Using $\sigma = \frac{dI}{db}$ and the chain rule, we can relate σ_{KS} to σ as

$$\sigma = \sigma_{KS}\left(1 + \frac{db_{xc}}{db}\right)_{b\to 0}, \tag{21}$$

a result originally derived in i-DFT [17], and discussed here in a slightly different form. While not operationally used in the rest of the paper, Eqs. (18-21) and the considerations above elucidate the importance of the b_{xc} correction. According to Eq. (21), we need to apply a different bias to the KS system than the one to the physical system, i.e. the conductance is the sum of the KS conductance and a dynamical correction [17, 39], which can be sizeable even if $b \to 0$. Correspondingly, a calculation which approximates v_{xc} but neglects b_{xc} (which is the same as to assume $\sigma = \sigma_{KS}$) will overestimate (assuming $b_{xc} < 0$) the calculated conductance, a scenario that occurs in NEGF+LDA calculations [39]. This is of paramount importance for strongly correlated systems. The latter have low conductance because of interaction effects, and b_{xc} can be on the order of $-b$, giving a vanishing physical conductance while keeping a large σ_{KS}.

3.2.1. Details regarding the reverse engineering algorithm The KS system should reproduce L lattice densities and the current, giving $L+1$ constraints. These are intertwined, in the sense that changing v_{KS} also affects the current and changing b_{KS} also affects the density in the system. Thus, v_{KS} and b_{KS} have to be adjusted simultaneously in a self-consistent scheme. We found that a method that seems to be quite stable is to first fix the bias, and then change v_{KS} until the densities match. At this point, we adjust b_{KS}, and then change v_{KS} again, and so on.

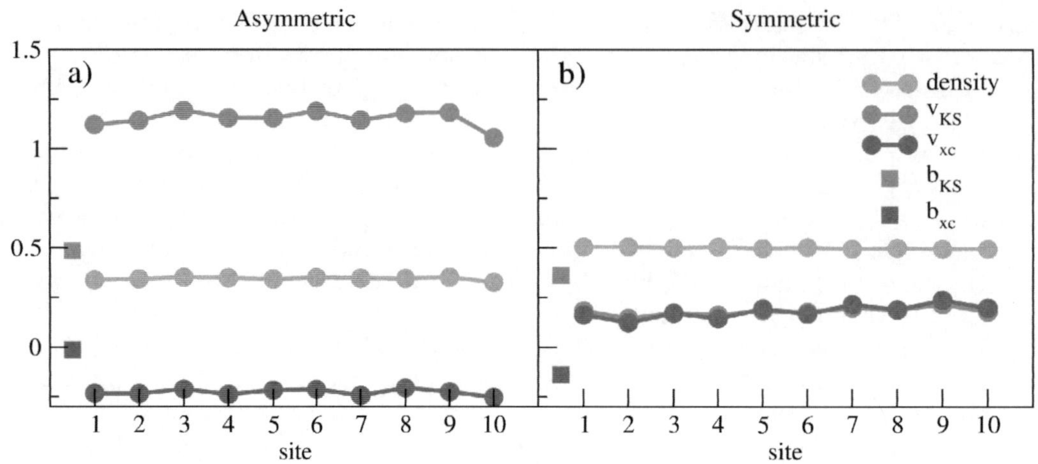

Figure 2. Density and effective potentials and biases for a $L = 10$-site chain. The applied bias is $b = 0.5$, while $U = 4$ and $\mu = 0$. In a), $b = b_L, b_R = 0$, and $v_i = 0$. In b), $b_L = -b_R = 0.25$, and $v_i = -U/2$, leading to a more homogeneous density. Since $b_{KS,R} = 0$, the zero potential is at 0.25 in b). The values of b_{KS} and b_{xc} are shown as squares.

More in detail, for a fixed bias b_{KS}, we start from an initial v_{KS}. We obtain the corresponding KS Green's functions, and then calculate the density via Eq. (15). We then adjust v_{KS} according to the following scheme

$$v_{KS}(i) \to v_{KS}(i) + \beta \frac{n_i^{KS} - n_i}{n_i}, \tag{22}$$

where n_i is the original density, and β is a numerical parameter that can be reduced to ensure a more stable convergence. When the density is converged, we change b_{KS} via

$$b_{KS} \to b_{KS} - \gamma \frac{I^{KS} - I}{I}, \tag{23}$$

where γ plays a role similar to β. With this new b_{KS}, we perform another density self-consistency, and then repeat the cycle until the densities and current are converged.

There is no guarantee that the converged solution is unique. Since the current is a non-monotonic function of the bias if we have finite leads, we expect to find at least two solutions, one for small bias, and one for high bias. In order to circumvent this problem, we start the self-consistent calculations with a low bias as initial guess.

4. Results

The shapes of v_{KS} and b_{KS} are, of course, heavily dependent on the physical parameters in the system. These quantities also depend in a non-local way on the densities and the currents. In the unbiased case, we must have $b_{KS} = 0$ since $I = 0$. Out of equilibrium, based on our many-body approximations, it seems that $b_{KS} < b$ ($b_{xc} < 0$), that is, the effective bias is lessened by exchange-correlation effects. To illustrate this point, we consider an interacting chain with $L = 10$ sites subjected to two different biases, shown in Figure 2. Electronic correlations are included at the 2nd Born level. In panel a), we consider an asymmetric bias with $b = b_L = 0.5, b_R = 0$. The interaction keeps the density quite uniform in the chain. In b), we consider a symmetric bias $b_L = -b_R = 0.25$, and put $v_i = -U/2$ in the chain, which makes the density even more uniform.

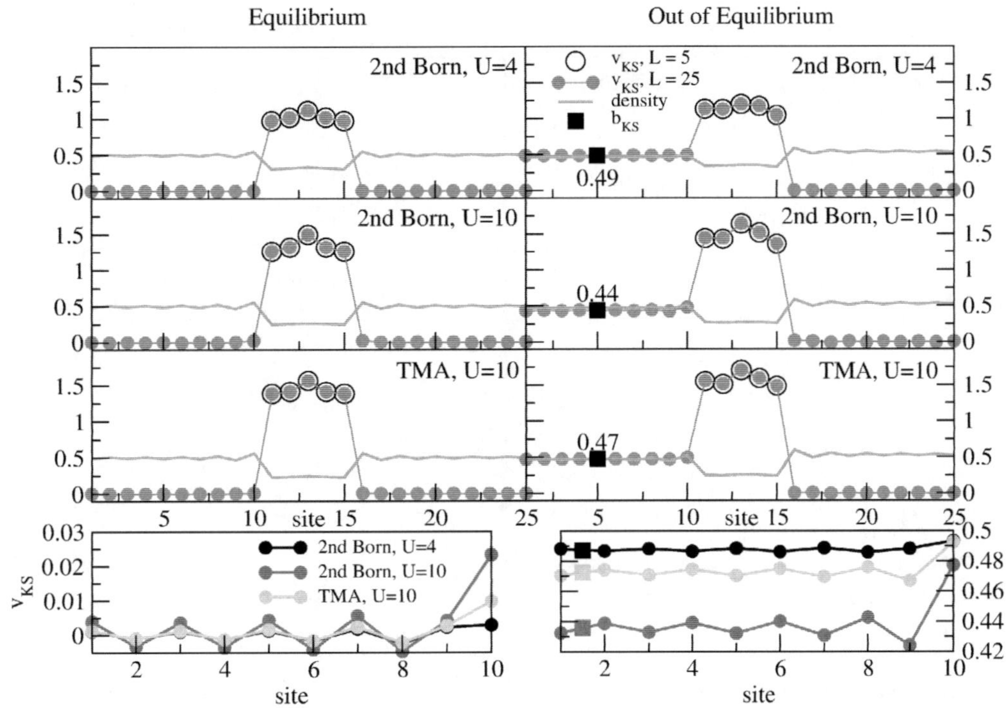

Figure 3. Reverse engineered v_{KS} and b_{KS} for a non-biased (left) and biased with $b_L = 0.5, b_R = 0$ (right) system for an interacting region of 5 sites. The leads are half-filled ($\mu = 0$). In the $L = 5$ case, we do not allow for spatial variations beyond the central region, and v_{KS} is shown with open circles. In the $L = 25$ case, we extend the central region to contain also 10 noninteracting sites of each lead. v_{KS} is converged already for $L = 5$. The density shown is the $L = 25$ density. For $U = 4$, 2nd Born and TMA are very similar, and thus we show TMA for $U = 10$ only. On this scale, b_{KS} does not change when extending the central region. The values of b_{KS} are shown as squares.

We see that v_{KS} shows considerable more structure (especially for the asymmetric case) than the density. This is because the application of a bias results in a potential step created at the interface between the lead(s) and the central region. In a non-interacting system, this in turn induces (Friedel-like) density oscillations, which are instead highly damped in the interacting system. For the symmetric case, Figure 2 b), the density is close to the lead density (half-filling). In this case, we find $v_{xc} \approx v_{KS}$ since the Hartree potential is very close to $U/2$. Since b_{KS} is reduced compared to b, there is still an asymmetry in the KS system, and to reproduce the almost uniform density within the non-interacting system, the corresponding v_{KS} must then oscillate in space. Clearly, $v_{KS}[n, I]$ depends non-locally on the density, an aspect missed by any local-density approximation.

It is interesting to see how allowing for v_{KS} to vary in the leads changes b_{KS}, and v_{KS} in the original central region. To illustrate the effect, we first consider a case in which b_{KS}, v_{KS} converge very quickly (Figure 3) and then one case where one needs to go to a considerably larger extended region (Figure 4). For these two situations, we take an original interacting central region of $L = 5$ sites in the presence of an asymmetric bias.

Starting with Figure 3, results for a 5-site KS region (i.e. of the same size of the physical interacting region) are compared to those from an extended KS central region with additional 10 sites from each lead ($L = 25$). For the chosen set of parameters, v_{KS} in the central region

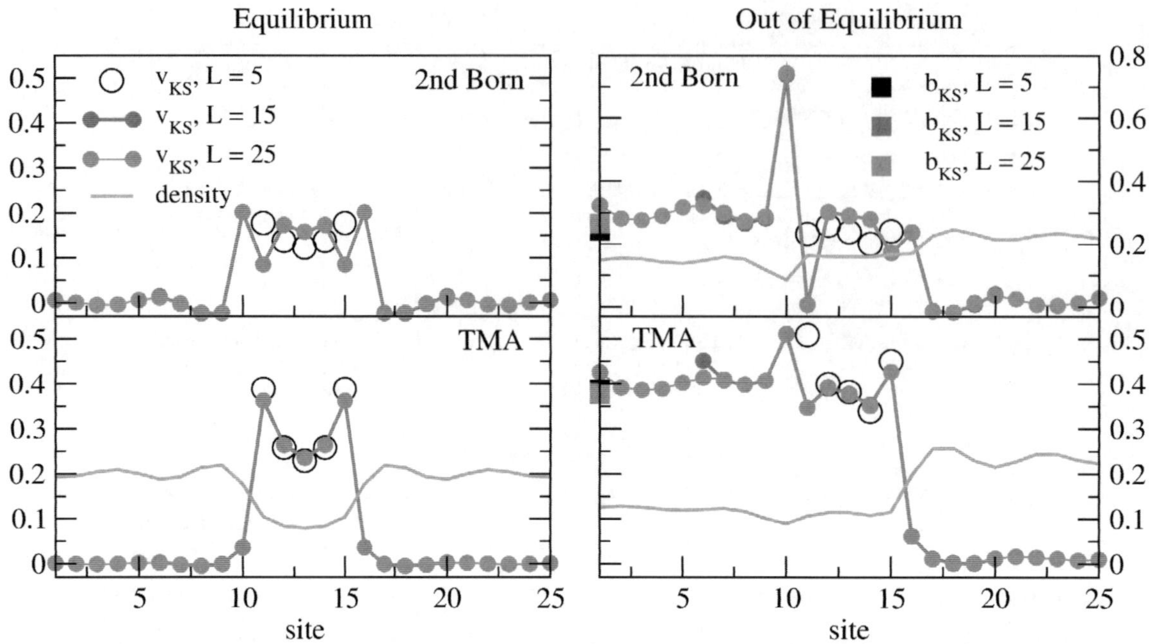

Figure 4. Reverse engineered v_{KS} and b_{KS} from 2nd Born and TMA for $U = 10$ in the low-density regime, with average lead filling $n = 0.2$ ($\mu = -2\cos(0.2\pi)$). Allowing for a potential varying in the leads changes v_{KS} in the central region. For the unbiased case, $L = 15$ is enough for v_{KS} to be converged, while $L = 25$ is not enough for the biased case. Furthermore, there is a difference between v_{KS} in the lead and b_{KS}. Also b_{KS} changes when increasing L. The density shown is for $L = 25$. The values of b_{KS} are shown as squares.

is quite well converged. In the leads, v_{KS} oscillates with a decaying amplitude, tending to b_{KS} deep in the leads. This is a clear display that the effective bias is simply the effective potential deep in the leads, as shown in [12]. Or, equivalently, that the exchange-correlation bias is the exchange-correlation potential deep in the leads. Finally, we note that in this parameter regime the treatment of correlations within TMA and 2nd Born yield the same qualitative behavior in the KS system.

We turn now to a set of parameters where the differences are larger (Figure 4). Here, we consider the same system as in Figure 3, but instead of half-filled leads, we consider the low-density regime, $n = 0.2$, where TMA is expected to perform well even for stronger interactions [31, 32, 34, 40].

In Figure 4, we see that the choice of partitioning changes v_{KS} in the physical central region, and that there is also a difference between b_{KS} and v_{KS} at the edges of the extended region. Clearly, the largest partition considered is enough for v_{KS} to be converged in the physical central region, but not enough to reach its asymptotic value b_{KS} at the edges. [3] These considerations seem to be much more relevant for the biased case, as the unbiased scenarios are already converged for $L = 15$.

We have seen that b_{KS} and v_{KS} depend strongly on the system parameters. We now try to quantify this dependence by considering chains of different lengths and different interaction

[3] As function of L, v_{KS} at the edges of the central region and b_{KS} converge at different rates to their true KS values (the first more slowly than the second). For low lead-density, this induces a small jump between v_{KS} at the edge and b_{KS}, which becomes vanishingly small only for very large L. However, except for such edge sites (and immediate neighbours), b_{KS} and v_{KS} reach convergence already for moderately extended central regions.

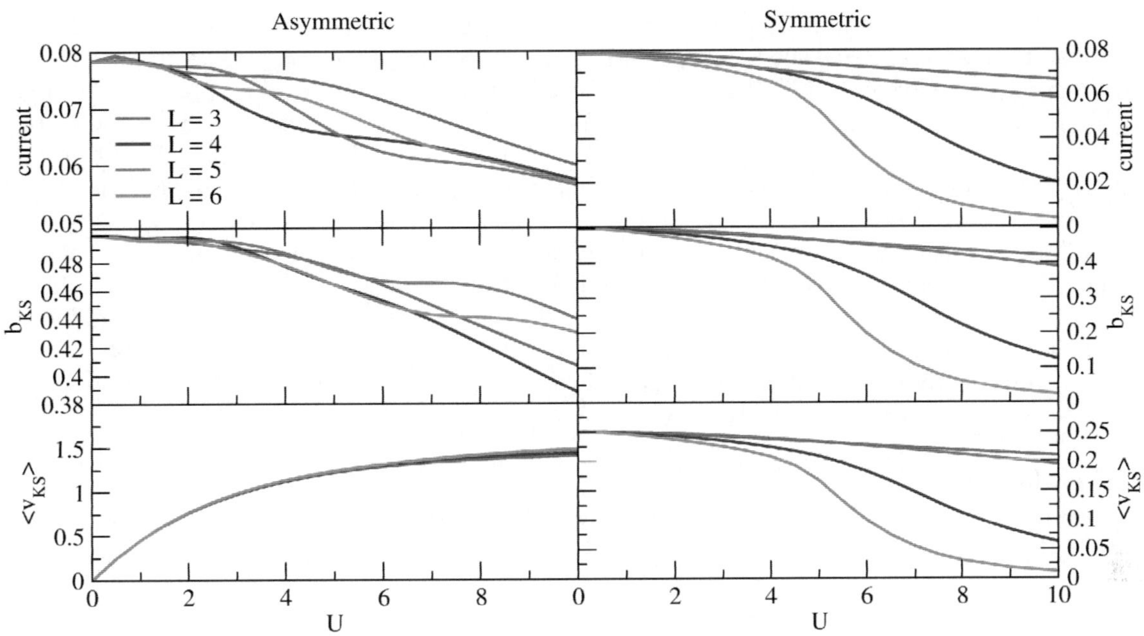

Figure 5. Currents, b_{KS} and average v_{KS} for a bias $b = 0.5$ for chain of various lengths, using 2nd Born. Left: $b_L = b, b_R = 0$. Right: $b_L = -b_R = b/2$ and $v_i = -U/2$.

strengths. The results are shown in Figure 5, where we also plot the v_{KS} site-average $\langle v_{KS} \rangle = \sum_i v_{KS,i}/L$. We consider two different cases: asymmetric bias, and symmetric bias with $v_i = -U/2$.

Due to interaction effects, the current generally decreases as we increase the interaction strength and/or the chain length. The KS system reproduces the same behavior by varying v_{KS} and b_{KS}. In the asymmetric case, the interplay between these two quantities induces a quite complex behavior for $b_{KS}(U, L)$.

In the symmetric case, v_{KS} plays a slightly smaller role, and thus $b_{KS}(U, L)$ is more similar to $I(U, L)$. Moreover, in some cases b_{KS} is very small ($b_{xc} \approx -b$), showing how crucial it is to have a good description of the effective bias/exchange-correlation bias. Specific to the case of symmetric bias, we note a different pattern for odd and even numbers of sites: for even (odd) L the current drops much faster (slower) as a function of U. These strong correlation effects are captured by the behavior of b_{KS} and v_{KS}, which show a similar trend.

Figure 5 shows results up to $L = 6$. For longer chains with high interaction strength, numerical convergence becomes much harder to attain. However, the short chains considered here appear to be appropriate to illustrate the important trends for b_{KS} just discussed.

In this work, we have considered transparent contacts. In this case, the results indicate that the smaller the density is, the slower v_{KS} approaches b_{KS} when moving away from the central region. On the other hand, for non-transparent contacts, more precisely for weak links and wide-band leads, the nonuniformity of v_{KS} close to the interface (on the leads side) most likely will diminish. Corroboration of this point is deferred to future work.

5. Conclusions

We have studied finite interacting chains strongly coupled to biased leads. We have shown that it is possible to reverse engineer effective potentials v_{KS} and biases b_{KS} that can reproduce densities and currents from many-body approximations in NEGF [16]. The non-interacting

Kohn-Sham system reproduces these quantities by varying v_{KS} and b_{KS} in a non-trivial way. Our results unambiguously show that effective potentials and biases depend strongly on the parameters in the system, and that it is unlikely that all correlation effects can be captured within a local-density approximation to v_{KS}. We also find that b_{KS} is reduced compared to the applied bias, in agreement with previous results in equilibrium [39, 41] and out of equilibrium regimes [17]. This further hints to the fact that any approximate KS scheme that describes transport through a strongly correlated system will need to also have a reliable approximation for b_{KS} (or b_{xc}), since the bias correction is sensitive to the nature of the approximation chosen.

The NEGF currents and densities were generated by using the standard many-body approximations 2nd Born and T-matrix. While being approximate, these two different treatments of correlations produce similar trends, reinforcing our conclusions and possibly paving the way to studies to further elucidate the role of the exchange-correlation bias, by e.g. extracting the exact v_{KS} and b_{KS} from steady-states obtained via time-dependent DMRG.

Acknowledgments

We acknowledge useful discussions with Miroslav Hopjan, and acknowledge the Royal Physiographic Society in Lund for support. D.K. thanks the Academy of Finland for support.

References

[1] L. P. Kadanoff and G. Baym. *Quantum Statistical Mechanics.* Benjamin, New York, (1962).
[2] G. Stefanucci and R. van Leeuwen. *Nonequilibrium Many-Body Theory of Quantum Systems: A Modern Introduction.* Cambridge University Press, (2013).
[3] K. Balzer and M. Bonitz. *Nonequilibrium Green's Functions Approach to Inhomogeneous Systems*, volume 867. (2013).
[4] M. Bonitz and A. Filinov. *Progress in Nonequilibrium Green's Functions III, Special issue of Journal of Physics: Conference Series 35.* (2006).
[5] P. Hohenberg and W. Kohn. *Physical Review*, **155**, (1964).
[6] W. Kohn and L. J. Sham. *Physical Review*, **140**, (1965).
[7] E. Runge and E. K. U. Gross. *Physical Review Letters*, **52**, (1984).
[8] G. Vignale and M. Rasolt. *Physical Review Letters*, **59** 2360–2363, (1987).
[9] I. V. Tokatly. *Physical Review Letters*, **110** 1–5, (2013).
[10] C. Caroli, R. Combescot, P. Nozieres, and D. S. James. *J. Phys. C*, **4** 916, (1971).
[11] M. Cini. *Physical Review B*, **22**, (1980).
[12] G. Stefanucci and C.-O. Almbladh. *Physical Review B*, **69** 195318, (2004).
[13] S. Kurth, G. Stefanucci, C.-O. Almbladh, A. Rubio, and E. K. U. Gross. *Physical Review B*, **72** 1–13, (2005).
[14] C. Verdozzi, G. Stefanucci, and C.-O. Almbladh. *Physical Review Letters*, **97** 046603, (2006).
[15] P. Myöhänen, A. Stan, G. Stefanucci, and R. van Leeuwen. *EPL (Europhysics Letters)*, **84** 67001, (2008).
[16] Preliminary reverse-engineering results for b_{KS}, v_{KS} were presented at the workshop *Time-Dependent Density-Functional Theory meets Quantum Information Theory*, December 15-18 (2014), Araraquara (BR).
[17] G. Stefanucci and S. Kurth. *Nano Letters*, **15** 8020–8025, (2015).
[18] C.-O. Almbladh and A. C. Pedroza. *Physical Review A*, **29** 2322–2330, (1984).
[19] M. Lein and S. Kümmel. *Physical review letters*, **94** 143003, (2005).
[20] R. Baer. *The Journal of chemical physics*, **128** 044103, (2008).
[21] Y. Li and C. A. Ullrich. *The Journal of chemical physics*, **129** 044105, (2008).
[22] C. Verdozzi. *Physical Review Letters*, **101** 166401, (2008).
[23] P. Schmitteckert, M. Dzierzawa, and P. Schwab. *Physical chemistry chemical physics : PCCP*, **15** 5477–81, (2013).
[24] P. Schmitteckert and F. Evers. *Physical Review Letters*, **100** 086401, (2008).
[25] J. P. Bergfield, Z.-F. Liu, K. Burke, and C. A. Stafford. *Physical Review Letters*, **108** 066801, (2012).
[26] E. Boström, A. Mikkelsen, and C. Verdozzi. *arXiv:1507.06975*, pages 1–5, (2015).
[27] M. Puig von Friesen, C. Verdozzi, and C.-O. Almbladh. *Physical Review B*, **82** 1–19, (2010).
[28] M. Farzanehpour and I. Tokatly. *Physical Review B*, **86** 125130, (2012).
[29] D. Karlsson and C. Verdozzi. *Physical Review B*, **90** 201109(R), (2014).
[30] K. S. Thygesen and A. Rubio. *Physical Review B*, **77** 115333, (2008).
[31] V. Galitski. *Sov. Phys. JETP*, **34** 104, (1958).
[32] M. Puig von Friesen, C. Verdozzi, and C.-O. Almbladh. *Physical Review Letters*, **103** 176404, (2009).

[33] M. Puig von Friesen, C. Verdozzi, and C.-O. Almbladh. *EPL (Europhysics Letters)*, **95** 27005, (2011).
[34] N. Schlünzen, S. Hermanns, M. Bonitz, and C. Verdozzi. *Physical Review B*, **93** 035107, (2016).
[35] G. Baym and L. Kadanoff. *Physical Review*, **124**, (1961).
[36] G. Baym. *Physical Review*, **127**, (1962).
[37] P. Pulay. *Chemical Physics Letters*, **73**, (1980).
[38] Y. Meir and N. S. Wingreen. *Physical Review Letters*, **68** 2512, (1992).
[39] S. Kurth and G. Stefanucci. *Physical Review Letters*, **111** 030601, (2013).
[40] M. Cini and C. Verdozzi. *Solid State Communications*, **57** 657–660, (1986).
[41] G. Stefanucci and S. Kurth. *Physical Review Letters*, **107** 1–5, (2011).

General Retarded Contact Self-energies in and beyond the Non-equilibrium Green's Functions Method

Tillmann Kubis[1] **Yu He, Robert Andrawis, Gerhard Klimeck**
Network for Computational Nanotechnology, Purdue University, Indiana 47907, USA

Abstract. Retarded contact self-energies in the framework of nonequilibrium Green's functions allow to model the impact of lead structures on the device without explicitly including the leads in the actual device calculation. Most of the contact self-energy algorithms are limited to homogeneous or periodic, semi-infinite lead structures. In this work, the complex absorbing potential method is extended to solve retarded contact self-energies for arbitrary lead structures, including irregular and randomly disordered leads. This method is verified for regular leads against common approaches and on physically equivalent, but numerically different irregular leads. Transmission results on randomly alloyed $In_{0.5}Ga_{0.5}As$ structures show the importance of disorder in the leads. The concept of retarded contact self-energies is expanded to model passivation of atomically resolved surfaces without explicitly increasing the device's Hamiltonian.

1. Introduction

The characteristic length scale of modern semiconductor nanodevices is typically smaller than the coherence length of electrons. In this length domain, details of the lead bandstructure and geometry influence the device transport characteristics [1]-[5]. The non-equilibrium Green's function method (NEGF) [6] is accepted to model transport scenarios most consistently that face coherent and incoherent effects. The NEGF method allows to model devices with open boundary conditions and it describes coherent quantum effects such as confinement, interferences and tunnelling as well as incoherent scattering on phonons, and device imperfections (such as roughness, defects and impurities). In the NEGF framework, open boundary conditions of the device are described in terms of contact self-energies. These self-energies represent the charge injection and extraction effects of the leads and allow modelling specific density of states in the leads. The distinction of contacts and central device allows limiting the explicit NEGF calculation to the central device area and thereby reducing the numerical load [6] significantly. The majority of algorithms that solve the contact self-energies rely on ideal (periodic or homogeneous) leads [7][8]. There are well-known algorithms to solve self-energies for ideal leads, such as the Sancho-Rubio method [9] and the transfer matrix method [10][11]. However, both methods are based on the assumption of semi-infinite, periodic leads. Realistic leads are not ideal and often have non-linear geometries [3] or imperfections like alloy disorder or defects and impurities. Instead, the complex absorbing potential (CAP) [12]-[17] can be extended to irregular leads including e.g. disorder effects or nontrivial geometries. In this work, the CAP algorithm is described and applied on several nonideal leads as they commonly appear in experiments.

Contact self-energies can also be used for numerical passivation of dangling bonds: An atomistic resolution of nanoscale devices is inevitable to realistically describe bandstructures and transport

[1] tkubis@purdue.edu

Journal of Physics: Conference Series **696** (2016) 012019 doi:10.1088/1742-6596/696/1/012019

properties. Atomistic representations of electronic structures typically face dangling bonds at atoms with insufficient bonding neighbours. This typically happens for atoms located at surfaces. This is particularly true for the wide spread empirical tight binding method. Likewise in experiments, those dangling bonds require saturation with Hydrogen atoms. Often, this saturation is modelled by explicitly including passivation atoms in the numerical discretization of the device. Depending on the ratio of dangling bonds versus actual device atoms, this explicit treatment of Hydrogen atoms can significantly increase the rank of the device Hamiltonian. This is insofar problematic since the computational load of NEGF equations scales cubically with the number of degrees of freedom in the Hamiltonian. In this work, it is shown how self-energies can be used to model the saturation of dangling bonds without increasing the number of explicitly discretized degrees of freedom.

In the following method section, the general retarded contact self-energy and the passivation self-energy are presented. The contact self-energies are applied within the NEGF method on regular leads to compare with the established transfer matrix method. The application to trumpet shaped leads proofs its applicability to non-periodic leads. Results of a $Ga_{0.5}In_{0.5}As$ random alloy device show the relevance of non-ideal leads for quantitative device performance predictions. The passivation self-energy is used throughout those examples, but its effect is also explicitly discussed with an electronic bandstructure calculation of a GaAs quantum well at the end of the results section. The paper ends with a summarizing conclusion.

2. Method

The electronic transport results shown in the next section are produced with the recursive non-equilibrium Green's function implementation of the multipurpose nanodevice simulation tool NEMO5 [18][19]. Electronic bandstructure data that illustrate the impact of passivation are results of NEMO5's Schrödinger equation solver. Electrons are represented within atomistic tight binding with a varying number of orbital representations [20][21]. For random alloys, the atomic structure is relaxed following the Keating strain model [22]. Strain induced changes of the electronic Hamiltonian are considered following the model of Boykin et al. [23]. All IV characteristics are solved including charge self-consistency of the NEGF equations with the Poisson equation. For transmission and bandstructure results only, a constant potential profile was assumed.

If not stated otherwise, the lead self-energies are solved with the CAP method [12]-[17]: The last few (~10) nanometres of the concrete lead structure that are closest to the device are considered in the lead calculation [24]. This lead portion is partitioned into segments to allow for a recursive solution of the surface Green's functions g_n of each segment n

$$\begin{aligned} g_n &= (E + i\eta_0 e^{-\lambda(n-1)} - H_{n,n} - \Sigma_n)^{-1} \\ \Sigma_n &= H_{n,n-1} g_{n-1} H_{n-1,n} \end{aligned} \tag{1}$$

Here, Σ_n is the corresponding contact self-energy and $H_{n,n}$ is the Hamiltonian of the lead segment n. The first segment's contact self-energy is assumed to vanish (Σ_0=0). An imaginary damping potential η_n with exponential shape is chosen as the CAP. It decays due to the decay parameter λ with increasing n smoothly to 0 when getting closer to the device. The initial damping potential η_0 is comparable with the energy E and therefore large enough to blur the impact of the vanishing initial guess of Σ_0. The purpose of the CAP η_n is mainly to improve the convergence behaviour [12]. A direct iteration of the equations (1) without the CAP suffers severely from bad convergence with n. Depending on the actual lead structure and geometry, the optimal η_0 and λ can vary. For the devices in this work, η_0 = -0.5eV and λ = 0.1 provide good convergence. For comparison and in the case of ideal leads only, the Sancho-Rubio method and transfer matrix method of NEMO5 are used to solve the contact self-energies. In these cases, the leads are considered periodic repetitions of a given unit cell.

If not stated otherwise, the electronic tight binding Hamiltonian $H_{SS'}$ between different surface atoms S and S' as well as the on-site one (i.e. S'=S) is given as

Journal of Physics: Conference Series **696** (2016) 012019 doi:10.1088/1742-6596/696/1/012019

$$H_{SS'} = H_0 + \sum_{i=1}^{N_H} \Sigma_{SS',i} \tag{2}$$

Here, H_0 is the original empirical tight binding Hamiltonian [25,26] with N_H unsaturated surface bonds. The sum in equation (2) that models the saturation of dangling bonds has first been introduced by Lee et al. in Ref.[27] for sp3-hybridized semiconductors. One can extend the passivation model beyond the sp3-symmetry when the $\Sigma_{SS',i}$ are given as

$$\Sigma_{SS',i} = H_{SH_i}(\varepsilon - H_{H_i})^{-1} H_{H_iS'} \tag{3}$$

Here, no transformation of the Hamiltonian into a sp3-hybridized basis is needed (in contrast to Ref.[27]). Equation (3) agrees with the contact self-energy formula of the NEGF method in equation (1). However, the inverted term in equation (3) represents a single Hydrogen atom with a fixed energy parameter ε and a Hydrogen Hamiltonian H_H instead of the energy dependent Green's functions of semi-infinite leads in equation (1). The passivation Hamiltonian of Ref.[27] can be identically reproduced with the parameters $\varepsilon = 1\text{eV}$, $V_{ss\sigma} = -2.739\text{eV}$, $V_{sp\sigma} = 4.743\text{eV}$, $E_s = V_{sd\sigma} = V_{ss^*\sigma} = 0$ which is done within this work. It is worth to mention other passivation parameter sets can be used to mimic the saturation with various type of atoms (such as with SiO_2 as in Ref.[28]).

3. Results

The presented CAP lead method is first verified on a 3x3x10nm homogeneous Si nanowire in the sp3d5s* tight binding representation [20]. Since the lead consists of pure Si and is considered to be periodic, the common boundary self-energy methods are applicable here and can serve as a verification baseline. The electronic transmission is solved with the NEGF method and the contact self-energies calculated by the transfer matrix method [10] and the CAP method. In the CAP case, the numerical lead structure for the calculation includes 156 slabs in total. Figure 1 shows that the transmission of the CAP case agrees well with the reference. All observed deviations are very close to van-Hove singularities where convergence of the presented method is indeed the worst. These deviations are typically within a very small energy interval and become irrelevant in charge self-consistent calculations when the solution of the Poisson equation creates inhomogeneous potential profiles. In particular, for charge self-consistent IV characteristics, such deviations have been found to be negligible.

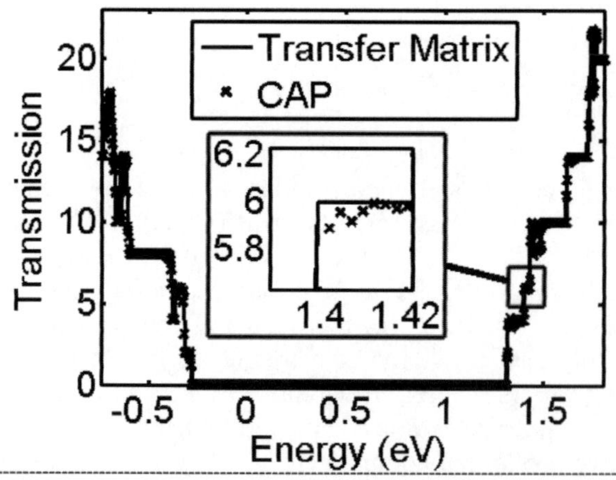

Figure 1. Transmission in valence and conduction band of a Si nanowire described in the text solved with contact self-energies of the transfer matrix method (solid line) and self-energies of the CAP method (symbols).

While the conventional lead methods are limited to periodic leads, the presented method allows solving contact self-energies of nonperiodic leads. In the following the CAP method is applied on a

20nm long, 10nm thick graphene nanoribbon (GNR) illustrated in Fig.2 (a). Here, electrons are represented within the pd tight-binding model [21]. The leads are GNRs with a width that increases constantly with increasing distance to the nanoribbon channel. To verify the modelling, the transmission of this structure is solved in two ways, i.e. with two numerically different devices: In the case of device1, the active region where the NEGF equations are solved is limited to the central nanoribbon with constant width of 10nm. In the case of device2, the region where all NEGF equations are solved extends 5nm beyond the central ribbon into the trumpet shaped leads as shown in Fig.2 (a). Obviously, the lead/device interface width differs in the two cases. However, the ballistic transmission describes electronic propagation from minus to plus infinity along the transport direction. Since the devices 1 and 2 are physically equivalent, their transmissions have to agree – as confirmed in Fig.2 (b).

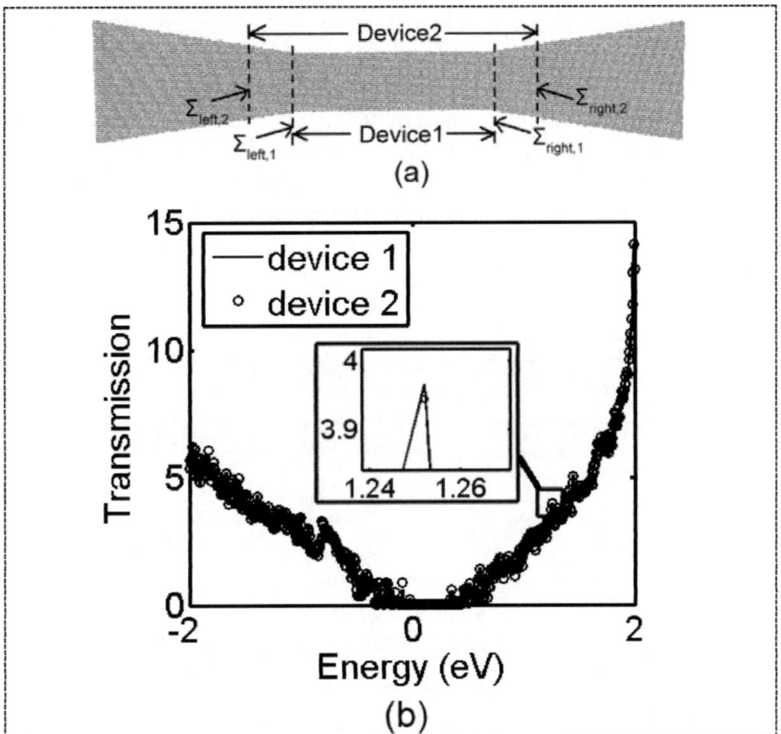

Figure 2. (a) Illustration of the trumpet shaped graphene nanoribbon and the two considered numerically different devices. (b) Calculated transmission of device1 (dashed line) and device2 (symbols).

Random disorder as it appears in random alloys can also be covered by the presented method. A common treatment of alloy disorder scattering in the NEGF framework involves a perturbative treatment [19][8]. The unperturbed configuration in that case is then given by the virtual crystal approximation that was shown to potentially miss some band bowing effects [29]. A full 3D atomistic modelling capability of a non-perturbative alloy disorder scattering treatment in the lead is available with the presented CAP method.

Approximate disordered nanowire bandstructures have been compared for AlGaAs wires [30] indicating the relationship between tunneling turn-on and the approximate bandstructure. These results also indicated that the atomistically disordered wire confines localized states which can result in resonant-like transport. In those simulations, however it cannot be concluded if the localization is induced by the centrally disordered device or the attached homogenous leads that form a

Journal of Physics: Conference Series **696** (2016) 012019 doi:10.1088/1742-6596/696/1/012019

hererostructure with the central device. With the approach described in this work one can treat quantum transport with disordered contacts.

Figure 3 (a) shows two $In_{0.5}Ga_{0.5}As$ nanowires with squared cross-sections of 3x3nm. The electrons are represented in sp3d5s* tight binding model [20]. The first device consists of $In_{0.5}Ga_{0.5}As$ in the central region only and ideal InAs leads surrounding it while the second device consists of $In_{0.5}Ga_{0.5}As$ in both device and leads. The In and Ga atoms in all $In_{0.5}Ga_{0.5}As$ areas are randomly distributed within the atomically resolved structure. All results that are subject to random atom distributions are averaged over 50 samples.

The transmission coefficients for the alloy cases and their error bars are shown in Fig.3(b) for devices with lengths of 18nm and 6nm. Figure 3(b) illustrates for the case of pure InAs leads that the transmission through a 6nm long device is significantly larger than that of a 18nm long device. In contrast, leads with random alloy yield virtually device length independent transmission coefficients. This indicates the tunneling of pure InAs lead electrons through the random alloy $In_{0.5}Ga_{0.5}As$ devices: the longer the device, the more pronounced is the dispersion mismatch between alloyed devices and ideal leads and the less likely is tunneling through the device. Similar lead/device reflections were reported when incoherent scattering was included in the active device, but ballistic leads were attached [31]. In contrast, when leads and device face the same effective random alloy the density of states between the leads and the device match and lead electrons face less reflection. As a result, the transmission remains the same when the device length changes. The small reduction of the device transmission with alloyed leads in the 18nm long device compared to 6nm long one (Fig. 3b) is due to a more pronounced weak localization in longer randomly alloyed devices.

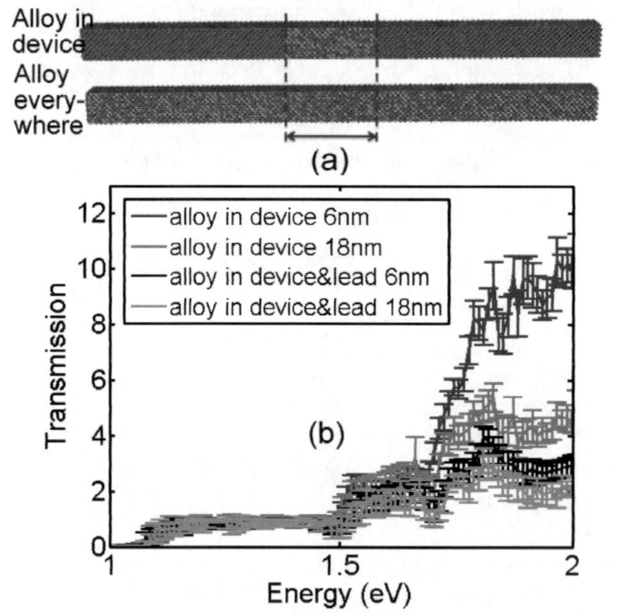

Figure 3. (a) Schematic of randomly alloyed $In_{0.5}Ga_{0.5}As$ nanowires with pure InAs leads and leads with $In_{0.5}Ga_{0.5}As$ random alloy. (b) Calculated transmission coefficients for the alloy randomness structures as shown in (a), with device lengths of 18nm and 6nm, respectively.

Contact self-energies can be used to model the impact of dangling bond passivation without increasing the numerical load. This is illustrated in the following with solutions of the Schrödinger equation. All results and statements are transferable to the NEGF method as well.

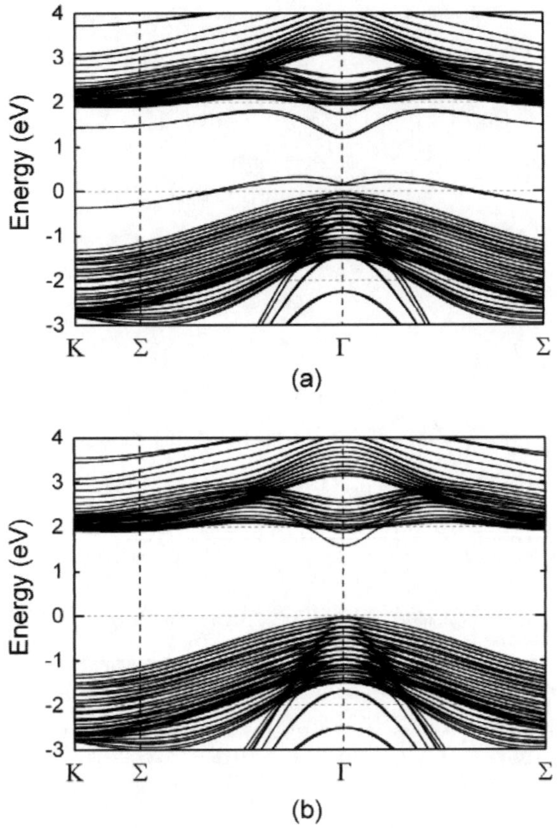

Figure 4. (a) Bandstructure of a 4.3nm GaAs quantum well grown in (111) direction in 10band atomistic tight binding representation. Without any special treatment of the surface atoms the bandstructure shows states within the bandgap of GaAs (0-1.4eV). (b) Bandstructure of the quantum well in (a) including the passivation self-energy of equation (3). The band gap is free of states.

Figure 4 (a) shows the bandstructure of a 4.3 nm thick GaAs quantum well without any special treatment of the surface atoms. The quantum well band gap should be larger than the GaAs material band gap, since the conduction (valence) band edge gets increased (lowered) by the confinement energy of the first confined valence and conduction band states. The effective (indirect) band gap of the unpassivated GaAs quantum well in Fig. 4(a), however, is smaller than the one of GaAs (~1.4eV). For comparison, Fig. 4 (b) shows the same bandstructure when the passivation self-energies of equation (3) are added to the Hamiltonian. In that case, no states lie within the GaAs band gap.

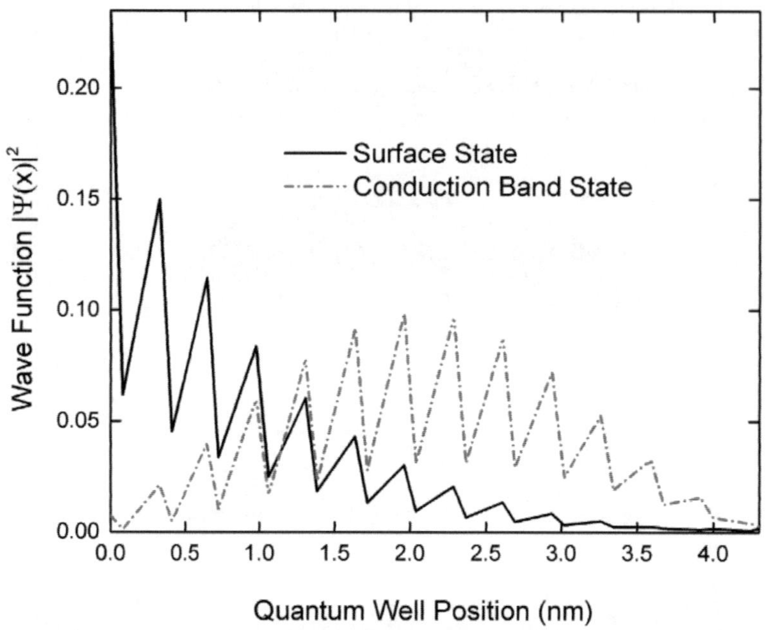

Figure 5. Absolute squared of the wave functions at the Gamma point of Figs. 4. A typical surface state at energy E=0.15eV in Fig.4(a) is shown with a solid line and a typical volume state (at E=-0.02eV in Fig.4(b)) is given by a dash-dotted line. Oscillations between anions and cations that superpose the exponential decay of the surface state and the first confined state envelope of the volume state are common in atomistic representations.

Figure 5 compares the absolute squared wave function of one of the dangling bond states in Fig. 4 (a) at the Gamma point at an energy of 0.15eV. This state is clearly located at the quantum well surface. In contrast Figure 5 also shows a typical state of the passivated quantum well of Fig. 4 (b) at the Gamma point and with an energy of -0.02eV.

4. Conclusion

It is demonstrated in this work how contact self-energies of irregular lead structures can be solved with the CAP method. It is also shown how contact self-energies can be used to efficiently model surface atom passivation. The CAP method is verified against conventional approaches for regular leads. The transmission of two numerically different, but physically equivalent irregular lead-structures agrees identically and verifies this method for irregular leads as well. Random alloy disorder in the leads is successfully modelled. It is shown that the typical assumption of ideal leads gives significant deviations from the disordered lead setup.

Atomistic tight binding calculations require to saturate dangling bonds of surface atoms. In this work it is shown how the concept of contact self-energies can be used to include passivation with Hydrogen atoms without increasing the system's electronic Hamiltonian.

5. Acknowledgment

The use of nanoHUB.org computational resources operated by the Network for Computational Nanotechnology funded by the US National Science Foundation under Grant Nos. EEC-0228390, EEC-1227110, EEC-0228390, EEC-0634750, OCI-0438246, OCI-0832623 and OCI-0721680 is gratefully acknowledged. This work was supported in part by funding from the Semiconductor Research Corporation's Global Research Collaboration (GRC) (2141.001, 2273.001), and Member Specific Research – Intel (MSR-Intel) (2434.001). Additional funding was provided by the Semiconductor Research Corporation membership in the Network for Computational Nanotechnology. NEMO5 developments were critically supported by an NSF Peta-Apps award OCI-0749140 and by Intel Corporation.

6. References

[1] Schwamb T, Burg B R, Schirmer N C and Poulikakos D 2008 *Appl. Phys. Lett.* **92** 243106
[2] Torricelli F, Smits E C P, Meijboom J R, Tripathi A K, Gelinck G H, Colalongo L, Kovacs-Vajna Z M, de Leeuw D M and Cantatore E 2011 *IEEE Trans. Electr. Dev.* **58** 3025
[3] Hwang W S, Tahy K, Li X, Xing H, Seabaugh A, Sung C Y and Jena D 2012 *Appl. Phys. Lett.* **100** 203107
[4] Dolcini F, Iotti R C and Rossi F 2013 *Phys. Rev. B* **88** 115421
[5] Pop E 2010 *Nano Res.* **3** 147
[6] Datta S 2005 *Quantum Transport: Atom to Transistor* (Cambridge University Press, Cambridge)
[7] Klimeck G, Lake R, Bowen R, Frensley W and Moise T 1995 *Appl. Phys. Lett.* **67** 2539
[8] Kubis T, Yeh C, Vogl P, Benz A, Fasching G and Deutsch C 2009 *Phys. Rev. B* **79** 195323
[9] Sancho M, Sancho J and Rubio J 1985 *J. Phys. F: Met. Phys.* **15** 851
[10] Luisier M, Schenk A, Fichtner W and Klimeck G 2006 *Phys. Rev. B* **74** 205323
[11] Bowen R C 1996 *Full Bandstructure Modeling of Quantum Transport in Nano-Scaled Devices* (PhD. Thesis, University of Texas at Dallas)
[12] Driscoll J A and Varga K 2008 *Phys. Rev. B* **78** 245118
[13] Varga K 2010 *Phys. Rev. B* **81** 045109
[14] Varga K 2011 *Phys. Rev. B* **83** 195130
[15] Cook B G and Varga K 2011 *Appl. Phys. Lett.* **98** 052104
[16] Zhang L, Chen J and Wang J 2013 *Phys. Rev. B* **87** 205401
[17] Yu Z, Zhang L, Xing Y and Wang J 2014 *Phys. Rev. B* **90** 115428
[18] Steiger S, Povolotskyi M, Park H H, Kubis T and Klimeck G 2011 *IEEE. Trans. Nanotechnol.* **10** 1464
[19] Lake R, Klimeck G, Bowen R C and Jovanovic D 1997 *J. Appl. Phys.* **81** 7845
[20] Boykin T B, Klimeck G and Oyafuso F 2004 *Phys. Rev. B* **69** 115201
[21] Boykin T B, Luisier M, Klimeck G, Jiang X, Kharche N, Zhou Y and Nayak S K 2011 *J. Appl. Phys.* **109** 104304
[22] Keating P N 1966 *Phys. Rev.* **145** 637
[23] Boykin T B, Klimeck G, Bowen R C and Oyafuso F 2002 *Phys. Rev. B* **66** 125207
[24] He Y, Wang Y, Klimeck G and Kubis T 2014 *Appl. Phys. Lett.* **105** 213502
[25] Tan Y, Povolotskyi M, Kubis T, Boykin T B and Klimeck G 2015 *Phys. Rev. B* **92** 08530
[26] Podolskiy A V, Vogl P 2004 *Phys. Rev. B* **69** 233101
[27] Lee S, Oyafuso F,von Allmen P and Klimeck G 2004 *Phys. Rev. B* **69** 045316
[28] He Y, Tan Y P, Jiang Z, Povolotskyi M, Klimeck G and Kubis T 2015 arXiv:1507.08712
[29] Boykin T, Kharche N and Klimeck G 2007 *Phys. Rev. B* **76** 035310
[30] Boykin T, Luisier M, Schenk A, Kharche N and Klimeck G 2007 *IEEE Trans. Nanotechnol.* **6** 43
[31] Kubis T and Vogl P 2011 *Phys. Rev. B* **83** 195304

Journal of Physics: Conference Series **696** (2016) 012020 doi:10.1088/1742-6596/696/1/012020

Development of non-equilibrium Green's functions for use with full interaction in complex systems

Richard A Lynn[1,2], **Robert van Leeuwen**[1,2]

[1] Department of Physics, Nanoscience Center, FIN 40014, University of Jyväskylä, Jyväskylä, Finland

[2] European Theoretical Spectroscopy Facility (ETSF)

E-mail: `richard.a.lynn@jyu.fi`

Abstract. We present an ongoing development of an existing code for calculating ground-state, steady-state, and transient properties of many-particle systems. The development involves the addition of the full four-index two electron integrals, which allows for the calculation of transport systems, as well as the extension to multi-level electronic systems, such as atomic and molecular systems and other applications. The necessary derivations are shown, along with some preliminary results and a summary of future plans for the code.

1. Introduction

The development of molecular electronics has in recent years made it desirable to be able to simulate time-dependent many-body phenomena. However, computational limitations have a significant effect, with any realistic system being impossible to calculate with current computing power. Therefore, we are forced to adopt approximate techniques. These can include mean-field theories such as Hartree–Fock[1–3], or methods such as density functional theory[4–7], which rely on an approximate potential. We look here at non-equilibrium Green's function theory[8], which is in principle exact. However, it requires a truncation of an infinite sum term known as the self-energy.

When considering a many-particle system in the ground-state, it is well-known that properties can be calculated using many-body Green's functions, containing a suitable approximation for the self-energy, Σ. However, in order to study time-dependent properties of these systems, it is necessary to propagate these Green's functions in time by solving their equations of motion. The equations of motion for the many-particle Green's functions are given by the coupled integro-differential Kadanoff–Baym equations[9–12]. The solution to these equations was calculated numerically by Dahlen and van Leeuwen in 2007[13], leading to the analysis of many time-dependent systems, including inhomogeneous systems[14], and strongly correlated few-electron quantum dots[15].

In 2009, this was extended to consider an embedded system, with a number of leads coupled to a central system of interest[16]. When propagated through time, both the transient and steady-state properties could be investigated, as well as the initial ground-state properties. This allowed for further investigations, such as image charge dynamics[17], quantum transport in AC and DC fields[18], and real-time switching between different steady states[19].

Journal of Physics: Conference Series **696** (2016) 012020 doi:10.1088/1742-6596/696/1/012020

In these earlier investigations, it was possible to use a simplification of the interaction term in the Hamiltonian, allowing for much faster computation. However, it has become desirable to include a fully described interaction, for the purposes of investigating transport and other applications using four-index two electron integrals, rather than the simplified two-index integrals used so far. The theory behind this expansion and the initial results and plans for the code in its new form are presented here.

2. Theory

2.1. Quantum Transport Model

When considering a two terminal quantum transport model, with a central system of interest connected to leads, the model Hamiltonian is described by:

$$\hat{H}(t) = \hat{H}_C(t) + \hat{H}_{\text{leads}}(t) + \hat{H}_T - \mu\hat{N} \tag{1}$$

where the Hamiltonian of the central system, $\hat{H}_C$, the Hamiltonians of the leads, $\hat{H}_{leads}$, and the tunneling Hamiltonian, $\hat{H}_T$, are given by Equation 2:

$$\hat{H}_C = \sum_{ij}\sum_{\sigma} h_{ij}(t)\hat{a}^{\dagger}_{i\sigma}\hat{a}_{j\sigma} + \frac{1}{2}\sum_{ijkl}\sum_{\sigma\sigma'} w_{ijkl}\hat{a}^{\dagger}_{i\sigma}\hat{a}^{\dagger}_{j\sigma'}\hat{a}_{k\sigma'}\hat{a}_{l\sigma} \tag{2a}$$

$$\hat{H}_{\text{leads}}(t) = \sum_{\alpha=L,R}\sum_{\substack{i\in C\\ j\in\alpha}}^{\infty}\sum_{\sigma}\left[h^{\alpha}_{ij} + W^{\alpha}(t)\delta_{ij}\right]\hat{c}^{\dagger}_{i\sigma\alpha}\hat{c}_{j\sigma\alpha} \tag{2b}$$

$$\hat{H}_T = \sum_{\alpha=L,R}\sum_{\substack{i\in C\\ j\in\alpha}}\sum_{\sigma} V_{i,j\alpha}\left[\hat{d}^{\dagger}_{i\sigma}\hat{c}_{j\sigma\alpha} + \hat{c}^{\dagger}_{j\sigma\alpha}\hat{d}_{i\sigma}\right] \tag{2c}$$

In Equation 2b the Hamiltonian for the leads, $\alpha = L, R$ is described, with h^{α}_{ij} being the nearest neighbour Hamiltonian, $\hat{c}^{\dagger}_{i\sigma\alpha}$ and $\hat{c}_{j\sigma\alpha}$ are the creation and annihilation operators respectively for the lead, and $W^{\alpha}(t)$ is the local potential dependent on time. Typically this local potential is modelled as a bias voltage generated by applying an electric field over the system.

Equation 2c describes the couplings between the central system and the leads, with $V_{i,j\alpha}$ being the matrix elements of the coupling Hamiltonian. The final term in Equation 1 couples the chemical potential, μ, to the total particle number operator, $\hat{N}$.

Of particular interest to this study are the one-body and two-body parts of Equation 2a, h_{ij} and w_{ijkl}. In previous studies, typically only the main local and non-local terms of these two terms were considered. In particular, this leads to the reduction of the interaction term $w_{ijkl} = \delta_{il}\delta_{kl}w_{ij}$, where $w_{ij} = 1/|i-j|$ is the long-range behaviour of the matrix elements of the interaction. However, it has now become desirable to remove this assumption, using the full descriptions as given by:

$$h_{ij} = -\frac{1}{2}\int d\mathbf{x}\phi^*_i(\mathbf{x})\hat{h}(\mathbf{x},t)\phi_j(\mathbf{x}) \tag{3a}$$

$$w_{ijkl} = \int d\mathbf{x}d\mathbf{y}\phi^*_i(\mathbf{x})\phi^*_j(\mathbf{y})\hat{w}(\mathbf{x},\mathbf{y})\phi_k(\mathbf{y})\phi_l(\mathbf{x}) \tag{3b}$$

where the standard position-spin coordinate description, $\mathbf{x} = (\mathbf{r},\sigma)$, has been used.

2.2. The Kadanoff-Baym Equations

For the purposes of transport calculations, the system is initially assumed to be in thermal equilibrium, before an external perturbation is applied at time t_0, after which time-evolution

Journal of Physics: Conference Series **696** (2016) 012020 doi:10.1088/1742-6596/696/1/012020

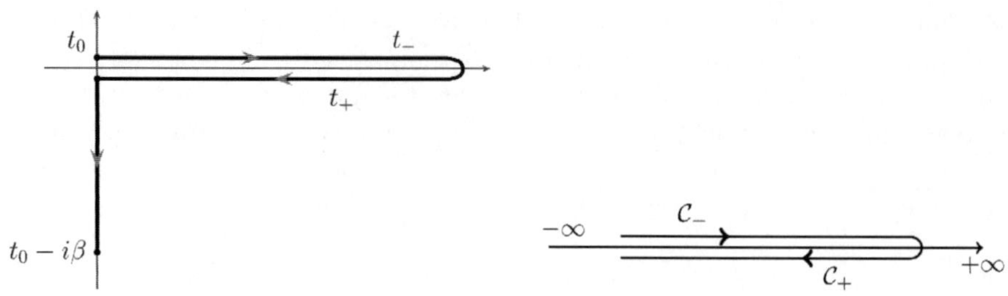

Figure 1. Keldysh Contours. The left gives the Keldysh contour showing the initial correlations in grand canonical ensemble form. The right gives a closed time-loop contour, $\mathcal{C}$. The forward time-ordered branch and backward anti-time-ordered branch are denoted by "−" and "+" respectively.

is carried out. The whole process is described by the Keldysh contour, as seen in Figure 1, with the ground-state described by the vertical imaginary track, and the dynamics described by the time-loop contour. Starting from the equations of motion for the single particle Green's function:

$$\left[i\partial_{z_1} - \hat{h}(1)\right] G(1,2) = \delta(1,2) + \int d3\Sigma_{\text{MB}}(1,3)G(3,2) \tag{4a}$$

$$G(1,2)\left[-i\overleftarrow{\partial}_{z_2} - \hat{h}(2)\right] = \delta(1,2) + \int d3G(1,3)\bar{\Sigma}_{\text{MB}}(3,2) \tag{4b}$$

different components can be distinguished on the contour $\mathcal{C}$ by utilising the Langreth rules. These components are called the Kadanoff-Baym equations, and are given as:

$$\left[i\partial_{t_1} - \hat{h}(t_1)\right] G^{\gtrless}(t_1,t_2) = \left[\Sigma^R \cdot G^{\gtrless} + \Sigma^{\gtrless} \cdot G^A + \Sigma^{\rceil} \star G^{\lceil}\right](t_1,t_2) = I_1^{\gtrless}(t_1,t_2) \tag{5a}$$

$$G^{\gtrless}(t_1,t_2)\left[-i\overleftarrow{\partial}_{t_2} - \hat{h}(t_2)\right] = \left[G^R \cdot \Sigma^{\gtrless} + G^{\gtrless} \cdot \Sigma^A + G^{\rceil} \star \Sigma^{\lceil}\right](t_1,t_2) = I_2^{\gtrless}(t_1,t_2) \tag{5b}$$

$$\left[i\partial_{t_1} - \hat{h}(t_1)\right] G^{\rceil}(t_1,\tau_2) = \left[\Sigma^R \cdot G^{\rceil} + \Sigma^{\rceil} \star G^M\right](t_1,\tau_2) = I^{\rceil}(t_1,\tau_2) \tag{5c}$$

$$G^{\lceil}(\tau_1,t_2)\left[-i\overleftarrow{\partial}_{t_2} - \hat{h}(t_2)\right] = \left[G^{\lceil} \cdot \Sigma^A + G^M \star \Sigma^{\lceil}\right](\tau_1,t_2) = I^{\lceil}(\tau_1,t_2) \tag{5d}$$

$$\left[-\partial_{\tau_1} - \hat{h}^M\right] G^M(\tau_1,\tau_2) = i\delta(\tau_1-\tau_2) + \left[\Sigma^M \star G^M\right](\tau_1,\tau_2) = I^M(\tau_1,\tau_2) \tag{5e}$$

$$G^M(\tau_1,\tau_2)\left[\delta_{\tau_2} - \hat{h}^M\right] = i\delta(\tau_1-\tau_2) + \left[G^M \star \Sigma^M\right](\tau_1,\tau_2) = I^M(\tau_1,\tau_2) \tag{5f}$$

The short-hand notation, $I^{\gtrless,\rceil,\lceil}$, refers to collision integrals, and the convolution integrals are defined as:

$$\left[a \cdot b\right](t_1,t_2) = \int_{t_0}^{\infty} a(t_1,t)b(t,t_2)dt \tag{6a}$$

$$\left[a \star b\right](t_1,t_2) = -\int_{0}^{\beta} a(t_1,\tau)b(\tau,t_2)d\tau \tag{6b}$$

Solving the Kadanoff–Baym equations numerically is far from trivial. The initial step is to solve the system along the imaginary time axis, that is the Matsubara component of the Green's function. The equation of motion of the Matsubara component is written as:

$$\sum_k \left(-\delta_{ik}\partial_{\tau_1} - h^0_{ik}\right) G^M_{kj}(\tau_1) = \delta_{ij}\delta(\tau_1) + \int_0^\beta d\tau_3 \sum_k \Sigma^M_{ik}(\tau_1 - \tau_3) G^M_{kj}(\tau_3) \tag{7}$$

Equation 7 can then be cast in the form of a Dyson equation:

$$G^M_{ij}(\tau_1) = G_{0,ij}(\tau_1) + \int_{-\beta}^0 d\tau_4 \int_0^\beta d\tau_3 \sum_{kl} G_{0,ik}(\tau_1 - (\tau_3 - \beta))\Sigma^{M,c}_{kl}(\tau_3 - (\tau_4 + \beta)) G^M_{lj}(\tau_4) \tag{8}$$

which can be solved iteratively to self-consistency using, e.g., Hartree–Fock. Using the solution to Equation 8, the two-time propagation can be started. By making use of symmetries, the subset of equations requiring propagation can be reduced to:

$$i\partial_{t_1} G^>(t_1,t_2) = \hat{h}(t_1)G^>(t_1,t_2) + I^>_1(t_1,t_2) \tag{9a}$$
$$-i\partial_{t_1} G^<(t_2,t_1) = G^<(t_2,t_1)\hat{h}(t_1) + I^<_2(t_2,t_1) \tag{9b}$$
$$i\partial_{t_1} G^\rceil(t_1,-i\tau) = \hat{h}(t_1)G^\rceil(t_1,-i\tau) + I^\rceil(t_1,-i\tau) \tag{9c}$$
$$-i\partial_{t_1} G^\lceil(-i\tau,t_1) = G^\lceil(-i\tau,t_1)\hat{h}(t_1) + I^\lceil(-i\tau,t_1) \tag{9d}$$

These equations can finally be propagated by using a Hartree–Fock Hamiltonian with small time-steps Δ from $t \to t + \Delta$.

2.3. Self Energy Approximations

The most important approximation used in the theory is that of the self-energy. This term is expanded in terms of the interaction, and we here consider two approximations. The first is the well-known Hartree–Fock approximation, described by Equation 10. The first term in this is the Hartree term, and the second is the Fock, or exchange, term. The other expansion used is the second-order Born approximation, given by Equation 11.[1] The first of the terms beyond Hartree–Fock is commonly referred to as the first order bubble term, while the final term is simply the second order correction to the exchange term. Both of these self-energies are presented diagrammatically in Figure 2.

$$\Sigma^{\mathrm{HF}}_{ij}(t) = -i\sum_{kl} G_{kl}(t,t^+)(2w_{ilkj} - w_{iljk}) \tag{10}$$

$$\Sigma^{(2)}_{ij}(t,t') = \sum_{klmnpq} G_{kl}(t,t')G_{mn}(t,t')G_{pq}(t,t') \times w_{iqmk}(2w_{lnpj} - w_{nlpj}) \tag{11}$$

2.4. Basis Sets

The one- and two-body parts of the Hamiltonian, described by Equation 3, depend on a chosen basis set. Any basis set can be used for this purpose, but for this current investigation, an atom-centred Slater basis set has been used. The definition of a Slater type orbital is given by:

$$S^\zeta_{nlm}(r,\theta,\phi) = N r^{n-1} e^{-\zeta r} Y^m_l(\theta,\phi) \tag{12}$$

[1] The second-order Born self-energy is not directly used in this work, but when time-dependence is introduced, it is intended that it shall be utilised substantially.

Journal of Physics: Conference Series **696** (2016) 012020 doi:10.1088/1742-6596/696/1/012020

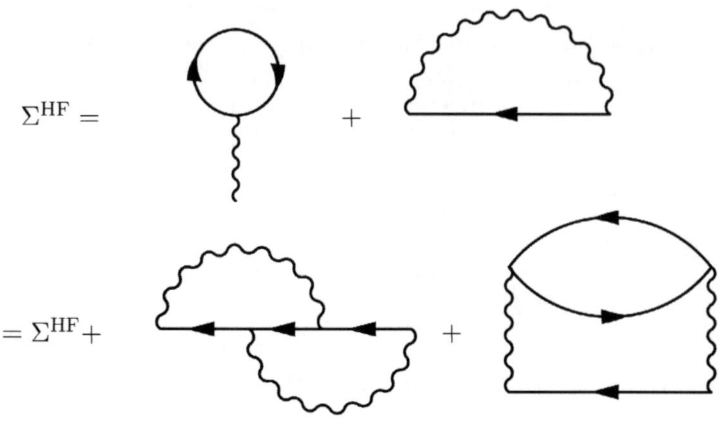

Figure 2. Diagrammatic representations of the self-energy approximations. The Hartree–Fock approximation, top, consists of the Hartree term, and the Fock, or exchange term. The second-order Born approximation consists of the Hartree–Fock self-energy, added to the first order bubble diagram, and the second-order correction to the exchange term.

where n, l, and m are the principle quantum numbers, which follow the standard rule $n>l \geq |m| \geq 0$. N is a normalisation constant, and Y_l^m are spherical harmonics, given by:

$$Y_l^m(\theta, \phi) = Ne^{im\phi}P_l^m(\cos\theta) \tag{13a}$$

$$P_l^m(x) = \frac{(-1)^m}{2^l l!}(1-x^2)^{m/2}\frac{d^{l+m}}{dx^{l+m}}(x^2-1)^l \tag{13b}$$

where P_l^m are the associated Legendre polynomials.

As the Slater type orbitals are based on the eigenvalues of the Hydrogen atom, they are beneficial in reducing the time taken to convergence in the calculations of molecular orbitals. When calculating these bases, it is possible to use ready prepared tables with values of ζ for all combinations of n, l, and m for different atoms. Using a seperate basis set for each atom, the Hartree–Fock orbitals are found by solving a generalised eigenvalue problem, and these orbitals then form a new orthonormal basis set.[2]

3. Initial Results

While the complete implementation has yet to be completed, there have already been some preliminary ground-state results obtained as a test of the reliability of the code. These initial tests were carried out on the hydrides of the first row elements, Lithium to Fluorine. These molecules were chosen as exact theoretical values for many of their properties are known, so can be directly compared. The size of the basis set used for the calculations was dependent on the atoms involved, i.e. there was one basis function for all different values of n, l, and m for each of the two atoms.

The first test was to find the ground-state equilibrium energy of the molecule. The theoretical values for comparison were calculated by summing the individual ground-state energies of the individual atoms, and adding the experimental dissociation energy. The numerical results were achieved by finding the ground-state energy of the molecule at the experimental equilibrium

[2] For more details, the full basis sets and a file of Slater functions and exponents are available for the reader upon request.

Molecule	Equilibrium Bond Length	Theoretical Energy	Numerical Energy
LiH	3.0094	-8.0701	-7.984791
BeH	2.5340	-15.2529	-14.881285
BH	2.2453	-25.2815	-24.920593
CH	2.0566	-38.4817	-37.916774
NH	1.9057	-55.2283	-54.555561
OH	1.8113	-75.7694	-75.385111
HF	1.7302	-100.5197	-100.063868

Table 1. The equilibrium bond lengths of diatomic molecules, together with the theoretical and numerical ground-state energy. All figures are in atomic units.

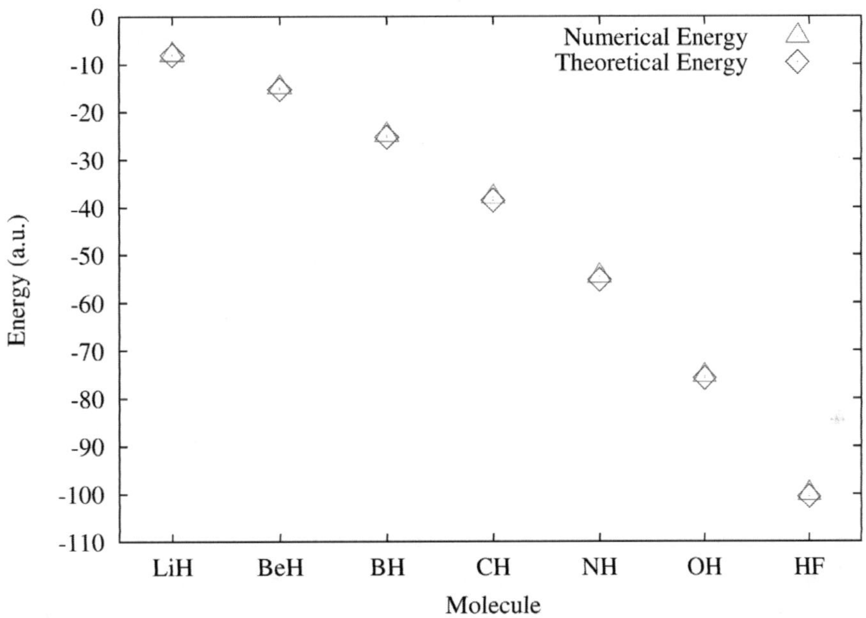

Figure 3. The ground-state energies of diatomic molecules calculated at the equilibrium inter-atomic bond length. The theoretical results are shown as blue diamonds, while the Hartree–Fock results are displayed as red triangles.

bond distance. Table 1 shows the two energies, and they are plotted for comparison in Figure 3. It can be seen that the agreement is very good between the two figures, with the numerical results slightly overestimating the equilibrium energy, as would be expected.

Having found the ground-state equilibrium results, the other significant property that was investigated was the dissociation energy. This was found by finding the numerical energy for various different bond lengths, and then fitting a common inter-atomic potential curve to the data. There were three potentials used to provide a thorough test, as well as give a range of energies. The first potential used was the Morse potential, given by:

$$V(r) = E + D_e\left(1 - e^{-b(r-r_e)}\right)^2 \tag{14}$$

with fitting parameters E, D_e, and b. The second potential was the Rydberg potential, given

Molecule	Experimental D_e	Morse D_e	Rydberg D_e	Varshni D_e
LiH	0.0926	0.0839041	0.0751182	0.0417132
BeH	0.0847	0.0462033	0.0417274	0.0298187
BH	0.1243	0.224196	0.204235	0.123077
CH	0.1270	0.313989	0.281394	0.138535
NH	0.1182	0.0182241	0.0164693	0.0142616
OH	0.1616	0.110868	0.106774	0.0979529
HF	0.2151	0.375695	0.340266	0.18733

Table 2. The experimental dissociation energies and the energies calculated by fitting data to the Morse, Rydberg, and Varshni inter-atomic potentials.

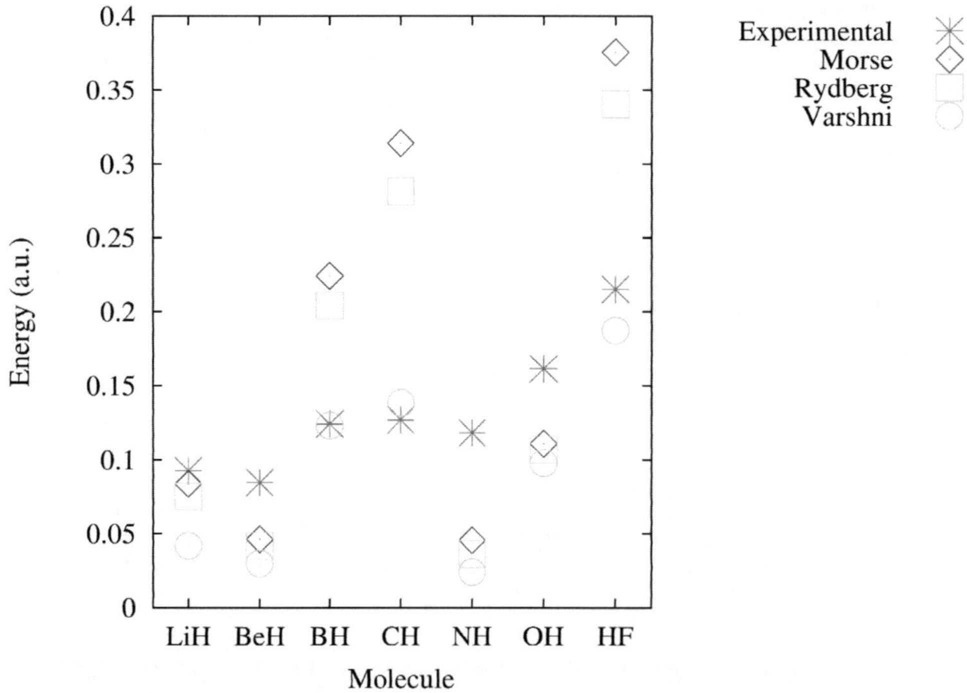

Figure 4. The dissociation energies of diatomic molecules calculated by fitting three different potentials to numerical data. The experimental energies are shown as red stars, blue diamonds show the Morse potential fit results, cyan squares show the results from the Rydberg potential, and the green circles show the Varshni potential.

by:

$$V(r) = E + D_e\left(1 - \left(1 + \frac{k}{D_e}\right)^{\frac{1}{2}}(r - r_e)\right)e^{-\left(\frac{k}{D_e}\right)^{\frac{1}{2}}(r-r_e)} \tag{15}$$

with fitting parameters E, D_e, and k. The final potential used was the Varshni potential, given by:

$$V(r) = E + D_e\left(1 - e^{-b\left(r^2-r_e^2\right)}\right)^2 \tag{16}$$

with fitting parameters E, D_e, and b. For all three potentials, the value of the fitting parameter

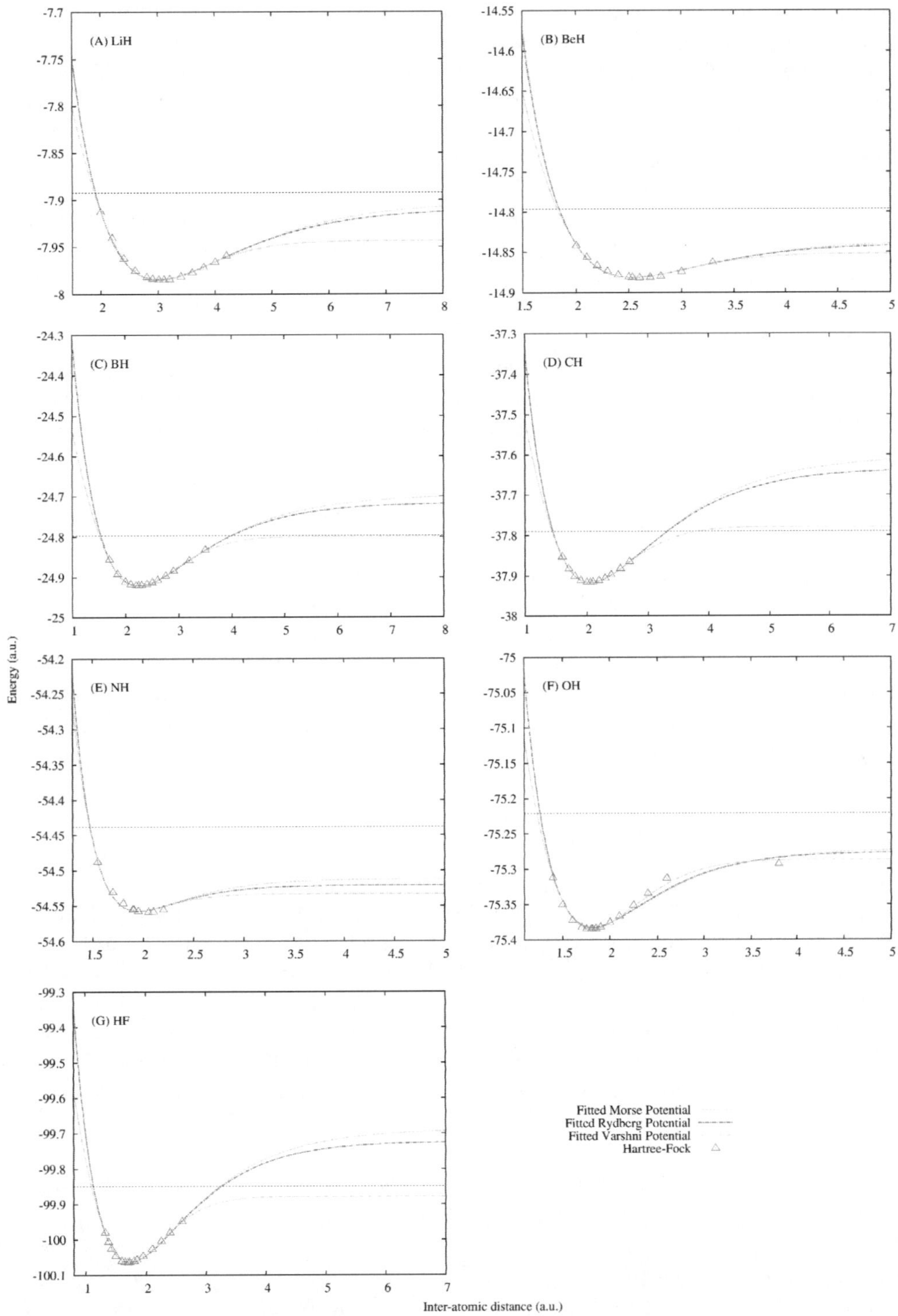

Figure 5. The dissociation curves for the first row hydrides. The Hartree–Fock results for various inter-atomic distances are shown as purple triangles, and various approximate atomic potentials are fitted to the data. A Morse potential is shown as a dot-dash-dotted green line, Rydberg potential as a dot-dashed red line, and Varshni potential as a dashed cyan line. The black dashed line shows the experimental dissociation energy of the molecule for comparison.

D_e gives the approximate dissociation energy. The values of D_e for each potential and molecule are shown in Table 2 with the experimental values for comparison, and the results are plotted in Figure 4. The individual dissociation curves for each molecule are shown in Figure 5.

The results shown here suggest that the calculation of dissociation energy is less accurate than the equilibrium energy value. However, this is in part due to the limited amount of data that has been used to fit the potential curves, and an inherent inaccuracy within the potentials themselves. The Varshni potential is the most consistent approximation, and follows roughly the correct trend. As these results were obtained primarily as a test of the functionality of the code, it can be said that the relative proximity between experimental and numerical results prove that the code is working as expected.

4. Conclusion and Future Plans

The initial results presented here show that the development of the code has already been successful for the calculation of ground-state properties, and implementation and testing of the time propagation have already been started for finding transient and steady-state properties. Once this has been completed, several specific applications have been planned for the future, which shall be briefly detailed here.

The topic of shot noise in quantum systems is an active area of research, and the more complex form of the interactions between particles being used in this new implementation make it possible to simulate. The system itself is a typical transport setup, with a central system of interest between two leads. Initially, the current over this system is measured as a function of time. If this is plotted, a fluctuation around the mean current can be seen. It is known that these fluctuations are correlated, and the current-current correlation function is defined as $\Delta I_\alpha(t,t') = I_\alpha(t) - \langle I_\alpha(t) \rangle$, where the current operator is defined as the time derivative of the total number of particles, given by $N_\alpha(t) = \sum_k c^\dagger_{k\alpha}(t)c_{k\alpha}(t)$. The property of interest is the time-dependent shot noise, which is defined as:

$$S_{\alpha\beta}(t,t') \equiv \frac{1}{2}\langle \Delta I_\alpha(t)\Delta I_\beta(t') + \Delta I_\beta(t')\Delta I_\alpha(t)\rangle \tag{17}$$

The other investigation planned using the new implementation is with the well-known Anderson impurity model. Typically, the simplified interaction used in the Anderson model removes the possibility of multiple energy-levels on a single site. By using the full interaction term, it is hoped that more can be understood about the simple one-site model, as well as providing the first stage in more complex systems. An isolated central site will first be studied, before the contacted case is considered and compared. It would be expected that a broadening of the peaks from the isolated case would be observed, along with the addition of a Kondo peak, as in the simplified model, but it is possible that additional physics may be observed.

Acknowledgments

We would like to acknowledge the Finnish Academy of Science for providing the funding for this project to be carried out.

Appendix A. Numerical Implementation

The calculation of the two-electron integrals, given by Equation 3b, makes up the majority of the additional code in this new implementation. As the integrand is made up of a product of four basis functions, with form described by Equations 12 and 13, along with the interaction $w(\mathbf{x}, \mathbf{y})$, this is a non-trivial integral to solve.

Before the integration itself, the variables within the bases are first read in from an input file. At this stage the input provides capability for modelling single atoms and diatomic molecules,

Journal of Physics: Conference Series **696** (2016) 012020 doi:10.1088/1742-6596/696/1/012020

but this can later be expanded to an N-particle system. As an initial simplification, those integrals where all four bases are centred on the same atomic site are considered separately.

The approach used considers the integral in spherical polar coordinates, with the separation of variables where possible simplifying the process considerably. The resulting multiple integrals are solved using a set of different numerical integration techniques.

The first main method used is Gauss-Legendre quadrature. This effectively reduces an integral to a sum of weightings applied to the function being integrated at fixed points. In general in Gaussian quadrature, the fixed points are determined by the roots of an associated polynomial. In the case of Gauss-Legendre integration, the associated polynomials are Legendre polynomials. This method is applied to integrands such as:

$$\int_1^\infty dx x^j \left(x^2 - 1\right)^m e^{-cx} \tag{A.1}$$

and

$$\int_{-1}^1 dx x^j \left(1 - x^2\right) e^{-cx} \tag{A.2}$$

The other numerical method used significantly is downwards recursion. This method can be applied, e.g. when the solution to an integral can depend on the solution to another integral with the same form of the integrand, but lower polynomial powers. This is applied in this code to situations such as:

$$\int_1^\infty dx x^j e^{-cx} \tag{A.3}$$

and

$$\frac{-1^l}{2} \int_{-1}^1 dt P_l(t) e^{-\beta t} \tag{A.4}$$

In addition to this, there are some integrals with simple known solutions, which are calculated directly, before the final value of w_{ijkl} is formed by combining these results.

For storing the two electron integrals, they must be kept in a two-dimensional array. As they are a four-dimensional object, this means some adjustment of the indices. In practice, they are stored as $v[p][q]$, where $i = 1 + (p-1)/nb$ and $j = p - nb * (i-1)$, with similar relations for k,l,and q.

References

[1] Hartree D R 1928 *Math. Proc. Cambridge* **24** 89–110

[2] Slater J C 1930 *Phys. Rev.* **35** 210–211

[3] Fock V A 1930 *Z. Physik* **61** 126–148

[4] Hohenberg P and Kohn W 1964 *Phys. Rev.* **136** B864–B871

[5] Kohn W and Sham L J 1965 *Phys. Rev.* **140** A1133–A1138

[6] Runge E and Gross E K U 1984 *Phys. Rev. Lett.* **52** 997–1000

[7] van Leeuwen R 1999 *Phys. Rev. Lett.* **82** 3863–3866

[8] Stefanucci G and van Leeuwen R 2013 *Nonequilibrium Many-Body Theory of Quantum systems: A Modern Introduction* (Cambridge University Press)

[9] Kadanoff L P and Baym G 1962 *Quantum Statistical Mechanics* (W. A. Benjamin, Inc., New York)

[10] Schäfer W 1996 *J. Opt. Soc. Am. B* **13** 1291

[11] Bonitz M, Kremp D, Scott D C, Binder R, Kraeft W K and Köhler H S 1996 *J. Phys. Condens. Matter* **8** 6057

[12] Binder R, Köhler H S, Bonitz M and Kwong N 1997 *Phys. Rev. B* **55** 5110

[13] Dahlen N E and van Leeuwen R 2007 *Phys. Rev. Lett* **98** 153004

[14] Stan A, Dahlen N E and van Leeuwen R 2009 *J. Chem. Phys.* **130** 24101

[15] Balzer K, Bonitz M, van Leeuwen R, Stan A and Dahlen N E 2009 *Phys. Rev. B* **79** 245306

[16] Myöhänen P, Stan A, Stefanucci G and van Leeuwen R 2009 *Phys. Rev. B* **80** 115107

[17] Myöhänen P, Tuovinen R, Korhonen T, Stefanucci G and van Leeuwen R 2012 *Phys. Rev. B* **85** 075105

[18] Myöhänen P, Stan A, Stefanucci G and van Leeuwen R 2010 *J. Phys.: Conf. Ser.* **220** 012017

[19] Uimonen A M, Khosravi E, Stefanucci G, Kurth S, van Leeuwen R and Gross E K U 2010 *J. Phys.: Conf. Ser.* **220** 012018